Partial Reconfiguration on FPGAs

Lecture Notes in Electrical Engineering

Volume 153

For further volumes:
http://www.springer.com/series/7818

Dirk Koch

Partial Reconfiguration on FPGAs

Architectures, Tools and Applications

Springer

Dirk Koch
University of Oslo
Oslo, Norway

ISBN 978-1-4614-1224-3 ISBN 978-1-4614-1225-0 (eBook)
DOI 10.1007/978-1-4614-1225-0
Springer New York Heidelberg Dordrecht London

Library of Congress Control Number: 2012932628

Printed on acid-free paper

Springer is part of Springer Science+Business Media (www.springer.com)

Preface

The research on partial runtime reconfiguration has a history that is as long as the great history of FPGAs. However, despite a time frame of more than two decades, the work on partial reconfiguration has not found its way into commercial systems yet, and is still considered very exotic. This is an interesting phenomenon as the benefits in saving costly FPGA resources and power consumption can be substantial. One reason for this phenomenon is that implementing partially reconfigurable systems is by far more difficult with ASIC-like design tools when compared with simply building static only systems that omit this nice feature.

There is a long list of system constraints, FPGA specific issues, problems with the design tools, documentation secrets, and missing techniques for resource management at runtime, that have to be processed to get a reconfigurable system running. Moreover, all these obstacles have to be understood in order to be able to build reconfigurable systems which *gain a significant improvement* over conventional static implementations, and which are *reliable*. In other words, it is typically not sufficient to have a simplified proof of concept implementation in order to build reconfigurable systems on a large scale.

Throughout this book, all major obstacles will be explained together with solutions. The goal of this book is to tackle the problems that come along with partial runtime reconfiguration, in a systematic way. This is done by proposing tools that are aware of the issues related to partial runtime reconfiguration, hiding most low level implementation details from a design engineer. By doing this, an error prone design flow is omitted and reproducible results can be guaranteed.

This book does not provide a general introduction into FPGA design techniques, which is already nicely covered in numerous textbooks. Here, the focus is put, in particular, on the effects of partial runtime reconfiguration that are not deeply covered by other textbooks as yet. However, readers of this book are not expected to have deep knowledge about hardware design and FPGA technology; all necessary technical basics have been provided.

This book has been written for design engineers who are interested in investigating runtime reconfiguration. Moreover, the book also targets graduate students, lecturers and researchers working in the field. The book provides theoretical and practical background, as well as design experience gathered through a decade of working with reconfigurable devices and systems.

Finally, this is the first issue of this book, and despite the review effort, there are probably some remaining mistakes. Therefore, I will be grateful to receive your comments in this regard. Moreover, this book reflects many years of personal experience, and the knowledge shared by many great colleagues and students whom I met since starting my work on FPGAs. However, many readers have gathered their own experience when using partial reconfiguration, and I am very pleased whenever they share their experiences with me.

Oslo, Norway Dirk Koch

Contents

Acronyms

ALU	Arithmetic Logic Unit[1]
API	Application Programming Interface
ASIC	Application Specific Integrated Circuit
BE	Basic Logic Element
BIST	Built-In Self-Test
BRAM	Block RAM
CAN	Controller Area Network
CIF	Common Intermediate Format
CLB	Configurable Logic Block
CPU	Central Processing Unit
CRC	Cyclic Redundancy Check
CCITT	Comitè Consultatif Internationale de Tèlègraphie et Tèlèphonie
DMA	Direct Memory Access
DSP	Digital Signal Processing
ECU	Electronic Control Unit
ESM	Erlangen Slot Machine
FIFO	First In, First Out memory
FPGA	Field Programmable Gate Array
GNU	GNU's Not Unix
ICAP	Internal Configuration Access Port
IMEC	Interuniversity Microelectronics Centre
I/O	Input/Output
ISP	Instruction Set Processor
IP	Intellectual Property
JTAG	Joint Test Action Group
LUT	Look-Up Table
MAC	Media Access Control

[1] Stan Kelly-Bootle gave in *The Computer Contradictionary* the following definition for an ALU: "A random-number generator supplied as standard with all computer systems".

NoC	Network on Chip
OS	Operating System
OS4RS	Operating System for Reconfigurable Systems
PCB	Printed Circuit Board
PCIe	Peripheral Component Interconnect Express
RTL	Register Transfer Level
SEU	Single Event Upset
SGC	Secure Group Communication
SPDIF	Sony Philips Digital Interface
SRAM	Static Random Access Memory
SSI	Stacked Silicon Interconnect

Symbols

b_{CLB}	bit offset within an CLB
b_{init}	initial configuration data offset for the first CLB
B	buswidth in bit
BE	basic element
c	communication cost in look-up tables
c_r	configuration readback data rate
c_w	configuration write data rate
c_{const}	constant cost for providing a configuration interface in look-up tables
C^S	channel of span S
CLB_h	configuration data per CLB in one frame
CLB_x	horizontal CLB position
CLB_y	vertical CLB position
Δ	vertical module displacement
η	compression ratio
$\widetilde{\eta}$	compression ratio for random data
$\widehat{\eta}$	worst case compression ratio
$\widehat{\eta}_{Huff}$	$\widehat{\eta}$ of a Huffman compressor
η_{LZSS}	compression ratio of a LZSS compressor
$\widehat{\eta}_{LZSS}$	$\widehat{\eta}$ of a LZSS compressor
η_{RLE}	compression ratio of a run length compressor
$\widehat{\eta}_{RLE}$	$\widehat{\eta}$ of a run length compressor
d_{FPGA}	FPGA peak configuration data rate
d_{MEM}	memory data rate
Δ_{AB}	vertical displacement between two modules in terms of BEs
f	clock frequency
f_b	size of the configuration frame buffer in frames
f_h	size of a single frame
f_m	number of frames of a particular module
F_{max}	maximal clock frequency of a module
γ	memory bandwidth limitation factor
h	module height in resource slots

h^{BE}	module height in basic elements
$\|h_{max}\|$	maximal Huffman code word size
$\|h_{min}\|$	minimal Huffman code word size
H	height of a resource slot in basic elements
i	interface requirement
I	finite set of module interface requirements (subset of $\mathcal{I}$)
$I_{\cup}$	finite set of all reconfigurable module interface requirements (subset of $\mathcal{I}$)
$\mathcal{I}$	finite set of interface requirements
IO_h	configuration data per I/O cell in one frame
j	number of routing multiplexer inputs
k	number of look-up table inputs
l	length in data items
λ	number of attached resource slots
L	logic overhead in look-up tables
L_{DR}	dedicated read signal logic overhead (refinement of L)
L_{DW}	dedicated write signal logic overhead (refinement of L)
L_{SR}	shared read signal logic overhead (refinement of L)
L_{SW}	shared write signal logic overhead (refinement of L)
m	module $m \in \mathcal{M}$
m_h	size of the module configuration data per frame
M	maximal number of simultaneously integrated modules
$\mathcal{M}$	finite set of modules
n	node $n \in \mathcal{N}$
n_c	number of configuration bits
N	interleaving factor
$\mathcal{N}$	finite set of nodes
o	data offset
o_r	overhead for configuration readback operations
o_w	overhead for configuration write operations
O	module configuration overhead in look-up tables
$\overline{O}$	average of O over all modules
$\widehat{O}$	peak value of O over all modules
p	number of ports receiving a burst
$\widetilde{p}$	probability
p_m	bitstream manipulation processing speed
R	total number of resource slots
R_h^{BE}	horizontal amount of basic elements
R_H	number of horizontal resource slots
R_V	number of vertical resource slots
s	module reference (start) resource slot
σ	number of look-up tables per resource slot
S	wire span
τ	token (data item emitted by a decompressor)
$\|\tau\|$	the number of bits to code a token

t_b	full buffer-based module configuration time
t_{conf}	delay of a configuration clock cycle
t_{defrag}	time for moving a module for a defragmentation step
t_{opt}	optimal configuration speed
t_{prop}	propagation delay
t_{wc}	worst case end-to-end propagation delay
$t_{wc^{pipe}}$	worst case propagation delay using pipelining
T	throughput
$\mathcal{V}$	finite set of tasks
$\mathcal{V}^{HW}$	finite set of hardware tasks (refinement of $\mathcal{V}$)
$\mathcal{V}^{SW}$	finite set of software tasks (refinement of $\mathcal{V}$)
w	module width in resource slots
w^{BE}	module width in basic elements
$\|w_{deco}\|$	decompressor output word size
$\|w_{FPGA}\|$	FPGA configuration port word size
W	resource slot width in basic elements

full buffer-based module configuration time

delay of a configuration clock cycle

Chapter 1
Introduction

More than two decades in FPGA evolution have emerged in a multi-billion US$ market and devices that are currently providing one thousand and more multipliers, megabytes of on chip memory, hundreds of thousands of logic cells, and clock speeds of a half gigahertz. With this progress, FPGAs went far beyond the glue logic stage from their early days and are now occupying central positions in highly complex systems. But not only the capabilities of the high-end FPGAs have expanded with the progress in silicon industry, also the cost per function has enormously decreased such that a one-US$ FPGA can nowadays host a microcontroller and a few peripherals. Hence, FPGAs can be found in many present systems and we all may probably have used FPGAs without being aware if this, like, for example in our home network equipment, flat screen televisions, or cars.

In addition to their commercial success, the introduction of reconfigurable devices has established a large area of research under the name *reconfigurable computing* which, in particular, comprises the investigation of *self-adaptive hardware systems*. These systems are capable to exchange, update, or extend hardware functionality after production at runtime. Instead of providing all hardware modules in parallel, a self-adaptive system may load or swap modules with respect to present requirements or environmental conditions. Consequently, such systems may be implemented more cost and power efficient on smaller devices or faster by allowing more area for exchangeable accelerators. This is highly relevant for FPGAs where the gap to application specific integrated circuits (ASICs) is monumental. For example, in the same 90-nm CMOS process technology, a standard cell ASIC requires roughly 18 times less area and 14 times less dynamic power, while being 3–5 times faster than an FPGA counterpart [KR07]. With the help of partial runtime reconfiguration, this gap can be materially narrowed. Despite these obvious advantages, runtime reconfiguration, as an integral part of a digital system, is still rarely exploited in commercial applications. The main reasons for this are open questions on the following topics:

D. Koch, *Partial Reconfiguration on FPGAs: Architectures, Tools and Applications*,
Lecture Notes in Electrical Engineering 153, DOI 10.1007/978-1-4614-1225-0_1,

1. **Methodologies:** How can hardware modules be efficiently integrated into a system at runtime? How can this be implemented with present FPGA technology? And, how can runtime reconfigurable systems be managed at runtime?
2. **Tools:** How to implement reconfigurable systems at a high level of abstraction for increasing design productivity?
3. **Applications:** What are the applications that can considerably benefit from runtime reconfiguration?

This book aims to answer these fundamental questions. An example of the right *application* will be given by an automotive application where a reconfigurable distributed network provides fault tolerance with a minimum of structural redundancy by self-adapting and self-healing the system according to the currently available resources. This includes migrating hardware and software functionality within the network, replacing and masking defective resources, and optimizing the hardware/software binding of the tasks on the particular network nodes. However, suitable applications for exploiting partial runtime reconfiguration are manifold and not limited to fault tolerant distributed control systems, but this application will demonstrate powerful capabilities for reconfiguring the hardware and software of a system autonomously at runtime. These abilities may then be used – all or in part – in various other domains including *networking* for adapting to changing protocols, *digital signal processing* for parametrizing filters or exchanging algorithms, *cryptography* by swapping sequentially required accelerators (e.g., the result of an asymmetric key exchange will typically be used by a following symmetric encryption) or FPGA-based systems in general. In these systems, mutually exclusive tasks may share the same resources (e.g., multimedia encoding versus decoding in a multimedia device). Or, to mention another possibility, a start-up processes could be accelerated (e.g., the software stored on a slow flash memory may be decompressed by a hardware accelerator that will be replaced with another hardware module after booting).

The answers on *methodologies required to efficiently implement runtime reconfigurable systems on FPGAs* are the most important and will consequently constitute the main topic in this book. For demonstrating the effectiveness of partial reconfiguration, it is not sufficient to prove that modules can be swapped at runtime on an FPGA, it is also necessary to achieve this with a substantial benefit as compared to a static only implementation. This implies that partial modules can be integrated into a system without a relevant overhead for providing an interface to the reconfigurable modules or by leaving considerable amounts of resources unused on the device. Furthermore, it must be possible to implement a partial system without dropping the clock speed of the modules. For this reason, a novel on-chip communication architecture for FPGAs has been developed, analyzed, and implemented that is capable to efficiently integrate dozens of partial modules into an FPGA at runtime. It is demonstrated that present FPGA technology provides sufficient capabilities to build corresponding systems for real world applications. In addition, focus is put on speeding up self-adaptation because of runtime reconfiguration is in many cases only feasible if the process can be accomplished within certain time boundaries. For example, if reconfiguration is utilized to mask faults in a control system, the time to repair is typically critical, or when accelerators are swapped over time in a signal processing application, this must often fulfill real time constraints. Consequently, all methodologies will be examined with respect to their temporal behavior and additional techniques have been introduced for reducing the reconfiguration latency, including configuration prefetching, configuration port overclocking, hardware accelerated configuration data decompression, and high-speed defragmentation.

Another important question, answered throughout this Book, is *how the implementation of self-adaptive systems can be automated with the help of tools.* The capability of runtime reconfiguration implies an increased design and test complexity and adequate tools are essential for enhancing design productivity. If, for example, a system may require several months of extra design and test time to save, let's say, half the cost of only an FPGA by exploiting partial runtime reconfiguration, it can be more economic to use a static only design for bringing the system faster to market at a lower risk. For bridging the design tool gap in implementing runtime reconfigurable systems, a novel and easily usable design flow – including powerful tools – will be introduced in this book. The flow allows to compose systems based on fully implemented modules that can be integrated to an initial system by merging configuration data or netlists while providing the option to swap modules at runtime without influencing other parts of the system that are not involved in the reconfiguration process. The proposed system composition allows to integrate modules among different sized FPGAs without an additional synthesis or place and route step. The modules can still be parameterized (e.g., adjusting the addresses or interrupts) while maintaining their timing. All this results in a highly modularized design flow that enormously simplifies the system integration phase, thus, making this flow attractive also for complex and fully static only systems. In other words, this fully component-based flow can gain more efficiency to traditional flows than the expected additional effort for implementing runtime reconfiguration capabilities in systems on FPGAs.

1.1 Basic Definitions on Reconfigurable Computing

The research in the reconfigurable computing domain is quite emerging and very active, thus many terms and definitions have been introduced from multiple researchers working on the same topic. The goal of this section is to clarify these definitions.

Definition: Reconfigurable Computing. *Reconfigurable computing* is defined as the study of computer systems embracing reconfigurable devices. This includes architecture aspects, models, algorithms, and applications. Furthermore, reconfigurable computing is a computer paradigm combining the flexibility of programmable software systems with the performance of pure hardware processing.

A *configurable device* is a chip allowing adoptions after or during production of a system, and, a *reconfigurable device* is respectively a chip allowing adoptions at any time in the system, even at runtime. A good definition for a reconfigurable hardware system was introduced with the Rammig Machine (by Franz Rammig 1977): "... a system, which, with no manual or mechanical interference, permits the building, changing, processing and destruction of real (not simulated) digital hardware." This definition has to be qualified such that the adoption must include a certain level of complexity. For example, exchanging an add operation with a shifter will not be considered as reconfiguring the device, while the change from a Gaussian window operator to a median noise canceler in an image processing system will be. Besides these larger adoptions in space on a chip, a configuration is commonly executed for a longer time in terms of clock cycles or iterations of an algorithm. Strictly following this definition, *a CPU is not a reconfigurable device* as the instruction stream causes only relatively small changes and because instructions change on a per cycle base.

Configurable devices can be distinguished in many ways and a fundamental classification of reconfigurable devices is their granularity. This includes (1) *fine-grained* reconfigurable devices such as FPGAs that are based on processing elements operating at bit-level and (2) *coarse-grained* devices that provide more complex functional blocks such as ALUs. Coarse-grained devices have demonstrated material performance and efficiency benefits (in terms of power or chip area) over traditional von Neumann or Harvard architectures [KSHT09], but these devices are commonly designed for a very specific application domain, e.g., digital signal processing (DSP) [BMN+01]. On the other side, fine-grained FPGAs can virtually implement any digital system or algorithm. Consequently, FPGAs may host a complete SoC while coarse-grained architectures are often embedded into an SoC, e.g., Sony's Virtual Mobile Engine, the picoChip array, or Tilera's TILE64 processor. For increasing area and power efficiency on FPGAs, multiple hard-IP cores have been integrated into FPGA architectures, hence, resulting in *mixed-grained* devices. This includes dedicated multipliers for accelerating DSP algorithms, CPU cores, or peripherals such as PCIe interface modules or memory controllers.

In this book, focus is put on distributed embedded control systems which comprises various different application domains, including networking, control systems, and signal processing. Furthermore, the individual nodes of such a network demand a CPU, hence, an FPGA platform is most suitable to meet all these requirements by one single device. However, many of the proposed methodologies throughout this book can also be applied to coarse-grained architectures.

FPGA architectures can be classified with respect to their configuration capabilities, as illustrated in Fig. 1.1. At the highest level, FPGAs can be separated into one-time configurable devices that can only be used as an ASIC substitute and configurable FPGAs. Configurable FPGAs can in turn be distinguished in globally and partially reconfigurable devices. When globally reconfiguring an FPGA, the complete device configuration is exchanged. As a consequence, all internal states get lost and the FPGA will have to restart its operation. This is appropriate for an in-field update of the FPGA but is unlikely for self-adapting reconfigurable systems.

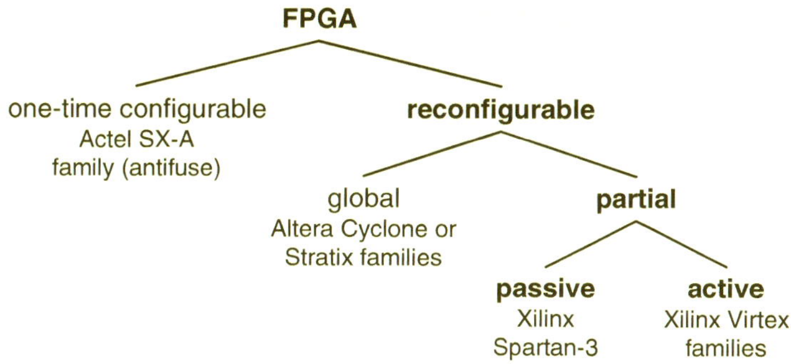

Fig. 1.1 Classification of FPGAs by their configuration capabilities. We will put focus on *passive and active partially reconfigurable FPGAs*

Consequently, focus will be put on *partially reconfigurable systems*, which allow to exchange just a fraction of the resources of an FPGA. Partial reconfiguration can be performed either *passive* by stopping the FPGA (e.g., by disabling all clocks) or *active* where the operation can seamlessly continue during the reconfiguration process.[1] When using active partial runtime reconfiguration, the system can be designed to perform the reconfiguration autonomously without additional off-chip control. This technique is called *self-reconfiguration* and constitutes a main topic in this book.

With putting focus not only the reconfigurable device but on building complete reconfigurable systems in general, the term *adaptive computing systems* is widely used. A straight and universal definition of this term is given by Teich in [Tei07]:

> **Definition: Adaptive Computing Systems.** Adaptive computing systems are computing systems that are able to adapt their behavior and structure to changing operating and environmental conditions, time-varying optimization objectives, and physical constraints like changing protocols, new standards, or dynamically changing operation conditions of technical systems.

With respect to distributed embedded control systems, this definition includes the network level as well as the node level. In other words, a distributed adaptive computing system can autonomously adapt the network, e.g., for compensating a link defect or adapt the nodes of a network, e.g., by reconfiguring a module on the FPGA.

[1] The differentiation in active and passive reconfiguration follows the documentation provided by the Xilinx Inc. that is the market leader in runtime reconfigurable FPGAs. However, there are other vendors that provide partially reconfigurable FPGAs. For example Abound Logic, Inc. sold large 750k 4-input LUT FPGAs before closing their business [Abo09]. Moreover, Altera Inc. recently announced to enable partial reconfiguration in their new Stratix-5 devices.

1.2 Reconfigurable FPGA Technology

This section gives an introduction into SRAM-based FPGA technology and how an FPGA implements the ability to change its structure and behavior at runtime. The next section introduces look-up tables as the basic building block to implement logic. Next, Sect. 1.2.2 looks into the routing. After this, we put these things together to an FPGA in Sect. 1.2.3.

1.2.1 Reconfigurable Logic on FPGAs

The majority of FPGA architectures implement Boolean logic with the help of *look-up tables* (LUTs). As shown in Fig. 1.2a, a LUT is basically a multiplexer where the select input is identical to the k LUT inputs. The multiplexer inputs are connected to 2^k configuration SRAM cells. A Boolean function is represented by its truth table that is stored in the SRAM cells, and *a LUT can implement any logic function with k inputs*. Including all permutations, there are 2^{2^k} different functions, because we have (2^k) configuration cells that can be configured with $2^{(2^k)}$ values (just assume we have a 2^k-bit counter we run once through). Figure 1.3 gives an example for the configuration of different logic functions. Note that not all inputs of a LUT have to be used (they will be tied to a constant value by the synthesis tool) and that the same function can be implemented even when inputs are swapped. The last property gives the router much more freedom in order to improve the routing (i.e., finding shorter paths). A look-up table can alternatively be configured in route through mode by configuring the truth table with the identity function of the input that shall be routed through. As shown by the last two examples in Fig. 1.3, this could be used to implement a reconfigurable multiplexer, where the multiplexer select state is controlled by the configuration bitstream.

The LUT output is connected to a state flip-flop in order to store the state of a synchronous circuit. In Fig. 1.2a, where the LUT is used as a function generator, only the state flip-flop might represent the state of a module. This implies that a

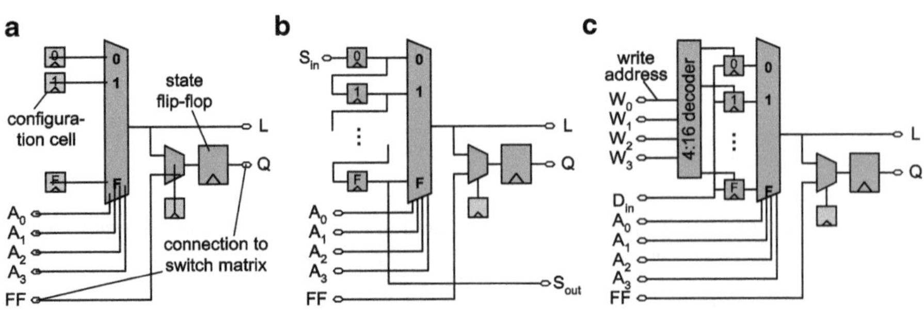

Fig. 1.2 (**a**) Implementation of a $k = 4$ input look-up table; (**b**) LUT in optional shift register mode; (**c**) LUT in optional memory mode

$A_3\,A_2\,A_1\,A_0$	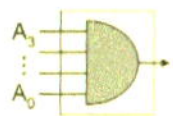	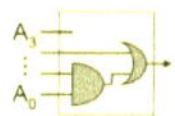	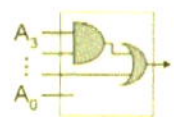	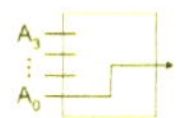	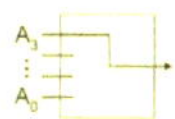
0: 0 0 0 0	0	0	0	0	0
1: 0 0 0 1	0	0	0	1	0
2: 0 0 1 0	0	0	1	0	0
3: 0 0 1 1	0	1	1	1	0
4: 0 1 0 0	0	1	0	0	0
5: 0 1 0 1	0	1	0	1	0
6: 0 1 1 0	0	1	1	0	0
7: 0 1 1 1	0	1	1	1	0
8: 1 0 0 0	0	0	0	0	1
9: 1 0 0 1	0	0	0	1	1
A: 1 0 1 0	0	0	1	0	1
B: 1 0 1 1	0	1	1	1	1
C: 1 1 0 0	0	1	1	0	1
D: 1 1 0 1	0	1	1	1	1
E: 1 1 1 0	0	1	1	0	1
F: 1 1 1 1	1	1	1	1	1

Fig. 1.3 Look-up table configuration examples

circuit uses no combinatorial latches to store a state, which is generally the case. However, most recent SRAM based FPGAs allow to configure some or all look-up tables as shift registers (Fig. 1.2b) or as tiny RAMs (Fig. 1.2c). This feature is called *distributed memory* and it improves efficiency for memory intense parts of a circuit, including register files, delay shift registers, or small FIFOs. In the case distributed memory is used, not only the state flip-flops are representing the state of a module but also the configuration SRAM cells (now part of the implemented circuit) may change during operation of a module (and not only by reconfiguration). This is important to remember, because on Xilinx FPGAs, the present value of state flip-flops cannot be directly overwritten, while for the 2^k LUT configuration SRAM cells there is no possibility to mask them from being configured. This has two consequences. (1) a module cannot be easily initialized with a predefined state due to the reconfiguration process,[2] and (2) reconfiguring the function of one LUT may destroy/corrupt the state of other LUTs that use the distributed memory option. This has to be considered during the RTL implementation of a reconfigurable module and when placing logic primitives on the FPGA fabric during the physical implementation of a module.

If we go from the logic level implementation to the transistor level, we see that the wide input look-up table multiplexer is actually implemented with pass transistors (or transmission gates), as shown in Fig. 1.4. For a k-input LUT, the multiplexer tree requires $2 \cdot 2^k$ transistors plus a level restorer to compensate the voltage drop of the pass transistors. However, the majority of the transistors is actually required for the configuration SRAM cells that need at least 5 transistors per cell or at least $5 \cdot 2^k$ transistors all together. Therefore, a basic 4-input (6-input) LUT itself is implemented with at least $(2 + 5)2^k = 112$ (448) transistors.

[2]This is only possible using a complex configuration sequence, as presented in Sect. 4.2.7.

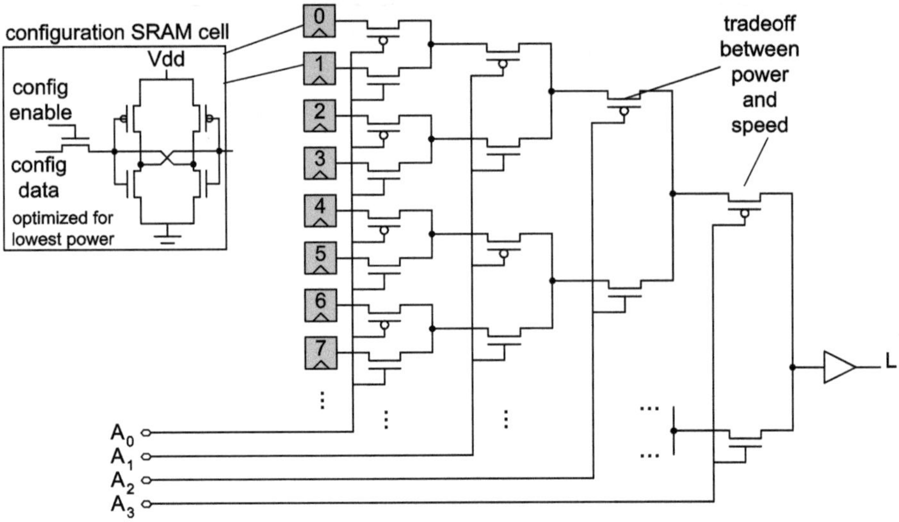

Fig. 1.4 LUT implementation on the transistor level ($k = 4$)

Writing to the configuration SRAM cells cannot be masked on Xilinx FPGAs (a feature that will be available for Stratix-5 FPGAs from Altera), but however, Xilinx guaranties that overwriting a configuration on the FPGA with its already stored configuration will cause no glitches. In other words, it is possible to manipulate parts of a logic function during operation without any interference if the involved configuration SRAM cells are not selected by the LUT inputs ($A_0 \ldots A_3$) during reconfiguration.

In real FPGAs, the look-up tables will provide some more features that vary among the different FPGA-families. This includes the already mentioned distributed memory modes, fast carry ripple chain logic between adjacent LUTs, the possibility to combine adjacent LUTs to larger LUTs with more inputs, or the possibility to split a LUT into smaller ones (e.g., a 6-input LUT into two 5-input LUTs). The latter feature allows the implementation of different smaller Boolean functions that possess an overlap in their input literals. Furthermore, also the shown state flip-flop will be more complex and many different configurations are commonly available. For example, the flip-flop might be configured to be synchronous or an asynchronous latch, or it can be configured if a reset is synchronous with a clock or not. Then, it might be possible to set an initial flip-flop state (the state after configuration) and independent to this, a reset value that is set in the state flip-flops when triggering a reset. For example, a modulo 10 (the reset value loaded after each round) down counter can start at 7 (the init value) after configuration. Note that on Xilinx FPGAs, loading the init value to the state flip-flops can only be applied globally to the whole device and as mentioned before in this section, there is no easy way to initialize the state flip-flops when using partial reconfiguration. This is in particular an issue if reconfiguration shall be used to transparently swap hardware

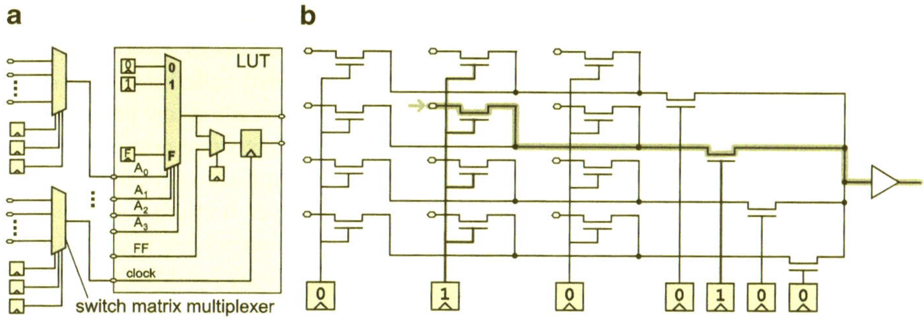

Fig. 1.5 Routing multiplexer. (**a**) Logic level; (**b**) pass transistor implementation

tasks similar to what we know from the software domain, because the state flip-flops will have to be reinitialized when resuming a module. This will be discussed in more detail in Sect. 4.2.7 on Page 216.

1.2.2 Reconfigurable Routing on FPGAs

With a reconfigurable look-up table, we can define the behavior of a digital circuit by loading truth tables of Boolean logic functions to configuration SRAM cells. The structure of a circuit is basically defined by how the different logic elements are linked together. This is carried out by switchable interconnects, that are again implemented with multiplexers, as shown in Fig. 1.5. Opposed the LUT multiplexers, the configuration SRAM cells for configuring the routing are connected to the select inputs of the routing multiplexers and the multiplexer inputs are the inputs to the interconnect network of the FPGA fabric. The routing multiplexers are typically relatively wide with 10–30 (or even more) inputs. They are again implemented with pass transistors that are arranged in different dimensions. The example in Fig. 1.5b is for two dimensions, where the left cluster of pass transistors selects an input column and the right cluster the row that is connected to the routing multiplexer output.

As we select exact one path per dimension, the corresponding control signal has to be one-hot encoded. On Xilinx FPGAs, there is no extra decode logic for this purpose[3] and the one-hot encoding is achieved by one-hot encodings for the different dimensions in the configuration bitstream. Depending on the FPGA architecture, some dimensions might be bypassed to allow faster routing of some

[3]The local interconnect in Spartan-3 FPGAs is implemented with three dimensional routing multiplexers. In that case, the circuit in Fig. 1.5b is duplicated and a third dimension selects which of the duplicated multiplexers is connected to the final output. This is controlled by a single configuration bit and the inverted output of the corresponding SRAM cell could be seen as a decoder.

paths [LAB+05]. With respect to Fig. 1.5b, this means that we could add an additional row that is directly connected to the multiplexer inputs without the first column dimension. This is useful as the slowest paths (also called *critical paths*) define the possible clock speed and in most cases, there are only a few paths within a circuit critical. It is then possible to assign faster inputs to critical nets.

We have learned that each dimension of a routing multiplexer is configured by a one-hot encoded part in the configuration bitstream. However, it might consequently by possible to activate more than one input per dimension which may result in short circuits when also applying input values of different logic level to the involved inputs. This may result in a power increase,[4] impact on the signal integrity, faster device aging (ware out), or even a damage of the device [BKT10]. Note that there is no check during configuration for short circuits and that it is easy to manipulate a bitstream such that it would even pass the CRC checksum test of the FPGA. This is a potential threat and somebody may develop an FPGA virus causing negative side effects to a system. This problem is not bound to a specific FPGA vendor and special considerations are necessary when opening a system to many configuration data suppliers.

Another important issue to remember is that the configuration of an FPGA is transferred sequentially. So there is no shadow configuration that allows to swap between modules in a single clock cycle. Different to SRAMs, FPGAs shift configuration data first into a shift register and from there the data is written to the configuration SRAM cells that are spread regularly over the whole fabric. So, rather than accessing the configuration in words of 8 or 16 bits, like in SRAMs,[5] the smallest atomic portion of configuration data is in the range of a kilobit or more. Xilinx named this portion of configuration data a *frame* and several frames have to be written to configure a LUT and the corresponding routing multiplexers. Even the configuration bits of a single routing multiplexer are typically spread over multiple frames. This has impact when partially reconfiguring the device. First of all, switching a reconfiguration multiplexer can take hundreds of configuration clock cycles to shift in all the required frames. Furthermore, as data is written sequentially, overwriting one configuration with another one (without blanking in-between) may cause short circuits. To explain this, we take another look on the example in Fig. 1.5b. Let us assume that the three configuration bits that select the input column are written in three consecutive frames. If the first configuration has selected the second column, as shown in the figure, there will be the situation of two selected columns if the new configuration will select the first column, and it takes one frame before the second column will be deselected. This short time will be not sufficient

---

[4]There is even an application for increasing power consumption due to short circuits: Several cryptographic algorithms can be attacked by analyzing the power consumption as a side channel to access secret data. By producing power noise with the help of short circuits, side channel attacks become far more difficult.

[5]For the sake of completeness, it has to be mentioned that the FPGAs of the AT40KAL family from Atmel provide an SRAM like interface.

to cause a damage to the device, but we can notice a considerable increase in the power consumption for reconfiguration [BKT10] and a system might not improve power efficiency by using a smaller FPGA when this situation occurs very frequent. Consequently, tools for building reconfigurable systems should provide bitstreams that blank a reconfigurable region before overwriting it with a new module.

The clustering has impact on the granularity in which we can reconfigure modules. While it can be useful to manipulate individual switch matrix multiplexers (e.g., for implementing a circuit switching network directly within the routing fabric), modules will typically not share the logic resources of a single CLB. This is because some resources are shared by the LUTs within a CLB and it would be difficult (even perhaps impossible) to implement separate modules sharing all common resources of a CLB.

We can take another look on Fig. 1.5b in order to count the transistors required to implement a routing multiplexer with j inputs. Assuming a two dimensional multiplexer implementation, there are at least $\lceil j/\lceil\sqrt{j}\,\rceil\rceil$ columns and $\lceil\sqrt{j}\,\rceil$ rows. This results in a minimum of $j + \lceil\sqrt{j}\,\rceil$ pass transistors plus $(\lceil j/\lceil\sqrt{j}\,\rceil\rceil + \lceil\sqrt{j}\,\rceil)\cdot 5$ transistors for the SRAM configuration cells (five transistors each):

$$\left[\, j + \lceil\sqrt{j}\,\rceil \,\right]_{\text{pass transistors}} + \left[\left(\left\lceil \frac{j}{\lceil\sqrt{j}\,\rceil} \right\rceil + \lceil\sqrt{j}\,\rceil\right)\cdot 5\,\right]_{\text{config. transistors}}$$

Note that we omit the transistors for the level shifting. A routing multiplexer implementation with $j = 20$ inputs costs consequently at least 70 transistors. And again, this number is dominated by the transistors for the SRAM configuration cells.

1.2.3 Putting Things Together: A Simple FPGA Architecture

With look-up tables and routing multiplexers, we can compose a complete FPGA fabric, as shown in Fig. 1.6. LUTs can be attached to a switch matrix to form a basic logic element, and we can arrange these elements in a regular mesh and connect them by wires. For connecting the fabric with the outside world, additional input/output primitives had been added. At this point, we are omitting a discussion about further dedicates primitives that commercial FPGAs provide. This includes memory or arithmetic blocks, clock generation primitives, or any specialized primitives such as memory controllers or embedded CPUs. These primitives will substantially improve implementation efficiency for specific applications and FPGAs can be tailored to certain application domains. For example, adding multipliers will improve in particular signal processing applications.

The configuration SRAM cells of our FPGA architecture can be accessed like in standard SRAMs or in a shift register fashion, as shown in Fig. 1.7. There are various possibilities that are basically a tradeoff between the wiring and logic cost on the die and the granularity to access the configuration bits. As mentioned for the column based Xilinx FPGAs, a courser grained granularity might result in

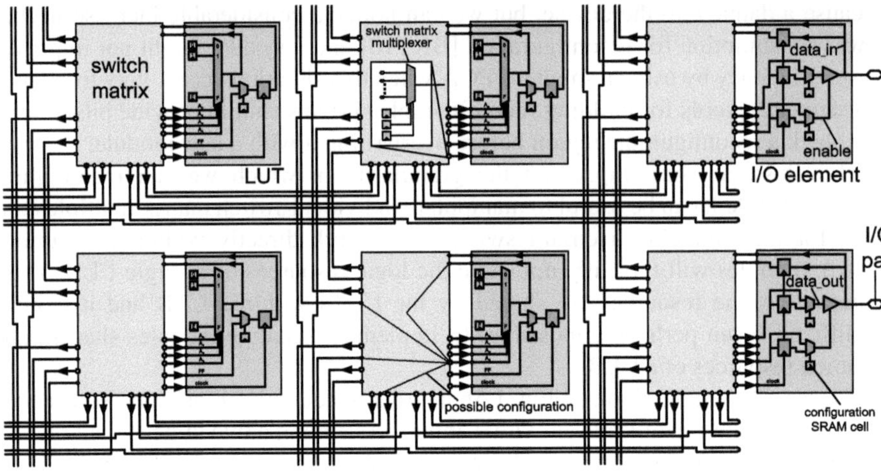

Fig. 1.6 Example of an FPGA architecture composed of look-up tables, I/O elements, and switch matrices for implementing the routing

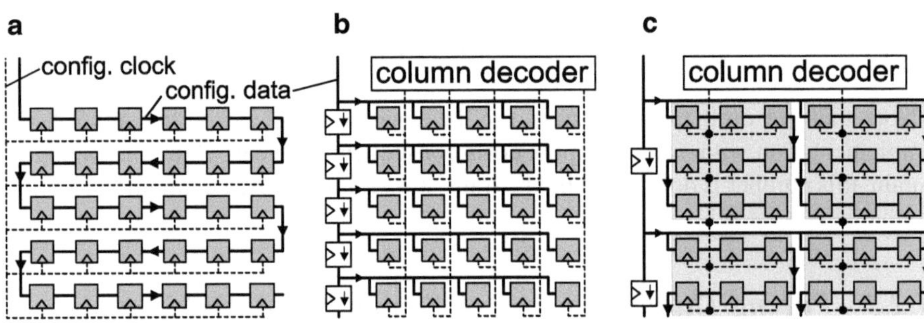

Fig. 1.7 Different possibilities to access the configuration SRAM cells. (**a**) Register chain. (**b**) Column based. (**c**) Mixed topology

side-effects when using partial reconfiguration. However, later in Sect. 3.5.6.1, we will see that the column based approach can be easily extended by a mask register to provide individual configuration bit access at relatively low implementation cost.

The configuration process itself is not trivial and the FPGA has to ensure that preinitialized states of the circuit are not corrupted during configuration. For example, when using a LUT as a shift register (see Fig. 1.2b), the FPGA has to ensure that an initial value in the shift registers is not destroyed. This might happen if the clock input routing multiplexer glitches during the configuration and when this multiplexer configuration is written before configuring the LUT table values. Or when using a LUT only as a function generator, it might still be possible that the LUT will be temporarily in a distributed memory mode during partial reconfiguration. This might require additional logic and/or a certain configuration order.[6]

It has been shown that it is beneficial to cluster multiple LUTs together to share a switch matrix [BRM99]. The idea behind clustering LUTs is to provide a dense routing among the LUTs inside a cluster (for efficiently implementing larger logic functions) and to share some resources among multiple LUTs (e.g., clock input multiplexers). Due to the enormous implementation cost, the switch matrices cannot provide all possible configurations from any input to any output. As a consequence, FPGA vendors optimize their switch matrix design and routing fabric layout to minimize the size for the routing multiplexers. In Sect. 2.6.2 we will provide more details in the switch matrix design of commercial FPGAs.

We can use our simple FPGA architecture model to estimate the transistor count of custom or even commercial FPGAs. For example, on Virtex-II FPGAs from Xilinx, eight 4-input LUTs have been clustered together to a block, called CLB (configurable logic block). In the case of Virtex-6 FPGAs, each CLB contains eight 6-input LUTs. For these two devices, the estimates for the different routing multiplexers of a CLB and the LUT implementations are listed in Table 1.1. The final transistor count will be a few hundred transistors more (for the level restorers, state flip-flop and internal slice options). However, the values represent the costly part of a configurable logic block on Xilinx FPGAs. The transistor count per LUT of this part is $\approx$1,600 for Virtex-II and $\approx$2,100[7] for Virtex-6 FPGAs. Thus, despite that Virtex-6 provide 6-input LUTs (which can roughly implement twice the amount of logic than the 4-input LUTs of a Virtex-II device), the transistor count rose only by less than 20%. In both cases, more than 60% of the transistors are used for the configuration SRAM cells. In other words, an FPGA is basically a large SRAM with embedded multiplexers that are clever arranged to implement any digital circuit.

It has to mentioned that the transistors in the SRAM cells take less space on the chip than the pass transistors. On most FPGAs, the total area spent on the configuration cells will be less than the area for the pass transistors and surrounding logic.

Table 1.1 points out the cost of the routing fabric: when only considering the CLB logic primitives, a Xilinx Virtex-II FPGA spends more than 90% of the transistors on the routing multiplexers. In case of the newer Virtex-6 family, it is still almost 80% of the transistors. This is one of the reasons why it is not good to put focus on a high utilization of the logic primitives. If we buy an FPGA, we pay by far more for

[6]This is used in Virtex-II FPGAs. When configuring this device, all LUT values are selectively written once again after a first configuration step.

[7]The implementation cost of a single look-up table (including its routing resources) on a modern FPGA is approximately the cost of a vintage Intel 4004 4-bit CPU.

Table 1.1 CLB implementation cost of Virtex-II/Virtex-6 FPGAs

Virtex-II			Transistors/mux		In total	
Routing resource	# muxes	# inputs	Pass	Conf.	Trans.	Conf. bits
LUT inputs	32	36	42	60	3,264	384
FF_in (BX/BY)	8	32	38	60	784	96
CLK	4	24	29	50	316	40
CE and SR	8	16	20	40	480	64
Double and hex	80	14	18	40	4,640	640
OMUX wirs	16	24	29	50	1,264	160
Longlines	8	8	11	30	328	48
TBUFs (I|EN)	4|4	25|17	30|22	50|45	588	76
Sum routing					11,664	1,472
LUT truth table		Trans.: 8×112	Conf. bits: 8×16		896	128
In total per CLB					12,560	1,600
Virtex-6			Transistors/mux		In total	
Routing resource	# muxes	# inputs	Pass	Conf.	Trans.	Conf. bits
LUT inputs	48	24	29	50	3,792	480
FF_in	8	28	34	55	712	88
D_in	4	22.5*	27.5	50	310	40
CLK and GFAN	4	14	18	40	232	32
WE|CR|SE	2|2|2	24|23|12	29|28|16	50|50|35	416	54
Local routing	96	20.5*	25.5	50	7,248	960
Longlines	4	12*	16	35	204	28
Sum routing					12,914	1,682
LUT truth table		Trans.: 8×448	Conf. bits: 8×64		3,584	512
In total per CLB					16,498	2,194

*Average value

the routing than for the logic.[8,9] Consequently, we should look on the utilization of the routing fabric [Deh99]. However, despite that the routing dominates the implementation cost and has substantial impact on the performance, its capabilities and structure is mostly hidden to the customers of an FPGA.

[8]In [SW07], reconfigurable nano wire interconnects are proposed. This is a technique to implement routing multiplexers directly in the metal layer of a chip with programmable interconnects between wire cross points. So far this technology is experimental but has the potential to revolutionize reconfigurable technology and semiconductor industry in general when being able to replace major parts of the expensive routing fabric.

[9]An interesting approach from the start-up company TierLogic is to remove all configuration SRAM cells for controlling the routing multiplexers from the die by shifting them on top of the metal layers. This is possible by using thin film transistors that are available in a special process developed by Toshiba. TierLogic claimed to build FPGAs 30% denser than conventional FPGA architectures by applying their trick. The higher density will not only allow for chips with more capacity, it will also improve performance, as denser packing moves logic cells closer together. And as shorter the routing distance, as faster a signal can be propagated. Despite their promising approach, TierLogic has not succeeded in the market.

There is an interesting observation to make in this section: despite that FPGAs are made of multiplexers they are weak in implementing multiplexers. For example, a 4-input multiplexer implementation using pass transistors costs a bit more than 30 transistors (see Fig. 1.4). However, when implementing such a multiplexer with 4-input LUTs, it requires two LUTs, or more than 3,000 transistors for the same task ($\approx$100 $\times$ more transistors). Anyway, this overhead is especially high for multiplexers and due to high volume production in latest process technology, this overhead does not result in a similar increased cost or power consumption in many cases.

With our knowledge on how to implement an FPGA fabric, we can think about to define our own FPGA. The room for design decisions and optimizations is virtually infinite and we can investigate different logic cells, different clustering, different specialized blocks, different possible connections within the switch matrices, or different wiring between the basic logic elements, just to mention a few parameters. There are tools that can be used to explore this design space. For example, VPR [BR97], is an open source place and route tool that can be targeted to custom FPGA architectures. With such a tool, we could examine how benchmark circuits can be implemented on various FPGA variations. This book is not further covering this topic and more information can be found in a book from Vaughn Betz [BRM99].

In this and the two preceding sections, we learned that the capability of reconfiguring an FPGA comes along with a massive area, power, and latency overhead; but runtime reconfiguration is an option to improve these factors. However, as compared to a static only design, implementing reconfigurable systems is more difficult and there exist a couple of additional restrictions that have to be followed for each specific FPGA family. This book will investigate these issues and propose solutions to automate the design of runtime reconfigurable systems.

1.3 Benefits of Partial Reconfiguration

In the early days of field programmable gate arrays, the available logic capacity was very limited and using runtime reconfiguration had been suggested to raise resource utilization or to squeeze larger circuits into the available logic. For example, the dynamic instruction set computer (DISC) [WH95] is capable to change its instruction set at runtime according to a running program. Consequently, only the currently used instructions are loaded on the FPGA which takes fewer resources than having a system providing all instructions at the same time. However, for decades implementing such systems required deep knowledge about the used FPGA architecture and has to follow an error prune difficult design flow. Furthermore, long configuration time was another crucial issue that detained the use of runtime reconfiguration (RC) in industrial applications.

With the progress in silicon process technology, logic capacity raised steadily while getting cheaper (and often more power efficient per logic cell) at the same time. The capacity growth for the FPGAs of the Virtex family from Xilinx and

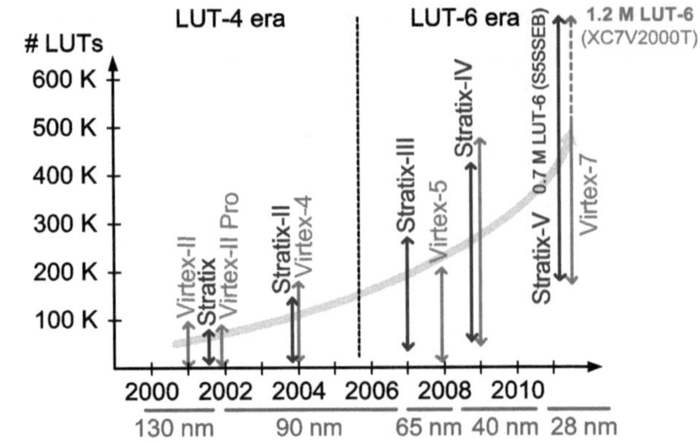

Fig. 1.8 Device capacity in terms of look-up tables of the Altera Stratix and Xilinx Virtex series FPGAs over the last decade

respectively the Stratix family from Altera throughout the last decade are listed in Fig. 1.8.[10,11] The explosion in capacity removed the pressure on the FPGA vendors to add better support for runtime reconfiguration in their tools and devices. However, by heading towards huge 1M LUT devices (million look-up table FPGAs), things are changing dramatically at the moment.

For present high capacity FPGAs, the *configuration time* required to write tens of megabytes of initial configuration data is too long for many applications. and partial reconfiguration can be used to speed up the system start as revealed further in Sect. 1.3.3.

A further consequence of having large high density FPGAs is their higher risk of failure due to *single event upsets* (SEU). However, SEUs can be detected and compensated with the help of partial runtime reconfiguration (e.g., using configuration scrubbing).

Another factor arising for current high capacity FPGAs is a strong relative increase in the *static power consumption*. The static power consumption is related directly to the device capacity. With the help of partial runtime reconfiguration, a system might be implemented on a smaller and consequently less power hungry device. This will further result in a *cheaper system*, as shown in Sect. 1.3.1.

The factors fast system start, SEU recovery, and static power consumption force FPGA vendors to enhance the support for runtime reconfiguration. As a consequence, partial reconfiguration will be available in the majority of future FPGA devices and it will be better supported by the corresponding tools.

---

[10]It is interesting that the entire smallest devices of the recent FPGA families provide roughly three times the amount of logic than the 10 year old flagship devices. Furthermore, the price of that flagship devices was about 10K US$ while the smallest Stratix-V or Virtex-7 devices will be about 100 US$ for larger volumes. In other words, regardless how huge we think today, it will be tiny tomorrow and we will get virtually infinite capacity at very low cost!

[11]The largest XC7V2000T FPGA consists of four chips mounted in one package using a technology called stacked silicon interconnect (SSI). For this device, the dies are arranged side by side with thousands of fast internal wires at the borders between the dies.

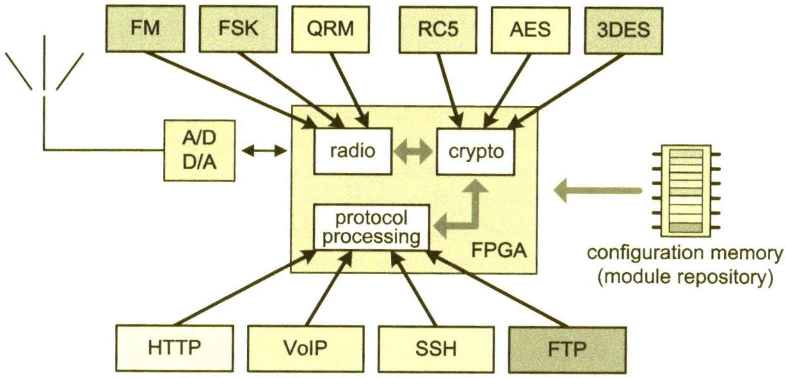

Fig. 1.9 Area saving by reconfiguring only the currently required accelerator module to the FPGA. Configurations are fetched from the module repository at runtime

Besides the mentioned functional aspects, the *design productivity* of complex multi million system gates designs can be substantially improved by closely integrating partial design methodologies into the standard FPGA design flow. While present tools support to render the logic synthesis process to a compute farm for fast parallel processing, the place and route phase is mostly implemented as a global optimization process. As a consequence, place and route can take many hours or even days to finish, hence making design respins very costly. However, throughout this book, it will be presented how the physical implementation of a design can be precisely constrained for individual modules. In other words, we can specify all the resources (including individual wires of the fabric) that are allowed for the place and route step of each module. This would permit to carry out the place and route process for multiple modules in parallel (see also Sect. 3.3). The following sections list only the main benefits and usecases for runtime reconfiguration and the potential applications are basically unbound.

1.3.1 Area and Power Reduction

Due to the increased leakage current in deep sub-micrometer CMOS process technology, static power consumption will dominate the total power consumption. By enhancing device utilization with the help of partial runtime reconfiguration, power and monetary cost could be saved. This is essential when following the trend towards more and more mobile devices. Furthermore, lower power consumption results in also lower cooling effort.

An example of a self-adaptive system featuring a mobile internet device is illustrated in Fig. 1.9. The system provides a software defined radio part (SDR), different decryption modules, and protocol processing accelerators for various protocols. Let us assume that the radio part will be adjusted according to the the

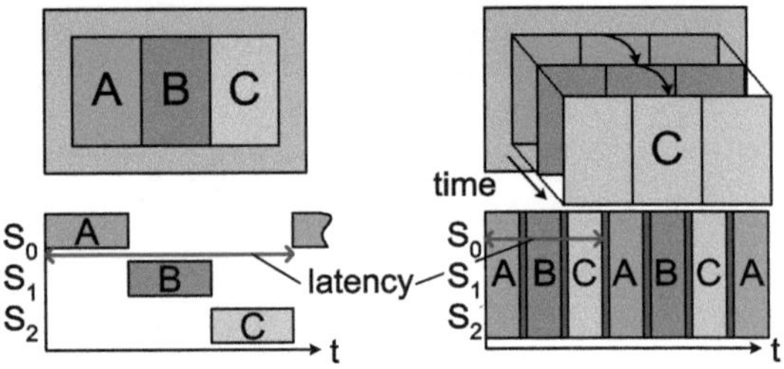

Fig. 1.10 System acceleration due to partial reconfiguration. By spending temporarily more area for each function, the overall latency is reduced

available bandwidth, and that the crypto and protocol processing accelerators are changed on-demand. We can then save substantial FPGA resources by not providing all variants in parallel, but by only loading the currently required modules to the device. This requires that the accelerator modules are either needed exclusively or that the system can time-multiplex the reconfigurable resources by sufficient fast reconfiguration. However, for low power operation, it should be mentioned that the configuration itself requires some power. This includes the power to fetch a configuration from the configuration memory and the power required by the FPGA itself for the configuration. Furthermore, we should note that the reconfigurable part will consume static power without providing useful work during the whole configuration process. If we assume that the system changes its operation modes on human interaction, the update rate will be sufficient low such that it easily amortizes the configuration power.

There are more potential benefits than only power and cost savings. If we can implement a system with a smaller FPGA, we might be able to use a smaller package which then permits to build the mobile device smaller. Or in the case of more complex systems that demand multiple FPGAs, runtime reconfiguration might be used to achieve a higher level of integration with the goal to save complete FPGAs. This might further simplify the PCB design and also help to save further power. The latter is possible because due to higher integration, we will typically have more on-chip and less off-chip communication (which saves power).

1.3.2 Performance Improvement

Beyond resource and power savings, partial reconfiguration can help to increase system performance. In most cases, this is achieved by spending more resources for specific accelerator modules. With more area, we can exploit a higher level of parallelism, which in turn permits a faster processing of the accelerator modules. If different accelerators are used sequentially (i.e., different accelerators are executed mutually exclusive in the time domain), we can exploit runtime reconfiguration in order to provide more resources per accelerator. As shown in Fig. 1.10, this results in faster processing of the individual algorithms as compared to a static only system. Even considering the configuration overhead before a new module can be started, this can result in a substantial lover overall latency. If we assume that the modules are executed in a periodic fashion, we see that the reduced latency allows a higher throughput of the system. A practical example for this method would be a hardware accelerated secure SSL connection. In this protocol, we firstly have to exchange a session key which is computed using asymmetric key exchange. After this, the session key is used by a symmetric (de-)cipher for the actual data transfer. Consequently, both steps are executed strictly sequential and we can apply the method that is shown in Fig. 1.10 for reducing latency.

Partial reconfiguration allows to trade-off computation arbitrary in time and space. However, there are algorithms that scale different with respect to these domains. For example, matrix multiplication of very large matrices scales superlinear in the space domain. This means that by spending more area, the speed-up will be typically proportionally higher than the relative area increase. For instance, if we spend twice the resources, we can typically compute a matrix multiplication in less than half the time. The reason for this is that very large matrices will not fit entirely into the on-FPGA memory. Consequently, the compute space has to be tiled into smaller portions and switching from one tile to another will introduce some overhead. By spending more area (= resources), we need less tiles and this will reduce some of that overhead. Note that a $N \times N$ matrix requires N^3 multiply-add (MAC) operations (i.e., the compute space). If we have only a single MAC unit, we have to spend $2 \times N^3$ I/O cycles to read the operands and the matrix multiplication would be I/O bound. As the other extreme, we can assume that the FPGA can host the full problem. Then only $2 \times N^2$ I/O cycles are required to read all operands and it would take $\mathcal{O}(N)$ MAC units and local memory for $2 \times N^2$ values. Note that in this case, allocating more MAC units will not help to further speed up the system. This is because we have to read all operands at least once and this implies a minimum linear time complexity. Anyway, for huge matrices, we can use runtime reconfiguration to compute larger tiles at lower latency. This can speed-up the overall system, as illustrated in Fig. 1.10. The trade-off between processing in time and space works nicely (within the throughput limits) for many compute intensive algorithms, including motion estimation, crypto accelerators, FFT, or FIR filters.

Runtime reconfiguration can also be used to sustain computation throughput for larger problem sizes. An example for this is shown for a sort accelerator in Fig. 1.11. When sorting large problems (i.e., problems that will not entirely fit into the on-FPGA memory), we have to sort in multiple runs. In the first initial step, we produce fully sorted blocks and store them in external memory attached to the FPGA. Depending on the problem size (the number of keys to be sorted), we need additional runs for merging blocks together to larger fully sorted blocks. This repeats until we have one final block. This is illustrated in Fig. 1.11b. In each run, the whole data is passed through the FPGA and the processing time is the number of runs multiplied by the time to read and write once all keys. While the latter time cannot be accelerated beyond the I/O throughput, we can still minimize the number of runs by performing more work per run.

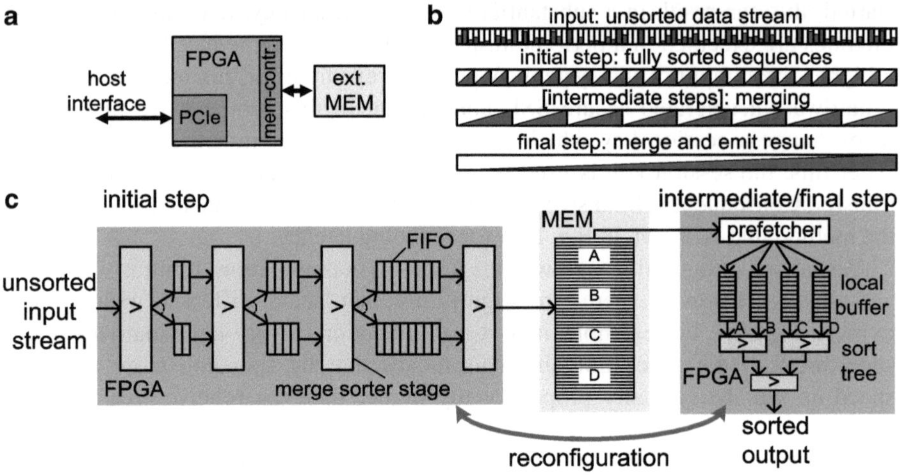

Fig. 1.11 Reconfigurable sort accelerator. (**a**) Hardware. (**b**) Multi-pass sorting; each run produces longer fully sorted blocks. (**c**) Sorter architecture and runtime reconfiguration

If we look closer on the accelerator in Fig. 1.11c, we see that the sorter for the initial run is different from the the sorter that is used in all succeeding steps in order to merge multiple blocks to a combined larger sorted block. The first sorter consists of multiple cascaded merge sorter stages whereof each stage merges two sorted input FIFOs together into a FIFO of double the size of the two input sequences. Each stage has two output FIFOs that will be filled once after the other in order to provide two FIFOs connected to the next stage. As a consequence, the sorted sequence will be doubled in each stage.[12] However, this is limited by the amount of on-FPGA memory for the FIFOs. If the problem is too large, we have to write intermediate blocks to external memory. These blocks are then merged by a sort tree in a second step. The tree sorter needs local buffers to make full use of the memory throughput (let us assume that we use external DDR memory where we have to read larger bursts in certain access patterns in order to achieve high throughput). Due to this buffering, the second sorter is also memory bound.

A sorter can only deliver results after receiving all sort keys. As both sorters for the initial step and the further merging can be designed to operate at wire speed [KT11b], the fastest sorting is achieved in one initial run and one final merge step. In this case, reading back the result can start almost directly after all values have been sent to the accelerator. If we compare a static two pass sorter without configuration with a reconfigurable counterpart, we see that we perform double the

[12]Fast software implementations of sorting algorithms (e.g., Quicksort) posses a time complexity of $\mathcal{O}(n\ log\ n)$ in order to sort n keys. When implementing a sort accelerator on an FPGA, the goal is to use $log\ n$ sort units in order to sort in linear time complexity (for fully utilizing the I/O bandwidth).

work in each step of the reconfigurable sorter, which is four times the work in total for the two assumed runs. In other words, partial runtime reconfiguration allows to sort a four times larger problem in only two runs than the static version when assuming the same FPGA. Note that the reconfiguration overhead is negligible for large problems with hundreds of megabytes of sort keys.

Partial reconfiguration can be exploited manifold and a performance improvement is not limited to the pure data processing part of a system. Let us take the example in Fig. 1.9 for another time and let us assume that we have to update or add some accelerator modules from time to time during the lifetime of the system. In this case, we only have to upload a smaller partial configuration bitfile rather than a full FPGA configuration file. Moreover, this permits the system to stay active during the whole process without a time consuming system restart after the update. Speeding-up system upgrades is in particular of interest for distributed embedded systems (e.g., automotive control units). In these applications, slow field buses are often used to transfer the update data. In some cases, the time to upgrade can be essential. For example, if a low orbit satellite has only a low speed upstream link and if the radio link can only be established during limited time intervals when passing a ground control station.

1.3.3 Fast System Start

A further use case for improving a system with the help of partial runtime reconfiguration is speeding up the start-up phase of a system. With larger system complexity in terms of FPGA size and software binaries, system start-up time becomes an important issue.

If we take the currently largest available FPGA, the Xilinx Virtex XC7V2000T, a full configuration bitstream consists of 55,917,152 bytes. Note that this device provides by far more SRAM configuration cells than the largest currently available SRAM memory devices. Configuring a XC7V2000T at full speed (400 MB/s) takes 140 ms for a complete configuration under optimal conditions. In this case, optimal means that the system can provide the configuration data at that high data rate. An overview of the reconfiguration bitstream sizes and corresponding optimal configuration times is listed in Fig. 1.12. In most systems, the initial configuration will be read from a Flash memory that provides a throughput that is typically an order of magnitude below the peak configuration speed of the FPGA. Consequently, we can expect an initial configuration time in the range of a second, which might be too long for some applications. For example, the PCIe host interface standard requires a device connected to the bus to answer within 100 ms after reset. This requires either a costly multi chip solution, changes in the software, or bootstrapping. In the latter option, only the time critical parts of the system (e.g., a PCIe interface core) will be configured at system start while leaving the configuration of the rest of the system to a second partial configuration step.

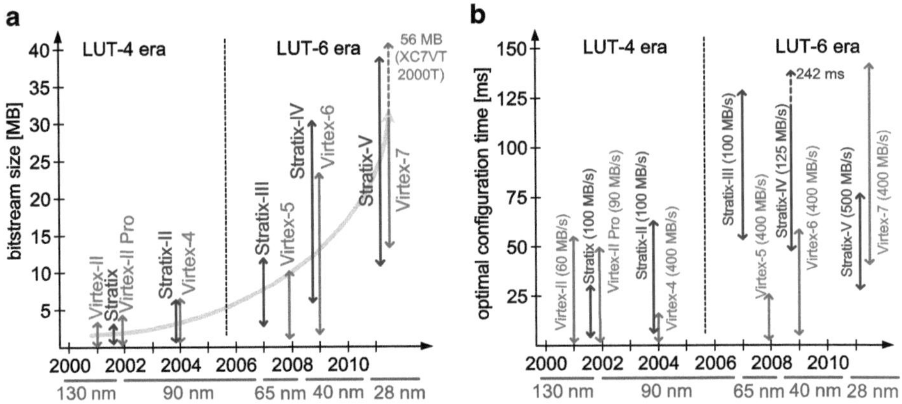

Fig. 1.12 (**a**) Configuration data size and (**b**) fastest specified configuration time of Altera Stratix and Xilinx Virtex series FPGAs over the last decade

Bootstrapping can be extended beyond an incremental system start. For example, we can include a hardware decompression accelerator in a very first configuration step. This accelerator can then be used to speed up the next time critical configuration step (by reading less data from the Flash memory due to bitstream compression). Next, we can continue configuring the remaining part of the hardware. After this, we might use the decompression accelerator to decompress the software binaries and memory file system. Finally, the decompression accelerator might be replaced with some other user logic, if the system is up running. More information on decompression accelerators is provided in Sect. 3.5.3. Fast system start-up and compression is an interesting option for power saving by fully powering down an FPGA during idle phases at runtime.

1.4 Basic Model of Partial Reconfiguration

The basic idea of partial reconfiguration is to share a part of the FPGA (i.e., a reconfigurable region) in order to share this part among multiple modules at runtime. The model behind this is basically a multiplexing between the different modules. This is illustrated in Fig. 1.13 for a mobile phone. As different functions are used exclusively to each other and never at the same time, it is applicable to share the same resources by partially reconfiguring the module to the device that is currently needed. By this, the FPGA will basically appear bigger than it physically is. The multiplexer is controlled by the configuration data that is written to the FPGA. Note that an FPGA might write the configuration on its own to one of its configuration ports which is called *self-reconfiguration*. During reconfiguration, the output of the multiplexer will be undefined. When finishing a module reconfiguration, the

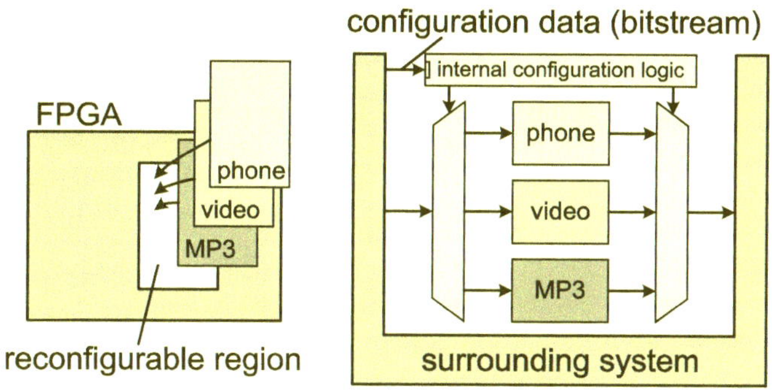

Fig. 1.13 Baseline model of reconfiguration, a multiplexer selects between different mutually exclusive modules based on the configuration data written to the FPGA

corresponding input and output signals will be connected to the rest of the system. We will call this *single island reconfiguration style* as one module can be loaded exclusively to a reconfigurable island of the FPGA. This configuration style is currently the only one that is supported by the tools from the FPGA vendors Xilnix and Altera.

The mobile phone example arises a couple of questions that are specific to the implementation of reconfigurable systems and that introduce additional complexity. For example, how do we guarantee that each of the modules gets the required resources? Note that different modules have different resource requirements (i.e., the number of LUTs, block RAMs, multipliers, etc.). In addition, the interfaces may be different as well. For instance, while the MP3 decode function might have a data input and audio output, a video decode module might require an additional video output. Furthermore, we have to think about internal module states. So, how can we continue to decode our favorite music at the same position when taking a phone call in-between? Has this to be cycle accurate or can we simply continue with the next MP3 frame? And who controls all this? Can this be done transparently by an operating system? A further aspect is the configuration time. So, can we change from the music decoder to the phone module within an acceptable latency. What is the impact on the power consumption of the system (reconfiguration is not for free)? Does the reconfiguration influence other parts of the system? This might be the case if configuration data (i.e., the module repository) is located in shared memories and is transferred over shared buses. In this scenario, the configuration process can cause a burst and result in delays for other parts of the system. And last but not least, how can we specify, test, and implement a reconfigurable system. In other words, the multiplexer model describes only the basic idea and abstracts away from several system properties and additional practical issues.

1.5 Partial Reconfiguration in Time

We can classify partial reconfiguration in the *time domain*, which means that we can distinguish in a system how fast or how frequent reconfiguration is applied. This is heavily related to the FPGA architecture that has to support the entire reconfiguration mode. The following paragraphs reveal this in more detail and we can classify if reconfiguration takes place rapidly within an operation clock cycle (*sub-cycle reconfiguration*), frequent as *single-cycle reconfiguration*, or seldom as *multi-cycle reconfiguration*.

1.5.1 Sub-Cycle Reconfiguration

This section is devoted to time-multiplexed FPGAs. The idea of time-multiplexing FPGA resources is to reuse costly look-up table logic and routing multiplexers by reconfiguring the fabric multiple times within a single clock cycle of the user circuit that is running on the FPGA. These devices are also called 3D-FPGAs because the two-dimensional fabric resources are re-used over time as a third dimension. However, the third dimensions is basically a logical view that is emulated by a two dimensional chip. Note that the overall goal of time-multiplexing FPGAs is to provide more logic capacity than a conventional FPGA on the same die size. This means that from a users point of view, the FPGA is not performing any reconfiguration and the whole device simply appears bigger at the cost of some performance penalty. Anyway, time multiplexed FPGAs might still be partially reconfigured as conventional FPGAs, but this is then not sub-cycle reconfiguration. However, the partial reconfiguration might take much longer time, but this would typically run transparently in the background and activating a new configuration might be performed in a sub-cycle.

Figure 1.14a gives an example of a 16-bit adder that is implemented on a conventional FPGA. In Fig. 1.14b, the same adder is time sliced into four chunks, called *microconfigurations*, that are executed in four consecutive *microcycles*. As shown in the timing diagram (Fig. 1.14c), configuration takes place in the beginning of each microcycle. In this example, the adder chunks seams to be identical; but however, this is not required on a time-multiplexed FPGA and all microconfigurations may differ.

In order to propagate signals from one microconfiguration to the next, time forwarding latches are required. These latches hold signal states stable during the reconfiguration of the fabric. If routing can be time-multiplexed over multiple microcycles, such forwarding latches will be required for each routing multiplexer. In the example in Fig. 1.14b, the carry ripple signal has to be propagated from one microconfiguration to the next with the help of time forwarding latches.

As compared to a conventional FPGA, a time-multiplexed FPGA needs an individual set of state flip-flops for each microconfiguration. If we think of our

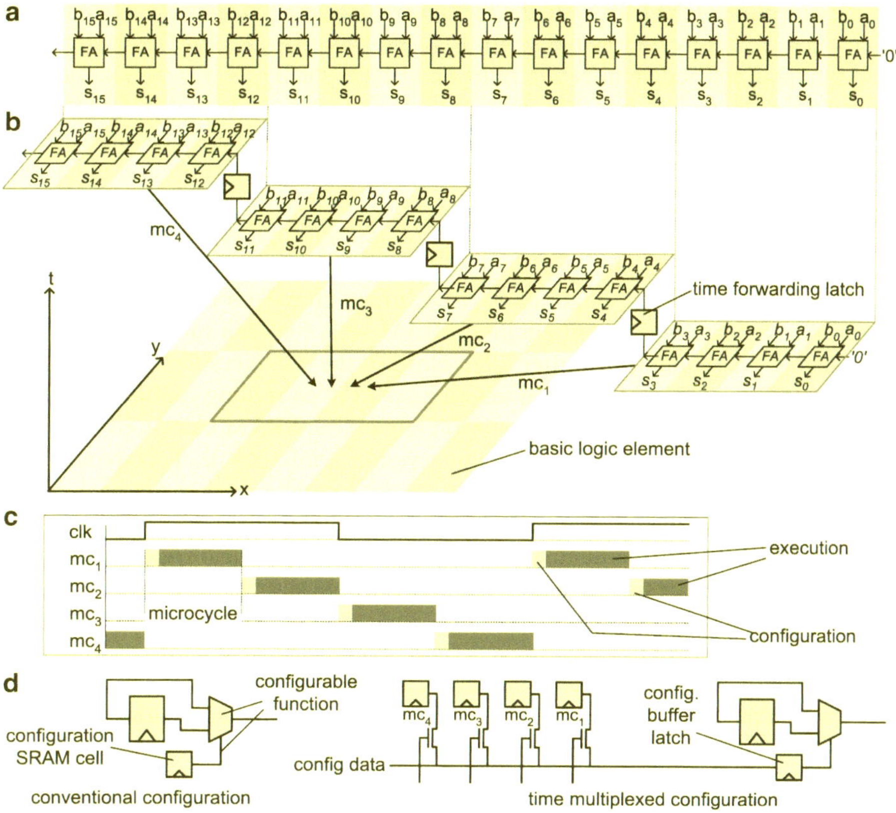

Fig. 1.14 Time multiplexed FPGA; (**a**) 16-bit adder on a conventional FPGA; (**b**) time multiplexed implementation using four microreconfigurations ($mc_1, \ldots, mc_4$); (**c**) timing; (**d**) configuration storage

16-bit adder example, it doesn't matter if we execute the circuit conventional or time-multiplexed; in both cases, we have to store the 16 result bits. Moreover, these bits might be accessed in arbitrary order over time, which results in additional multiplexers. This means that any state flip-flop from a preceding user clock cycle might be needed in any microcycle. However, due to the partitioning into microcycles, each microcycle netlist is smaller and it is unlikely for practical circuits that we really need all state flip-flops in each individual microcycle. This can be used to reduce the multiplexers to access the state-flip-flops.

One issue related to time-multiplexed FPGAs is the more difficult design and verification of these devices. During the technology mapping phase, logic has not only to be fitted into look-up tables, the look-up tables have also to be partitioned into the different microconfigurations. This has to consider that signals can only be propagated into the future and that there might be data dependencies between the different configurations. For example, when time-multiplexing an adder, we have to fit the execution of the LSBs into the first microconfiguration, then the next bits into a consecutive configuration and so forth, as shown in Fig. 1.14b. The design tools and time-multiplexed architecture must be of high quality as it is very difficult to debug errors that result from the time-multiplexing itself. Note that also EDA tools might have bugs. Furthermore, when monitoring internal signals of an implemented circuit through I/O pins, the routing of these signals will also be time multiplexed. As a consequence, transient events such as glitches might get filtered by the time forwarding latches (that are also required in the routing fabric) and these events cannot be seen at an output pin.

The different microconfigurations have to be stored on-chip. Consequently, there is a separate set of configuration SRAM cells for each microconfiguration. By placing a configuration buffer latch close to the configurable function that is controlled by the microconfigurations (e.g., a LUT or routing multiplexer), very fast configuration is possible from slow configuration SRAM cells. Consequently, more time can be spent on the execution within a microcycle. Furthermore, the buffer latch allows to change selectively a microconfiguration in the background without interfering with the execution of the currently executed microconfiguration. As stated in Sect. 1.2, the number of SRAM configuration bits is huge and rapidly reconfiguring will result in additional dynamic power consumption. This was evaluated by Xilinx in [TCJW97] for a time-multiplexed version of a XC4000 FPGA in 0.5 μm CMOS technology. A time-multiplexed FPGA with only 20 × 20 CLBs and eight possible microconfigurations took tens of watts when running at 40 MHz. In the simulation setup, 100,000 configuration bit lines are involved on the reconfiguration step within each microcycle. Due to this high power consumption, the device went never into production.

1.5.1.1 Tabulas' Spacetime FPGAs

With present deep sub-micrometre process technology, things have changed and static power consumption has started to dominate. So, if logic density is increased with the help of time-multiplexing, the relative amount of static power per logic cell decreases which in turn can amortize the dynamic power for fast reconfiguration. This is one of the technical motivations for the FPGA start-up Tabula that developed a time-multiplexed FPGA architecture. The currently available ABAX 3PLD family has a microcycle clock rate of up to 1.6 GHz (i.e., 625 ps per microconfiguration) and configuration memory for up to 8 microconfigurations (called "folds"). It is currently fabricated in 40 nm process technology. This results in a remarkable user clock speed of 200 MHz. Tabula claims faster user clocks, when using fewer folds. Therefore, users can tradeoff between performance and logic density on the same device. The largest device provides 630,000 time folded 4-input LUTs and 5.5 MB in dedicated memory blocks. The memories can be accessed in each microcycle which virtually provides up to eight rather than one port to the memory. However, if all the virtual ports can be used depends on the application. For example, if we have a long combinatory path, we might be bound to read all operands in the first microconfigurations. Or, if we assume that the operands in the adder example in Fig. 1.14b came from a memory block, we are reading a subword of the operands in each microcycle and there would be no microcycles left for additional accesses. This can be improved by not fitting single operations in multiple microconfigurations but by partitions where a single operation is executed in one microconfiguration and where different operations are time-multiplexed atomically over different microconfigurations. Note that parallel memory reads can be serialized with the help if the forwarding latches (called "time via" by Tabula), which however, takes more routing resources.

In Fig. 1.14d, it is shown that fast reconfiguration is possible even if latency to the configuration SRAM cells is slow. This is done with the help of pipelining the configuration by a buffer latch. This has been used when floorplanning the ABAX FPGAs and the SRAM cells are not spread fine grained across the fabric together with the look-up tables or routing multiplexers. Instead of this, configuration memory and the reconfigurable resources have been separated in densely packed clusters [Hal07]. This – together with the time-multiplexing – results in shorter routing distances inside a logic cluster, which in turn improves performance.

1.5.2 Single-Cycle Reconfiguration

As we learned from Sect. 1.2, FPGAs consist of an enormous amount of configuration memory cells. Consequently, reconfiguration takes a considerable amount of time to overwrite this memory. The configuration time might be too long in many applications. For example, if a video compression accelerator module should be implemented to use reconfiguration on a per-frame base, lets say to change between the motion estimation and the rest of a video encoder, there are only a few milliseconds available for reconfiguration and the reconfiguration would cause a significant idle time of the FPGA.

In order to provide fast reconfiguration, *multi-context* FPGAs have been proposed by Scalera and Trimberger [TCJW97, SV98]. These devices allow for swapping between two or more configurations, called *configuration planes* within a single cycle. Opposed to the sub-cycle reconfiguration for time-multiplexing an FPGA within a netlist, we assume here that the fabric is configured at maximum once in a user clock cycle. Trimbergers time-multiplexed FPGA architecture [TCJW97] supports sub-cycle and single cycle reconfiguration. As shown in Fig. 1.15, single clock cycle reconfiguration is basically done by duplicating all logic (e.g., the LUTs) as well as all routing multiplexers. Additional multiplexers are then used to switch between the duplicated planes. While it is possible to reconfigure a plane in the background (concurrently to the operation of the active other plane), swapping the active plane can be done in a single clock cycle. As an alternative, only the configuration bits can be duplicated. This removes the multiplexer for the plane selection from the logic path which improves timing. However this requires a

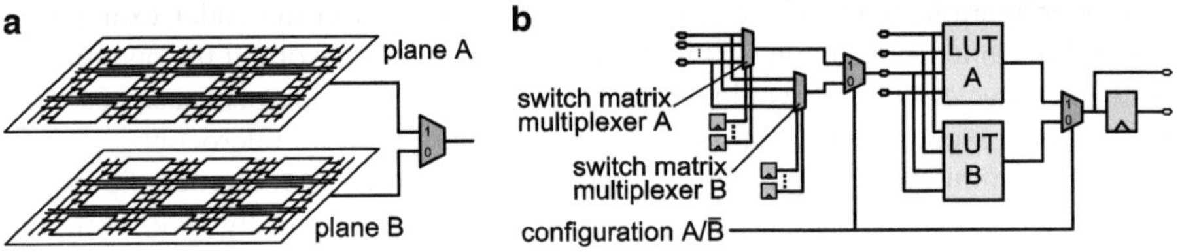

Fig. 1.15 (**a**) Multi context FPGA for switching between two configuration *planes A* and *B* in a single cycle; (**b**) implementation of a multicontext FPGA

separate multiplexer (or some other logic, like a shift register) for each configuration memory cell in order to activate a specific plane, which can be more expensive than duplicating a whole look-up table or routing multiplexer.

We can estimate, that regardless of which implementation alternative for a multicontext FPGA is used, the number of transistors will roughly double. This would not only make this approach very costly, it will also degrade performance. This is because the transistors for the different planes will have to share the same two-dimensional chip. And, as more area (= transistors) a single basic logic element occupies, as longer the routing distance from logic element to logic element, and as longer the latency. Moreover, a longer routing distance takes typically more power. In other words, despite that the model of a multi-context FPGA is a three dimensional stacking of logic planes (Fig. 1.15a), the implementation has to be done on a two-dimensional die.

The extra cost for the different planes will have to be spend regardless if reconfiguration can be actually used or not. For example, if the FPGA implements a memory controller or a control CPU that has to be active at any time, only one plane could be used for this or, as an alternative, all planes will have to contain these static parts of the system at the same position. We can summarize that removing the idle time of one plane during reconfiguration is costly bought by idling additional background planes all the time. For all these reasons, single cycle multi-context FPGAs haven't been successful in the market. However, multiple contexts is still an interesting option for coarse-grained architectures that provide complex functional units (e.g., ALUs) and wide configurable data paths (e.g., 32 bits wide) between the function blocks. For these devices, the relative amount of chip area spent for the configuration SRAM cells is far lower than in the case of fine-grained FPGAs. As a consequence, duplicating the configuration will have much lower impact on the transistor count.

Rather than having multiple planes, we could simply switch between active areas on a conventional single plane FPGA. For example, we can implement a function A in the left half of an FPGA and another function B in the right half of the device and use a multiplexer to switch between A and B in a single cycle. Note that we could still use partial reconfiguration to overwrite a function that is currently not selected. This approach provides all the advantages of a single clock multi context FPGA without a waste of resources for static parts of the system.

1.5.3 Multi-Cycle Reconfiguration

The standard option for partial reconfiguration will take many clock cycles to write in a new configuration. During the reconfiguration process, the involved modules and FPGA resources are not providing any service and the reconfiguration results in pure overhead. Consequently, updating smaller parts of the device less frequent reduces the overhead, but might not exploit the full potential of reconfiguration.

Regardless of how the configuration SRAM cells are actually accessed, the bottleneck for the configuration speed is typically the configuration port. However, while the CMOS processes has enormously improved throughout the last FPGA generations, the specified speed for the configuration has been kept pretty much the same for the last decade. For example, a Xilinx Virtex-II Pro FPGA, fabricated in a 130 nm process, has a configuration speed of 100 MHz or 400 MB/s, when using the wide 32-bit configuration port. This is the same maximal configuration speed than a modern 65 nm Virtex-5 or a 40 nm Virtex-6 FPGA is specified for. One of the reasons, why FPGA vendors haven't released much faster reconfiguration is that a higher guarantied property might result in a lower yield. This is because some devices might only fail to deliver a higher reconfiguration speed. In addition, there is currently not much customer demand, as most systems are configured from a relatively slow non-volatile memory (e.g., flash memory).

As mentioned in Sect. 1.2, the FPGAs from Xilinx provide an internal configuration port and configuration data is shifted internally into a long shift register. From there, the configuration is written in portions (called frames on Xilinx FPGAs) to the fabric. We can assume that the shift register can be filled faster than the specified 100 MHz when considering a 65 or 40 nm process. This has been explored by Claus [CMZS07] and Hanson [HKT11], and the latter work achieved a sustained configuration throughput of 2.2 GB/s. This work points out the headroom for fast reconfiguration; and partial configuration can be achieved in the range of micro or milliseconds. Let us consider the example of the configuration of a vintage 6502 processor (that was used in the Commodore 64 computer) on a modern FPGA as a softcore CPU. Then the reconfiguration could be done in 10 μs considering the configuration throughput of 2.2 GB/s. As a reference, this is the time the original Commodore 64 takes to perform two ADD assembly instructions. Or, as a more recent example, a fully featured Microblaze (a 32 CPU optimized for Xilinx FPGAs) can be loaded in less than 100 μs. However, this implies a system that can deliver this high throughput (see also Sect. 3.5.3 that is about this issue).

1.6 Partial Reconfiguration in Space and Time

In the last section, we looked into how frequent reconfiguration can be applied. In this section, we will in particular investigate and classify *how much* and in which spatial granularity partial reconfiguration can be applied. Partially reconfigurable system are partitioned into two parts. (1) The part of the system that is always present is commonly called *static region*. This part might include a memory controller, a soft CPU, or the configuration port interface logic.[13] (2) The part containing runtime reconfigurable modules is called *partial region*.

The amount of logic that is changed during reconfiguration could be as small as one or a few LUTs or routing multiplexers, as presented in Sect. 1.6.1. Of course, partial reconfiguration allows for changing much larger modules that will be loaded into the partial region. This is commonly called *module-based reconfiguration* [Inc02]. Here, we can distinguish if the partial region can only host one module exclusively (Sect. 1.6.2) or if there can be various modules be integrated more flexible at the same time. This could be done by tiling a reconfigurable region in one or two dimensions (Sects. 1.6.4 and 1.6.5).

1.6.1 Small Netlist Manipulations

In some cases, even small changes of the reconfiguration within a module (rather than exchanging a full module) can be quite beneficial. The main advantage of performing small changes is that this often requires only little configuration time. The use cases for small changes are manifold. For example, we could use this to change the clock input routing multiplexers in order to change to another global clock tree with another clock frequency.[14] Another example is the implementation of a circuit switching network directly within the switch matrices of the FPGA without using additional logic resources. In this case, only some routing multiplexers will be selectively reconfigured while keeping all surrounding modules active (see also Sect. 2.7.5.3 on Page 142).

There are also cases where it is useful to only change some look-up table functions while not touching the routing. An advanced concept using this is known as *tunable look-up table functions* (TLUT). As shown in the example in Fig. 1.16, tunable LUTs are used to remove the multiplexer and one branch of the datapath. In that example with 4-input LUTs, reconfiguration removed three of the original five LUTs with the help of reconfiguration. The approach is not bound to simple logic gates and just two different configurations. Tunable LUTs have been demonstrated

[13] All recent FPGAs from the vendor Xilinx provide an internal configuration access port (ICAP) for self-reconfiguration and no external pins are required to access the FPGA configuration. The same is announced for the new Stratix-V Series from Altera.

[14] Changing the clock input multiplexers can be done without glitches on Xilinx FPGAs. Taking the example in Fig. 1.5b on Page 9, it is possible to switch from one input within the same row to another one by temporarily enabling both inputs for the transistion. For instance, for a glitch free reconfiguration from the marked port in the middle column to the port right beside this port, we only have to change the column select configuration. Then, starting from the configuration "`0 1 0`", we firstly write "`0 1 1`" and then "`0 0 1`". If both inputs carry the same clock signal during the configuration process, there will be no unintended activation of a connected flip-flop.

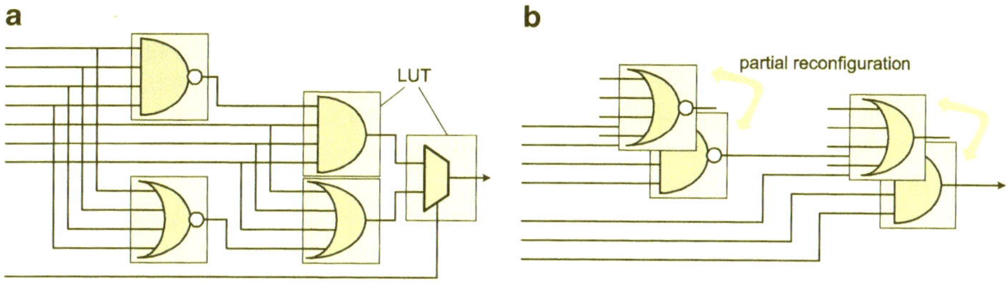

Fig. 1.16 Tunable LUT. (**a**) Standard look-up table implementation of a switchable wide input gate; (**b**) logic switching implemented by selectively reconfiguring the LUT table values (LUT tuning)

for constant coefficient multipliers and for key changing in crypto modules. The mapping from a standard LUT implementation to a TLUT implementation can be automated, as presented in [BMMAAS09]. However, the TLUT approach requires a strong similarity in the different data paths that are mapped together into the same LUTs. Otherwise, the routing will be significant slower and/or very congested and changing also the routing will in most cases result in a lower latency of the datapath. Furthermore, changing from one branch of the data path to another requires still a reconfiguration process that is thousands of cycles slower than switching the multiplexer in a conventional implementation. On recent FPGAs from Xilinx and Altera that provide 6-input look-up tables, roughly one fifth of the configuration data of the logic part is used by the LUT values. Consequently, the TLUT approach can potentially reconfigure up to five times faster by not writing the routing configuration[15] as compared to a full module swapping.

1.6.2 Island Style Reconfiguration

As shown in Fig. 1.17, the partial region can be arranged in different configuration styles. In the easiest case (Fig. 1.17a), the partial region follows the *island style* approach that is capable of hosting *one reconfigurable module exclusively per island*. A system might provide multiple islands and if a set of modules can only run in one specific island, this is called *single island style*. If modules can be *relocated* to different islands this is called *multi island style* .

While the island style can be ideal for systems where only a few modules are swapped it typically suffers *internal fragmentation* in most applications. This is a waste of logic resources that arises if modules with different resource requirements

[15]This implies that the configuration data for updating the LUTs are separated from the routing within the configuration data and that both parts can be accessed individually. This is the case on all recent Xilinx FPGAs.

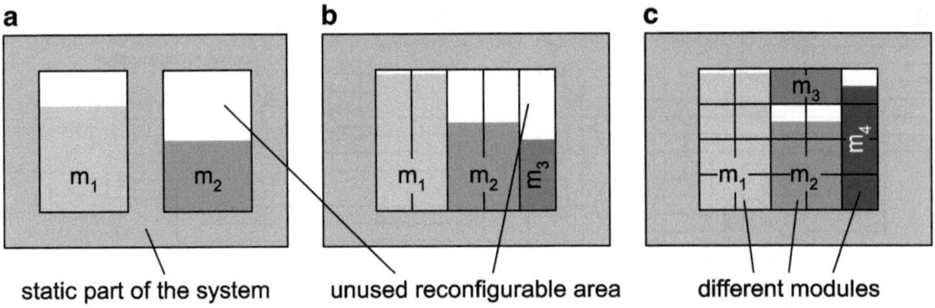

Fig. 1.17 Different placement styles of reconfigurable modules. In all examples, the total reconfigurable area and the individual module sizes are the same. With smaller tiles, internal fragmentation can be minimized and the number of simultaneously placed modules be maximized. (**a**) Island style. (**b**) Slot style. (**c**) Grid style

share the same island exclusively. This means, if a large module cannot be replaced by multiple smaller ones (to be hosted at the same time), the utilization of the reconfigurable region is getting weak. However, hosting only one module per island makes it simple to determine where to place a module.

In Sect. 1.4, we mentioned that we have to think about how to carry out the communication between the static part and the partial part of the system. In general, we have do define which routing resources of the FPGA will be used to cross the static to partial region border. However, for systems supporting multi island style reconfiguration, the used routing resources have to be constrained such that each island provides a compatible interface for the relocatable modules. In other words, the constrained routing resources act basically like a physical plug and the plug position within an island and the plug pinout has to be identical for different islands. Moreover, when using multi island style reconfiguration, the *resource footprint* that is the layout of the primitives (e.g., look-up tables or memory blocks), has to be identical for each reconfigurable island that should be able to host a relocatable module. More details on the footprint are listed in the next section.

In addition, we have to consider that the static system might require to cross a reconfigurable island on order to carry out the routing. In this case, the configuration data of a module has to include information about the static routing. This routing has than be part of the reconfigurable module and the corresponding partial configuration bitstream. This is difficult as the static routing that crosses the reconfigurable regions is in general different in each island. The vendor tools from Xilinx and Altera are currently not supporting multi island style with module relocation. Even if multiple identical instances of a module will be used in the system, it is required to implement an individual module for each possible island even if the resource footprints of the islands are identical. This also implies that the system has to provide an individual bitstream for each used placement position of a module instance. For example, a system with four reconfigurable islands and five different modules that shall run in each of the islands requires 20 individual routing steps and 20 partial bitstreams have to be provided at runtime. Moreover, as changes in the static design result in general in a different routing also for the paths that cross the reconfigurable islands, all 20 modules have to be rerouted on each change of the static system. Therefore, the vendor tools are bound to rather simple reconfigurable systems (e.g., for incremental system start via bootstrapping). At different places throughout this book, it will be shown how the static system and modules can be constrained to permit module relocation. This includes the possibility to cross a partial region with he routing of static signals as well as module relocation even among completely different systems (as long as the FPGA family is used).

1.6.3 FPGA Footprint and Module Footprint

As the footprint of the FPGA fabric and the reconfigurable modules has substantial impact on the reconfiguration style, a dedicated section is used to classify the term *footprint* in more detail.

1.6.3.1 Resource Footprint

As already mentioned, FPGAs are relatively complex and the fabric consists of various different resources. This includes logic resources, memory blocks, and dedicated arithmetic blocks that are commonly arranged in columns spanning the full height of the FPGA. On the more recent Virtex-series FPGAs from Xilinx, there are in addition columns of I/O cells and clock generation blocks fitted into the fabric. Furthermore, an FPGA might provide many further dedicated function blocks targeting domain specific applications. Examples for such blocks are high-speed transceiver blocks, PCIe interface cores, dedicated memory cores, or embedded CPUs, to mention only the most famous options. The situation is basically the same for all FPGA vendors. The different resources result in a device specific *FPGA resource footprint*. However each device within an FPGA family typically has its own individual resource footprint. Moreover, the footprint of each device might have an irregular pattern. For example, on Xilinx Virtex-II pro FPGAs, a BRAM column was located after six logic resource columns (CLBs) on all devices. This resulted in a very regular pattern of the resource footprint among the whole FGPA family. For recent devices, the resource footprint layout is more heterogeneous. However, for most FPGAs, there typically exist partial footprints that are identical at different positions within a single device or even among different devices. For example, due to the column-wise layout, the footprint is regular in vertical direction and it commonly differs in vertical direction of the FPGA fabric.

All these different resources available on an FPGA might be used by a reconfigurable module. This consequently results in a corresponding *module resource footprint*. If a module has to operate at different positions of the FPGA or even in completely different systems, the module resource footprint has to match the

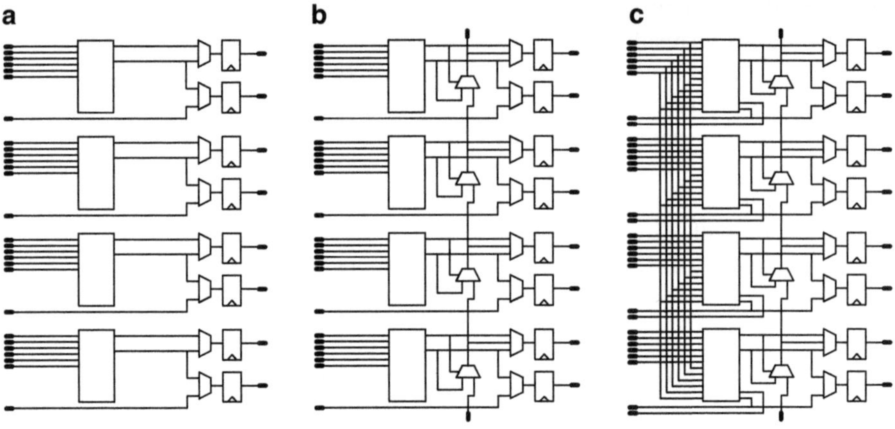

Fig. 1.18 Simplified view on the different logic resource types (slices) provided on Spartan-6 FPGAs. (**a**) SLICEX. (**b**) SLICEL. (**c**) SLICEM

FPGA resource footprint. This holds for all resources within the bounding box of a reconfigurable module (which corresponds to the reconfigurable island, in case of island style reconfiguration).

For a sake of completeness, it should be mentioned that, depending on the FPGA, there are possibilities to wildcard some primitives by simply not using resources that result in different footprints for different slots [BLC07]. This is possible, because the routing of the switch matrices and the wires between them are homogeneous throughout the whole FPGA fabric, independent of the resource type.

Furthermore, on some Xilinx FPGAs (e.g., Virtex-6 or Spartan-6 FPGAs), the logic primitives differ and there are simple LUT primitives, primitives with carry chain logic, and primitives with carry chain logic and distributed memory, as sketched in Fig. 1.18. As a more complex primitive can work also as a simpler one, module relocation is possible, even of the logic resource footprint varies among different reconfigurable islands. For example, if a module uses only SLICEX logic primitives (logic only), it could also work at positions that provide SLICEL (with carry chain logic) or SLICEM (distributed memory) slice primitives. Note that it is in this case possible to perform module relocation at the netlist level without rerunning place and route. However, this might not necessary allow to relocatable bitstreams (see also the section on the bitstream footprint).

1.6.3.2 Wire Footprint

Not only the logic resource primitive footprint has to compatible, the same holds for the footprint of the routing resources. For example, on many Altera FPGA families, there exist larger clusters of resources with denser routing inside a cluster; and despite the logic resource footprint looks homogeneous, the routing architecture varies between the center of a cluster and the cluster border. Examples of a wire footprint mismatches can also be found on Xilinx FPGAs. For instance, on the smallest Spartan-3 FPGAs, there are no hexline wires available that exist on all larger FPGAs within the Spartan-3 family. This has to be considered when migrating modules among different FPGAs (e.g., for a component-based design methodology).

1.6.3.3 Timing Footprint

There is also a timing footprint of the FPGA fabric that has to be considered when relocating modules. For example, the configuration logic will be located somewhere and result in longer routing delays at some points of the fabric (e.g., the center CLB columns on Xilinx FPGAs). Or the clock signals on Xilinx FPGAs are routed in multiple horizontal paths from where they branch into vertical splines up and down. If we now relocate a module from an upwards clock spline to a downwards spline, this impacts the timing. This is because a vertical routed signal within the module will once propagate with the clock wave and one against it.

In Chap. 2 special on-FPGA communication architectures for a more flexible module placement will be introduced. Such communication architectures have an internal latency which may impact the attached partial module. The latency in systems with such communication architectures can be placement position dependent (e.g., if no registers are used at the interface of the communication architecture). In this case, the system will typically have to follow a timing footprint, when using module relocation.

1.6.3.4 Bitstream Footprint

On Xilinx FPGAs, modules can be relocated by changing an address field in the configuration bitstream header. However, there are some device specific restrictions that have to be followed when relocating bitstreams. For example, the frame length of older Xilinx FPGAs is device specific. But also for recent FPGAs with an identical configuration frame format within a device family, configuration bitstream relocation might be restricted. For instance on Xilinx Spartan-6 FPGAs, modules using only the LUT function generator functionality (i.e., not using carry chains or distributed memory) can be relocated on a netlist level (as long as the resource footprint matches). This is possible without any changes to the internal logic and routing of such a module. However, the bitstream encoding differs for CLB columns providing distributed memory (CLEXM CLB columns) as compared to other logic columns (see Fig. 1.18 for an illustration of the different logic primitives). Consequently, it can require more than one partial module bitstream for supporting arbitrary module relocation.

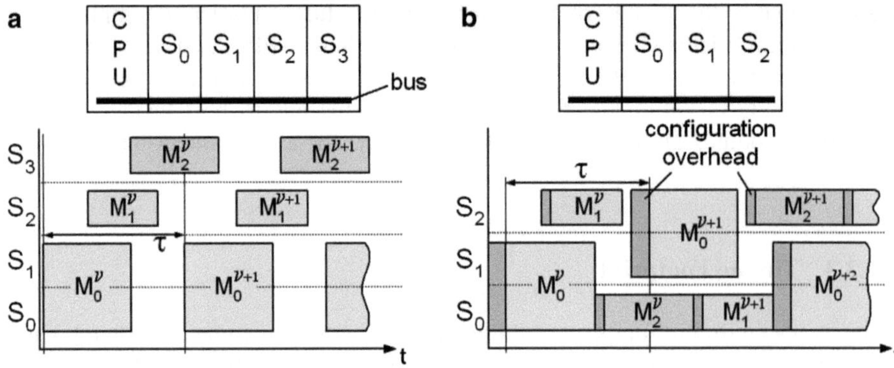

Fig. 1.19 Packing of modules into a system using slot-style reconfiguration. By allowing module relocation, one slot can be saved

1.6.4 One-dimensional Slot-style Reconfiguration

As mentioned in the last section, the island style reconfiguration suffers a considerable level of internal fragmentation. We can improve this by tiling reconfigurable regions into slots. This results in a one-dimensional *slot-style* reconfiguration style, as shown in Fig. 1.17b. In this approach, a module occupies a number of tiles according to its resource requirements and multiple modules can be hosted simultaneously in a reconfigurable region.

Figure 1.17 shows how the tiling influences the packing of modules into a reconfigurable region. However, that figure does not show the *time*. In general, partial reconfiguration always includes a packing of modules in the time domain. In order to fully improve the utilization of the reconfigurable resources by a slot reconfiguration style, modules have to be relocatable to different start slots. This has to follow pretty much the same rules that we discussed for the multi island reconfiguration style. Figure 1.19 gives an example of how module relocation helps to better fit modules into a reconfigurable region over time.

Tiling the reconfigurable region is considerable more complex as the system has to provide communication to and from the reconfigurable modules and to determine the placement for a module. The latter one has to consider that FPGA resources are in general heterogeneous. For example, there are different primitives like logic, memory, or arithmetic primitives on the fabric, which we summarized by the resource footprint in the last section. Moreover, depending on the present module layout, a tiled reconfigurable region might not provide all free tiles as one continuous area. If this results in slots that cannot be used, this overhead is called *external fragmentation*. External fragmentation can be removed by defragmenting the module layout which is called *compaction*. Note that some modules cannot be interrupted for compaction because a module would loose its internal state or it would not meet its throughput requirements.

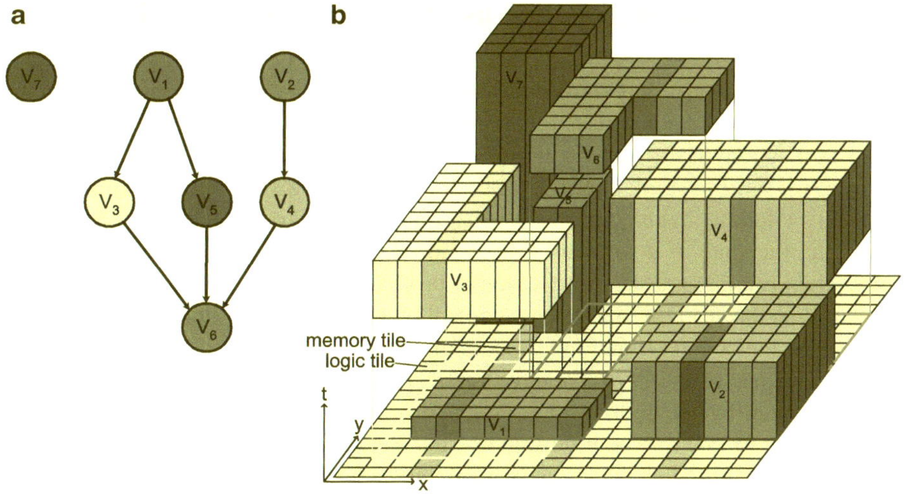

Fig. 1.20 Grid style module packing. (**a**) Task graph and (**b**) corresponding space-time packing. Note that the time runs top-down in the task graph and bottom-up for the space-time packing

1.6.5 Two-dimensional Grid-style Reconfiguration

The internal fragmentation of a reconfigurable region that is tiled into one-dimensional slots can still be large and in particular the dedicated multiplier and memory resources can be affected by this. The reason for this is that these resources are arranged in columns on the FPGA fabric. If a module needs only a few of these resources, it is beneficial if another module can use the remaining resources. Consequently, the vertical slots in Fig. 1.17b can be tiled another time in horizontal direction. This results in a two-dimensional grid-style reconfiguration style as shown in Fig. 1.17c.

The implementation and management of such a system is even more complex than the slot style reconfiguration approach. For example, we have to provide a communication architecture that can carry out the communication of the modules with the static system, and in general, also communication directly between reconfigurable modules. Note that the communication must be established within a system at runtime in absence of complex design tools. Together with the time domain, the two-dimensional grid style placement states a three-dimensional packing problem. The packing has to follow several constraints, including the schedule, the resource footprint or the communication between the modules. This packing, considering also fragmentation, has to be managed at runtime. The packing problem is visualized in Fig. 1.20. The shown packing problem is also known as *temporal module placement* or *FPGAtris*.

1.7 Partial Reconfiguration as FPGA Context Switching

Partial reconfiguration has similarities to context switching which is known from the software world and some researchers use the term *FPGA context switching/swapping* as a synonym for partial runtime reconfiguration. However, before using this phrase, we have to think about what the *context of an FPGA* actually is and what context switching means. A context is basically a state stored in some memory. For example, the state of a software task is stored in CPU registers, stack memory or files when running a task. In the case the task is preempted, the CPU registers will be saved on the stack or in a process control block. This saving of the CPU registers belongs to the context switching process of the software task.

With respect to an FPGA, we have not only the *internal state* of the modules (that is represented by the content of its state flip-flops and memory blocks), but also, independent to this, the configuration of the FPGA itself (i.e., the present module layout). Therefore we have to distinguish between context switching on the *FPGA technology level* and context switching on the *module logic level*. While the internal state of the modules is modified by the operation of the modules themselves, the configuration can only be accessed through a configuration port of the FPGA.

Figure 1.21 presents a classification scheme for FPGA context switching on the FPGA technology level by reconfiguration and FPGA context switching on the module logic level by considering the internal states of modules. Note that a runtime reconfigurable system typically uses more than one scheme. For example, a system might consist of a static system providing some basic infrastructure that is required at any time and that cannot be used for context switching at all (e.g., a memory controller). In addition to this, modules might be swapped in and out in a partially reconfigurable part dynamically at runtime.

The vast majority of work done on runtime reconfiguration does not consider internal module states. Here, FPGA context switching becomes synonymous to

	FPGA technology level: static	FPGA technology level: dynamic
Module logic level: static	• Module runns forever • Single configuration • Single module context "ASIC-like", for example: memory controllers, CPUs	• Run-to-completion model • *Partial reconfiguration* • No context after termination Example: secure SSL connection key exchange → data transfer
Module logic level: dynamic	• Module runns forever • Single configuration • *Multiple module contexts* Example: multi channel crypto	• Hardware module preemption • *Partial reconfiguration* • *Multiple module contexts* Example: load balancing in a ReCoNet (see Section 4)

Fig. 1.21 Classification of FPGA context switching

partial reconfiguration without module preemption or module contexts. The use of hardware preemption is discussed controversial as partial reconfiguration is time consuming, consuming power, and accessing the internal states of modules is difficult and not well supported by present FPGAs. However, context switching permits some interesting tasks, including the preemption of modules for full FPGA power down, checkpointing and rollback recovery for hardware tasks, defragmenting the module layout within a reconfigurable region, or an easier utilization of the FPGA resources over time.

The last task leads to the question: *what is FPGA utilization?* On a CPU, this is the relative amount of *time* the processor is performing useful work. With respect to an FPGA, we have to weight this by the used resources (i.e., the *area*[16]). However, in case of an FPGA, we have to clarify the term time. When considering a static system with no reconfiguration, all modules occupy the device at any time and the FPGA utilization is equal to the resource utilization. When also considering runtime reconfiguration, the utilization of modules has to be scaled down by the relative amount of time a module is actually needed to be present on the device (including possible reconfiguration overheads or overheads to access internal module states). With this definition, it does not matter whether a module is idle on the FPGA or if a module is not being configured to the FPGA. By time-sharing modules of low utilization with the help of context switching, we can enhance the total utilization of the FPGA.

However, the module utilization alone is not sufficient for determining if runtime reconfiguration can be applied or not. This is defined by additional system properties and constraints. For example, like two software tasks cannot share a CPU if they have to run at exactly the same time, regardless of their utilization, there are restrictions for using context switching on FPGAs. First of all, modules must be *mutually exclusive in space and time.* This means that a reconfigurable module must not occupy resources from other active modules during the time it is loaded and executed on the FPGA. Then, there could be restrictions in a system like limited throughput to transfer configuration data, too long latency for configuration, additional buffer requirements for the case a module cannot provide temporarily its service, and so on.

Note that implementation inefficiency results in higher utilization. For example, as an inefficient software implementation results in more CPU cycles to execute a program (and consequently in a higher CPU utilization), an inefficient hardware module implementation takes more resources and/or more time (and results again in a higher FPGA utilization). In other words, we can utilize resources heavily with inefficient modules.

---

[16]There is a project called "distributed.net" where it is possible to share *idle CPU time* to solve a large compute problem. In [KKT06] an FPGA version of a scalable compute client is proposed that can share *idle FPGA area* of a low utilized FPGA to compute the same problem, but much faster than a standard PC

1.8 Overview

This book comprises three key contributions: (1) Chap. 2 reveals methodologies required to integrate hardware modules into FPGA-based SoCs at runtime with the help of partial reconfiguration, which, in particular, comprises a novel communication architecture. Next, (2) Chap. 3 covers all design and tool aspects for implementing self-adaptive reconfigurable systems on FPGAs. After this, (3) Chap. 4 gives an application example for using partial reconfiguration for self-organizing fault-tolerant distributed control systems and Chap. 5 presents how to efficiently extend a CPU with reconfigurable instructions. Chapter 6 reveals concluding remarks. Finally, the appendix gives an outlook on an upcoming tool and design flow for all recent FPGAs from the vendor Xilinx. In the following, a brief overview of these chapters is given.

Partial runtime reconfiguration demands strategies for carrying out the communication between a static system, providing the basic hardware (e.g., a CPU, a memory controller, and some peripheral modules), and the reconfigurable modules, as deeply examined throughout Chap. 2. When designing FPGA-based runtime reconfigurable systems, the device resources have to be partitioned into a static region and a reconfigurable area tiled into a plurality of resource areas for hosting reconfigurable modules. In Sect. 2.1, the impact of such a tiling on the *placement flexibility* is discussed. After this, in Sect. 2.2, the *impact of the tiling on the resource overhead* is investigated. There, it is proven that existing approaches are in most practical cases not sufficient to achieve a material benefit when using runtime reconfiguration and that the tile grid has to be arranged much finer as known from existing approaches, with, in some cases, hundreds of individual tiles. Such a fine tile grid has to be physically mapped on an FPGA which requires an according *architecture model* as introduced in Sect. 2.3. Such a model has to reflect the geometrical aspects of the logic and routing resources of the underlying device architecture.

The research on reconfigurable computing systems has been quite active for more than a decade and various techniques for integrating reconfigurable modules have been proposed before. Section 2.4 provides a comprehensive survey on *related work* in the field of physically integrating modules into FPGA-based systems. This survey does not disclose techniques to efficiently provide the communication when tiling the reconfigurable area in a very fine grid. This issue is addressed in Sect. 2.5 with the introduction of a novel *bus-based communication architecture* that allows integrating modules by partial reconfiguration in a hot-plug manner. This architecture is capable to provide a bus connection in hundreds of individual tiles at high speed and low resource overhead. Besides buses, partial modules often require dedicated *point-to-point* links with other parts in the system. For this reason, communication macros, tailored to streaming ports, have been developed, as revealed in Sect. 2.6. The proposed communication architecture is *experimentally evaluated* in Sect. 2.7 and compared with state-of-the-art solutions in Sect. 2.8.

The techniques for integrating modules into an FPGA-based system at runtime require additional *methods and tools for building such reconfigurable systems*, as covered in Chap. 3. The communication architecture, for example, has to be implemented regularly structured and existing tools are not capable to build such structured circuits for FPGAs. This resulted in the development of a novel and easily usable *framework for component-based system design*. The framework will be presented in Sect. 3.1 and includes solutions for *simulating runtime reconfiguration* (Sect. 3.2.7), a *tool for the synthesis and mapping* of the proposed communication architecture which additionally comprises a *floorplanner* for defining and managing reconfigurable regions on an FPGA (Sect. 3.2), and a program for *assembling partial configuration files* as well as the initial configuration (Sect. 3.2.6). After these offline design aspects, Sect. 3.4 puts focus on the technical issues required to *load and parameterize partial module configurations* into a particular region on an FPGA at runtime. Loading configurations to an FPGA is in many cases time-critical and the configuration time dictates if runtime reconfiguration can be applied in a system or not. Hence, in Sect. 3.5, methods are proposed for *speeding-up the configuration process*. This includes configuration prefetching techniques, hardware accelerated bitstream decompressing, configuration port overclocking, and FPGA architecture extensions for fast defragmenting the module layout.

The introduced methodologies and tools permit novel opportunities to use partial reconfiguration in applications. For exploiting the potential flexibility of a distributed embedded control system, a model is introduced in Chap. 4 that separates the functionality, given by a set of communicating tasks, from the network topology. This allows, for example, to relocate tasks (including hardware modules) from one node within the network to another, as described in Sect. 4.1. It is demonstrated that this is possible completely transparent for the migrated tasks and operation can seamlessly continue on another node by using checkpointing techniques also for hardware modules, by including the internal states when migrating functionality. Moreover, with the operating system service of *hardware/software morphing* it is alternatively possible to change the implementation style of a task between software and hardware, or vice versa, as detailed in Sect. 4.3. Again, this is transparently possible without losing internal task states. These services have been integrated into a demonstrator running an automotive application in a distributed manner, as depicted in Sect. 4.4. Among other features, this demonstrator can detect and autonomously compensate node or link failures by migrating tasks and rerouting traffic in a small distributed embedded system.

Chapter 5 gives an example of how software programs running on a MIPS CPU can be considerable accelerated with custom instruction accelerator modules. Due to the efficient zero logic overhead integration technique, which is presented in Sect. 5.2, only very few resources are required to integrate multiple reconfigurable modules at the same time. It will be shown that only a few hundred calls of a custom instruction set extension often justifies the configuration overhead to load new instructions.

The last chapter, Chap. 6, concludes that with the introduced architectures, methods, and tools the key prerequisites are available to implement more advanced runtime reconfigurable systems very efficient. Moreover, in Sect. 6.1 future trends and research will be discussed.

These chapters constitute a reference on runtime reconfiguration using FPGAs. They introduce fundamentals, on-FPGA communication aspects, tools and design flows, as well as different practical examples. It is shown that building reconfigurable systems has to follow many obstacles in order to gain a substantial benefit by this technique. However, these obstacles can be systematically circumvented by tools that are aware of the special needs of reconfigurable system. This helps to hide most of the low-level implementation details from a design engineer and removes many error prune design steps known from other flows. The overall goal of this book is to provide a better understanding of the FPGA related effects during reconfiguration as well as a demonstration that partial reconfiguration is ready to be used for various industrial applications.

Chapter 2
Intra-FPGA Communication Architectures for Reconfigurable Systems

This chapter is devoted to communication architectures required to design and implement runtime reconfigurable systems. A special communication architecture is required whenever using partial runtime reconfiguration. It must provide all communication demands (e.g., bus connections or point-to-point links) between the static part of the system and the reconfigurable modules. The communication architecture has a large impact on the achievable performance in terms of throughput, the flexibility of the system, and the resource overhead. Moreover, it has impact on the whole design process. For example, can partial modules be implemented independent of the static system while guaranteeing timing closure. Finally, the communication architecture has large impact on the system management at runtime. For example, will the architecture mask effects resulting from the configuration process from the rest of the system.

The goal of this chapter is to propose and to analyze new on-chip communication architectures for FPGAs. This includes a novel bus-based architecture, called *ReCoBus*, for directly connecting runtime reconfigurable modules to a backplane bus and a further communication architecture, called *Connection Bars*, that can be used to provide point-to-point links to partially reconfigurable modules. The proposed architectures provide the following properties:

- *Direct interfacing* Reconfigurable modules can be directly hot-plugged into an on-chip backplane bus with the help of partial runtime reconfiguration. This includes integration of reconfigurable bus masters into the system. In addition, modules can connect to I/O pins, other on-chip resources or other reconfigurable modules within the system.
- *Module relocation* Instead of binding modules to a fixed reconfigurable island, they can be placed freely within the reconfigurable area. This area may be tiled into a one-dimensional slot style layout or a two-dimensional grid.
- *Variable module sizes* This allows integrating multiple modules with different resource requirements into the system without wasting large resource areas for smaller modules. This permits a system to replace a large module by multiple smaller ones or vice versa.

D. Koch, *Partial Reconfiguration on FPGAs: Architectures, Tools and Applications*, Lecture Notes in Electrical Engineering 153, DOI 10.1007/978-1-4614-1225-0_2,

- *Fine placement grid* The module placement grid is a quantity for the granularity of how fine a module bounding box can be adjusted to just fit the area demands of a particular module. In addition, the module grid defines how flexible these can be placed within a reconfigurable area. The module grid also defines the possible placement style of the modules within the reconfigurable resource area. The placement style ranges from an island style placement with fixed resource areas over one-dimensional slot based placement styles to free two-dimensional module placement techniques (see also Sect. 1.6).
- *Multiple instances* This attribute denotes that multiple instances of the same module could be integrated into the system at the same time. Using multiple instances allows a system to scale more flexibly on changing load demands at runtime.
- *Low logic overhead* to implement the communication architecture.
- *High performance*: The communication bandwidth and the latency of the reconfigurable communication architecture can compete with traditional static only systems.
- *Hot-swapping* Modules can be replaced without interfering with the rest of the system. The configuration process itself can be managed completely transparent to running communication cycles on the communication architecture.

This chapter puts focus on communication architectures for reconfigurable FPGA-based systems that optimize internal fragmentation as well as the communication logic overhead. The impact on the overhead will be discussed in Sect. 2.1. For clarifying all geometrical aspects, an FPGA architecture will be introduced in Sect. 2.3. After this, in Sect. 2.4, related work on such communication architectures will be summarized. Next, in Sect. 2.5, the novel ReCoBus architecture for reconfigurable on-chip buses will be presented that is in particular tailored to FPGA-based systems. This communication architecture can be efficiently implemented in one-dimensional slot based systems as well as for systems following a two-dimensional grid placement style, as revealed in Sect. 2.5.6. Afterwards, in Sect. 2.6, architectures for point-to-point communication will be proposed. These architectures provide dedicated reconfigurable connections of partial modules with I/O pins or other modules within the system. The introduced architectures will be experimentally evaluated in Sect. 2.7. Finally, the contributions and the progress against related approaches will be summarized in Sect. 2.8.

2.1 Preconsiderations on Flexibility and Module Placement

When utilizing partial runtime reconfiguration in a system, a structural partitioning step for dividing the FPGA resources into a static and one or more reconfigurable regions is required. The partitioning massively affects the placement flexibility and results into a specific *placement style* as discussed in Sect. 1.6.

As it is unlikely to manage the reconfigurable area in plenty of small look-up tables, it is common to group multiple look-up tables into one or more rectangular tiles. Such a tile is the smallest atomic area that can be assigned to a module, and, depending on the system, multiple adjacent tiles may be occupied by a module. In order to emphasize that these tiles provide the logic and routing resources for implementing some modules that are dynamically plugged into the system by partially reconfiguring an FPGA, such a tile will be named *resource slot*. As revealed in Sect. 1.6, the resource slot grid may follow (a) a simple *island style*, (b) a one-dimensional *tile style*, or (c) a two-dimensional *grid style*. The placement style is massively influencing the *configuration overhead* that consists of two factors: (1) the logic required to provide communication with other modules or the static part of the system and (2) the internal fragmentation that arises by fitting modules into resource slots. In this sense, a resource slot has some similarities with a sector on a hard disc drive. A sector is the smallest chunk of memory that is directly accessible and a file may occupy multiple sectors for storing its data, while one sector is not shared among multiple files.

In order to quantify the flexibility of different reconfigurable systems, the *placement flexibility* is defined. This value denotes the sum of all possible starting points to place a partially reconfigurable module on an FPGA-based system. This is regardless of the question whether a particular module could start in all the provided slots. With respect to the example in Fig. 1.17 on Page 32, the island style case provides a placement flexibility of 2 and the grid style case a flexibility of 20, respectively.

2.2 Efficiency Preconsiderations

When assuming that one resource slot provides σ look-up tables, then, if omitting communication cost, half of the amount of resources available in one slot ($\sigma/2$) will be wasted on average per module just because of internal fragmentation. This is because in the best case, all resource slots will be completely filled and in the worst case, a module will leave almost the amount of logic provided in one resource slot unused. Consequently, the resource slots should be small. On the other side, the cost for implementing the communication with the reconfigurable modules that are located inside the resource slots may unlikely increase. For the following, it is assumed that the communication cost for the inter-module communication within each resource slot is c look-up tables and that a resource slot provides altogether σ LUTs. Then, all modules $m_i \in \mathcal{M}, \forall i = \{1, \ldots, |\mathcal{M}|\}$ possess the following *average module overhead* $\overline{O}$ in terms of look-up tables:

$$\overline{O} = \frac{1}{|\mathcal{M}|} \cdot \sum_{i=1}^{|\mathcal{M}|} \left(\left\lceil \frac{|m_i|}{\sigma - c} \right\rceil \cdot \sigma - |m_i| \right). \qquad (2.1)$$

Where $|m_i|$ denotes the size of a module in terms of LUTs and where $\left\lceil \frac{|m_i|}{\sigma - c} \right\rceil$ states the amount of resource slots required to implement a reconfigurable module m_i.

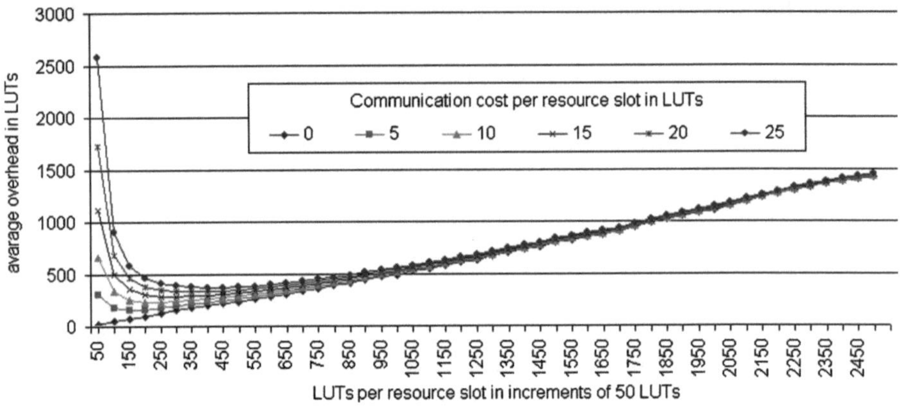

Fig. 2.1 The impact of the resource slot size σ and the communication cost c on the average module overhead $\overline{O}$ for a synthetic module distribution

The average module overhead $\overline{O}$ allows a first rough estimation of an optimal resource slot size as revealed in Fig. 2.1. Here, a synthetic scenario is assumed with modules ranging from 300 to 10,000 LUTs, which roughly corresponds to 10–300% of the logic required to implement an average 32-bit softcore CPU. In this example, the amount of modules is chosen in such a way that all modules sizes contribute with the same overall amount of look-up tables to the set of modules $\mathcal{M}$. At this point, a scenario is assumed with $|\mathcal{M}| = 9{,}700$ different modules ranging from 300 to 10,000 LUTs in increments of one LUT. As can be seen in Fig. 2.1, the communication cost is dominating for smaller resource slots, while for larger slots, the internal fragmentation will dominate. The curve points out that the size of a resource slot should be around a few hundred LUTs. In the case that $|m_i| \gg \sigma$, the placement overhead converges to $\frac{\sigma}{\sigma - c}$. In other words, for large modules requiring many resource slots, the internal fragmentation becomes negligible (as one slot more or less does not matter much for large modules) and the communication cost c will dominate (as most of the communication logic will be left unused).

For a given set of modules $\mathcal{M}$, the optimal resource slot size σ_{opt} against the communication cost can be computed by differentiating (2.1) with respect to the slot size σ. As the ceiling function is discontinuous, its bounds are considered: $\frac{|m_i|}{\sigma - c} \leq \left\lceil \frac{|m_i|}{\sigma - c} \right\rceil < \frac{|m_i|}{\sigma - c} + 1$. While the lower bound expresses that a module perfectly fits into the resource slot without internal fragmentation, the upper bound denotes the case of maximal fragmentation where one resource slot is almost unused. For the lower bound, internal fragmentation is optimized by reducing the resource slot size to $\sigma = 0$. However, in the case of a communication cost $c > 0$, this results in an explosion of the of the average overhead (see the results in Fig. 2.1 for small

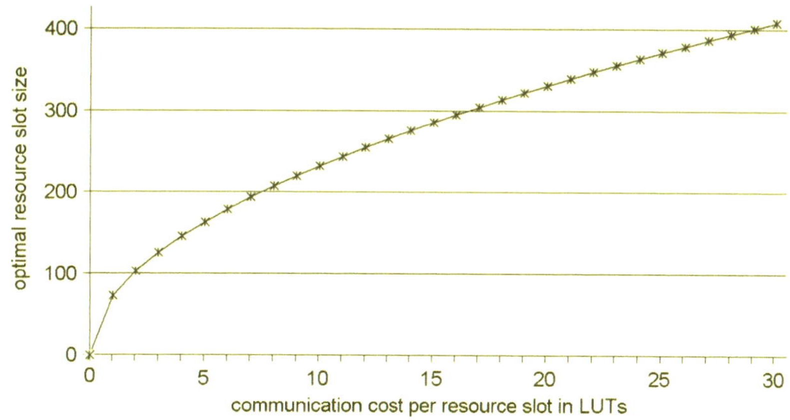

Fig. 2.2 Average optimal resource slot size σ_{opt} for a given communication cost c in each slot. The results are obtained for the same synthetic module distribution as assumed for the plot in Fig. 2.1

resource slots). For the upper bound, the average fragmentation over all modules $m_i \in \mathcal{M}$ can be minimized as follows:

$$\widehat{O} = \frac{1}{|\mathcal{M}|} \cdot \sum_{i=1}^{|\mathcal{M}|} \left(\left(\frac{|m_i|}{\sigma - c} + 1 \right) \cdot \sigma - |m_i| \right)$$

$$\frac{\partial \widehat{O}(\sigma)}{\partial \sigma} = 0 \Rightarrow \widehat{\sigma}_{opt} = c \pm \sqrt{|\mathcal{M}| \cdot c \cdot \sum_{i=1}^{|\mathcal{M}|} |m_i|}. \tag{2.2}$$

The average overall resource slot size is then approximately the mean value between the optimal slot sizes, determined for the two bounds: $\sigma_{opt} \approx \widehat{\sigma}_{opt}/2$. Figure 2.2 plots the relationship between the communication cost c and σ_{opt}. The results match the minima depicted in Fig. 2.1.

The geometrical view on a concrete system allows to state whenever runtime reconfiguration is suitable or not. The geometrical view is equivalent to a view on the partitioning of the FPGA resources. Let $\widehat{\mathcal{M}} \subseteq \mathcal{M}$ be the module subset requiring the largest amount of resource slots in a reconfigurable system at a point of time, then the following inequation must be met for achieving an overall benefit when applying partial runtime reconfiguration:

$$\sum_{\forall m_i \in \widehat{\mathcal{M}}} \left\lceil \frac{|m_i|}{\sigma - c} \right\rceil \cdot \sigma + c_{const} < \sum_{\forall m_i \in \mathcal{M}} |m_i|. \tag{2.3}$$

Here, c_{const} summarizes all additional resources required for allowing the system to exchange modules by partial reconfiguration at runtime, for example, a reconfiguration interface controller. Inequation (2.3) is a necessary condition but

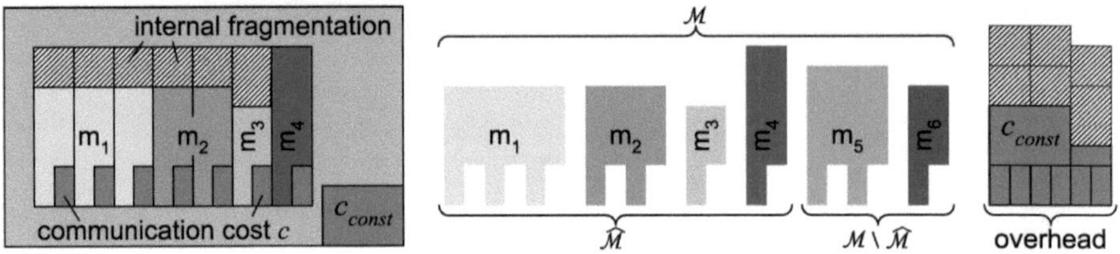

Fig. 2.3 Example of a system with seven resource slots ($R = 7$) and a set of reconfigurable modules $\mathcal{M}$. Each slot provides σ resources and comprises a communication logic overhead of c resources. At runtime, the peak resource requirement is demanded when the modules of subset $\hat{\mathcal{M}}$ are loaded to the FPGA. In this case, the total reconfiguration logic overhead is $R \cdot \sigma - \sum_{\forall m_i \in \hat{\mathcal{M}}} |m_i| + c_{const}$, which, in this particular example, is larger than the area that can be saved with the help of partial runtime reconfiguration that is $\sum_{\forall m_i \in (\mathcal{M} \setminus \hat{\mathcal{M}})} |m_i|$. Consequently, this example is more suitable for a static only implementation

not a sufficient one, because other important factors including reconfiguration time, external fragmentation, or design issues (e.g., tools) are omitted in this consideration. Figure 2.3 reveals an example where Inequation (2.3) is not met, because a static only implementation without using runtime reconfiguration would require less FPGA resources. In order to achieve an overall benefit by using runtime reconfiguration, the internal fragmentation could be reduced by (1) tiling the resource slots in a finer grid, by (2) reducing the communication overhead per resource slot, and by (3) optimizing c_{const}.

Inequation (2.3) and the example in Fig. 2.3 points out that runtime reconfigurable systems have to be designed with ultra low resource overhead in order to justify this approach. As this efficiency seems not to have been demonstrated so far, runtime reconfiguration has not been introduced widely in industrial applications and is still very exotic, despite its potential benefits.

2.3 FPGA Architecture Model

Independent from the utilized FPGA architecture, FPGAs are based on (1) functional units such as logic cells, I/O blocks or memories and on (2) an interconnect network for wiring these functional units. The goal of this section is to define a generic FPGA architecture model that reflects all commonly used FPGA architectures with respect to the implementation and analysis of runtime reconfigurable systems. The resulting model has to provide a geometrical view rather than a logical view on the functional units and the interconnect resources of different types of FPGAs. This is essential for examining the additional steps required to enhance a static only system to a reconfigurable system. This includes floorplanning and tiling an FPGA into resource slots, as well as all communication issues.

In common, one or more functional units and the according switching elements of the interconnect network can be combined into one *basic element* (BE).

Multiple BEs, each consisting of equivalent logic and routing resources, are placed on the FPGA in a highly regular fashion. On Xilinx FPGAs, a BE would be equivalent to a so-called *configurable logic block* (CLB) [Inc08b] that consists of multiple look-up tables (LUT) to implement logic functions. In the Xilinx Virtex-II architecture, for example, eight 4-bit LUTs share one so-called *switch matrix*. The equivalent to a CLB on FPGAs from Altera Inc. is called a *logic array block* (LAB) [Inc08a]. The architecture of a BE may vary even between different FPGA architectures from the same vendor, but in general, the routing within a BE is extensively faster when compared to the routing among different BEs. In summary, all commercially available FPGAs follow an FPGA architecture model with regular tiled BEs that are also connected by regular structured wires, as shown in Fig. 2.4.[1] Besides FPGAs, this model may also fit to other regular structured logic devices including CPLDs, PLDs, PLAs, or even structured ASICs. The geometrical parameters for partitioning the FPGA into a resource slot grid for hosting multiple reconfigurable modules are as follows:

W: resource slot width in BEs
H: resource slot height in BEs
R: resource slots in total as the product of horizontal and vertical resource slots: $R = R_H \cdot R_V$
M: maximum amount of simultaneously integrated modules
w: module width in resource slots
h: module height in resource slots

With respect to the example in Fig. 2.4, module 1 has a width of $w = 1$, a height of $h = 2$, and the placement position $S = (0, 3)$.

The BEs of a few currently available FPGA architectures are not identical over the complete device. For example, in Xilinx Virtex-5 or Spartan-6 FPGAs, only every second column of logic blocks can be configured as distributed memory. Similar irregularities also apply for the Stratix-4 [Inc08a] architecture from Altera. However, by grouping a set of CLBs (or LABs, respectively) together, it is still possible to construct a completely homogeneous BE layout on such FPGAs that fits the model in Fig. 2.4. For instance, it is possible to group in each case two horizontal adjacent CLBs of a Xilinx Virtex-5 or Spartan-6 FPGA together into one BE that will then possess the same logic and routing resources throughout the FPGA. A further concern are dedicated resources like embedded memory blocks or hardwired multipliers. By examining the Xilinx Spartan-3, Virtex-II, Virtex-4, Virtex-5, Spartan-6, and Virtex-6 FPGA architectures, it was discovered that the dedicated resources provide exactly the same routing resources as available within all logic tiles. It might be that these dedicated tiles cannot provide module connections as compared to the CLBs, but the routing architecture allows to implement homogeneous structures throughout the whole FPGA.

[1] Inhomogeneous elements, including distributed memory, dedicated memory blocks, or hardwired multipliers will be covered later in this section.

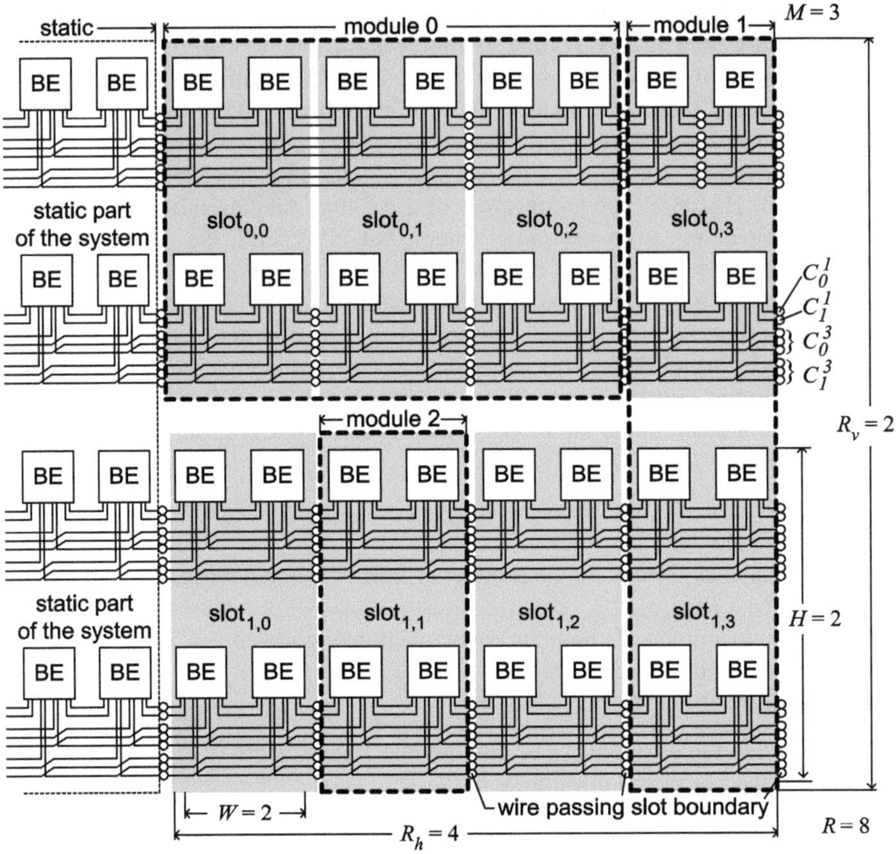

Fig. 2.4 Simplified FPGA architecture model. An FPGA consists of a two dimensional regular grid of *basic elements* (BE) containing the switch and logic resources. Wires of different span (S) between the BEs carry out the distribution of signals. For each span, there exists in each BE a number of output wires (counted by i) per span forming a *channel* C_i^S. To simplify the figure, only some horizontal routing wires are illustrated, although they may be arranged in any desired horizontal, vertical, or diagonal fashion. The reconfigurable system will be built onto such architecture. A system offers R_h horizontal and R_v vertical resource slots that are each W BEs wide and H BEs high. Modules can have different sizes that are a multiple of W BEs wide and a multiple of H BEs high. The total number of resource slots is $R = R_H \cdot R_V$. The maximum number of reconfigurable modules that may be integrated into the system at the same time is M. In addition to the resource slots, the reconfigurable SoC will contain a static part to control the system. The static part can communicate with the modules via a special communication architecture

Typically, FPGA architectures have routing resources that span different lengths. In Xilinx Virtex-II FPGAs, for example, there exist so-called *double lines*, spanning over a width of two CLBs and *hex lines* spanning over six CLBs, respectively. As revealed in Fig. 2.5, some of these wires can be tapped by multiple CLBs which is typically utilized for efficiently implementing multicast signals, such as enable or reset signals. On latest Xilinx FPGAs, most of the wires are simple

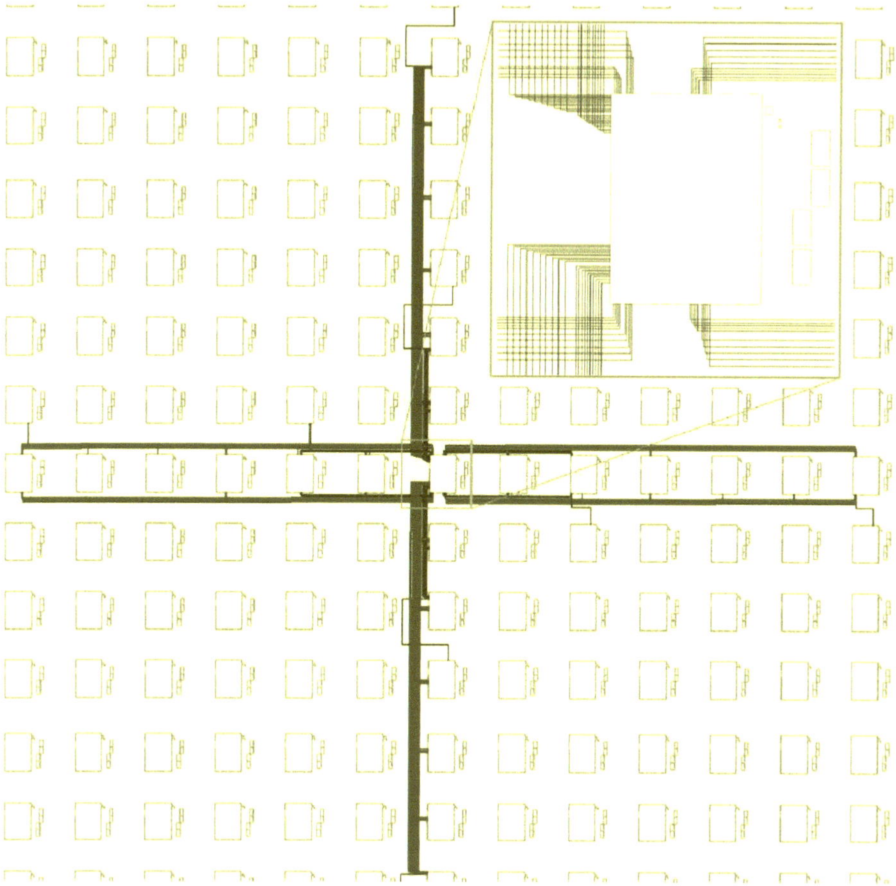

Fig. 2.5 FPGA Editor screenshot of a single configurable logic block (CLB) with highlighted hex (for routing 6 CLBs further) and double lines (for routing 2 CLBs further) on a Xilinx Virtex-II device. Only outgoing wires are marked. On this FPGA, all CLBs provide exactly the same interconnect network with 10 hex and 10 double lines starting in both, horizontal and vertical directions. Each of these wires can be tapped in the middle after three or one CLBs, respectively. As depicted enlarged, a CLB provides four so-called *slices*, connected to one switch matrix on this FPGA architecture. Each slice contains two 4-bit look-up tables, some further logic, and two flip-flops

point-to-point links as shown in Fig. 2.6. While on Virtex-II FPGAs the routing was strictly aligned in horizontal and vertical direction, newer devices provide routing in diagonal direction. Each output of a basic element forms a *channel* that contains as much wires as its *span* S. The different channels will be named by their span and their index: C_i^S. (See Fig. 2.4 that gives examples for triple and single wires.) The direction of a channel is obvious by its usage and will be consequently omitted. If, for example, a channel is used for a link from the static part of the system towards the reconfigurable area, it must drive a signal easterly, if the system is floorplanned as shown in Fig. 2.4.

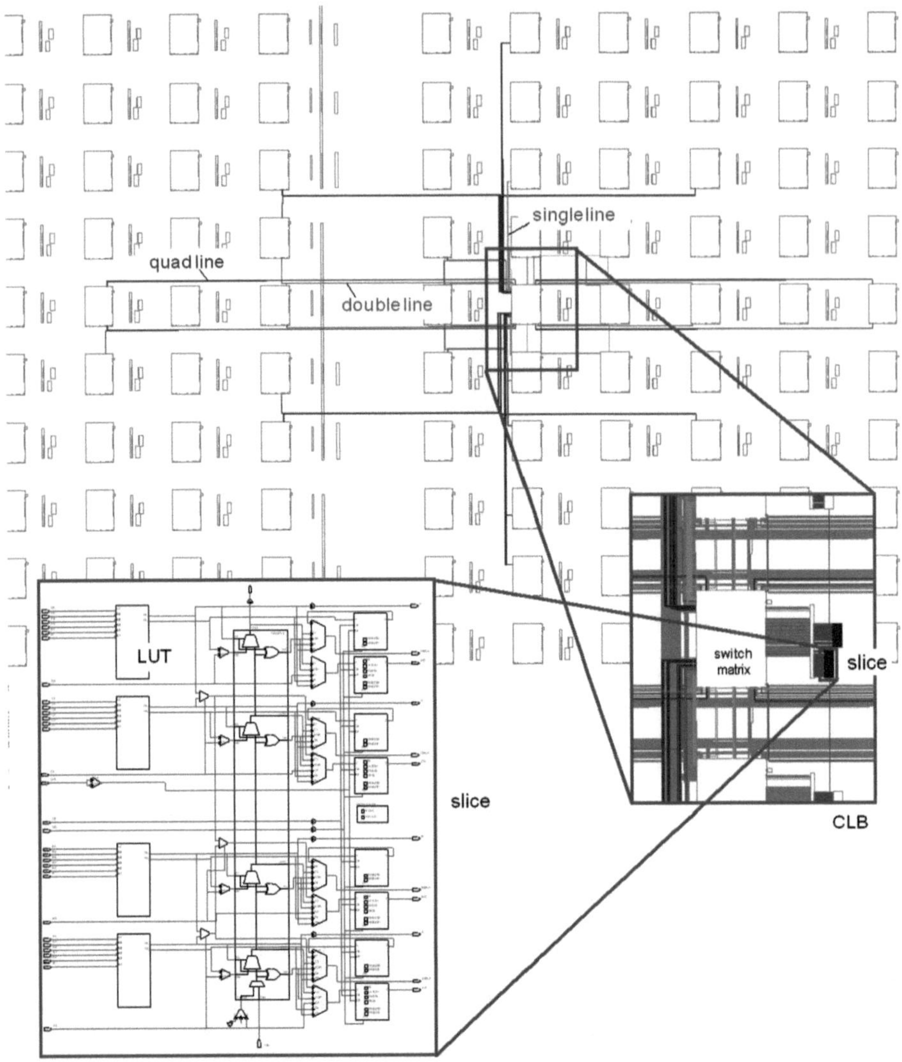

Fig. 2.6 FPGA Editor screenshot of a single configurable logic block (CLB) with highlighted quad lines, double lines, and single lines on a Xilinx Spartan-6 device. Only outgoing wires are marked. As opposed to the Virtex-II device, Spartan-6 routing is less regular and more, but closer located CLBs are directly connected. In total, there are roughly the same amount of wires leaving each CLB, but only half the number of wires end per CLB (because in Virtex-II, there is an access point in the middle of each wire). Each slice contains four 6-bit look-up tables, some further logic, and two flip-flops

The communication resources between the static part and the reconfigurable area have a preferential direction. This means that in the case of one dimensional, vertically aligned resource slots, the reconfigurable modules are located in a vertical manner beside the static part of the system. In such a system, the communication resources will be horizontally aligned and will cross all resource slots starting from the static part of the system. A resource slot may provide some logic and routing resources for the access to a bus and/or a point-to-point link. Such an access to the communication infrastructure is named a *module connection point*. The BEs of a module connection point must have the same relative position within all resource slots and must further contain the same internal logic functions. This allows connecting different modules to the communication architecture at the same position over time without interrupting the communication of other reconfigurable modules that may be carried out over the currently reconfigured resources. Among multiple different modules, the bus and point-to-point interface signals are allocated to fixed logic and routing resources within a resource slot.

All following figures that show some details of this communication architecture will be arranged in such a way that the static part is located at the left hand side of the system, while the resource slots will be horizontally aligned. This scheme matches best to the FPGAs offered by Xilinx Inc. that feature a column-based configuration scheme where the smallest piece of configuration data is a complete column spanning over the complete height of the device.[2] However, the presented techniques can be transferred easily to build reconfigurable on-chip buses towards any direction. At the end of Sects. 2.5 and 2.6, it will be described how the communication architecture for one-dimensional reconfigurable systems can be extended towards more flexible two dimensional systems.

2.4 Related Work on Communication for Reconfigurable Systems

All approaches for on-chip communication architectures can be classified into the following three classes: (1) *circuit switching* techniques (Sect. 2.4.1), (2) *packet switching* mechanisms (Sect. 2.4.2), and (3) on-chip *buses* (Sect. 2.4.3). Circuit switching is often referred to *point-to-point links*. Such a classification into different structures and protocols had been proposed by Mak et al. [MSCL06], as revealed in Fig. 2.7. The classification targets mainly static only systems.[3]

With respect to communication architectures for runtime reconfigurable systems, it is common to link the modules via a uniform, homogeneous interface that is equivalent in all resource slots for allowing module relocation. For the same

[2]The newer FPGA architectures Virtex-4, Virtex-5, Spartan-6 and Virtex-6 from Xilinx Inc. are configured sub-column wise. These devices possess still a preferable column wise module layout, because of restrictions in the local interconnect between the logic elements (e.g., interconnects for fast carry-chain logic is only available between vertically aligned logic elements).

[3]The classification is primary based on physical aspects. With respect to the OSI Reference Model, higher layers, e.g., the data link layer, are not considered at this point.

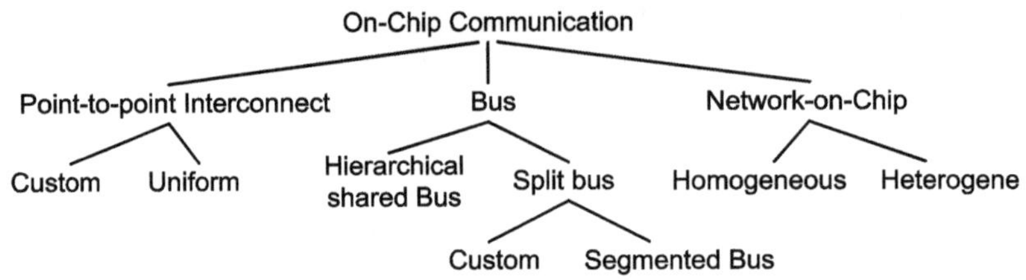

Fig. 2.7 Taxonomy of on-chip communication architectures (from [MSCL06]). From a system level perspective, the *ReCoBus* communication architecture is a hierarchical shared bus, because it connects the bus segment of the reconfigurable subsystem with the backplane bus of the static system. With respect to the reconfigurable subsystem, it is a fully homogeneous structure with identical interfaces to all reconfigurable modules. The *Connection Bar* architecture is a custom point-to-point interconnect network, because it allows to set connections when there is a necessity. This is sometimes also referred as ad-hoc interconnect

reason, hierarchical buses have not peen proposed for runtime reconfigurable systems, despite of the fact that such buses are often used in static only systems. A segmentation of routing wires is not only bound to buses (see Fig. 2.7) and has been demonstrated for circuit switched networks, as well [WJ06, ABD^{+}05]. In the following, the different communication architecture classes and associated protocols will be separately discussed in more detail. After this, an overview on physical implementation techniques will be outlined.

2.4.1 Circuit Switching

Circuit switching is a technique where physical links are established between two (what is also known as *point-to-point* communication) or more module *terminals* for a certain amount of time. To be more concise, after establishing a *channel* of fixed bandwidth between a source and one or more destinations, communication can take place with a fixed delay (as opposed to packet switching), and the channel is locked until all modules agree to release the channel. Circuit switching is advantageous if the channels are used at high bandwidth, because almost all wires of the channel may carry data, thus allowing a good interconnect utilization of the routing resources. This is typically the case for video or network processing systems. However, the interconnect utilization may be enhanced by time multiplexing different channels or by serializing the individual channel signals [BT93, DT01].

Typically, the module terminals are implemented by more or less complex crossbar switches. The switching state of the internal multiplexers can be controlled by the static system in a centralized manner or distributed by some logic in the crossbar switches itself.

Reconfigurable Modular Bus (RMB) In [ABD^{+}05] a circuit switching architecture is presented where the switching state is determined by the crossbar

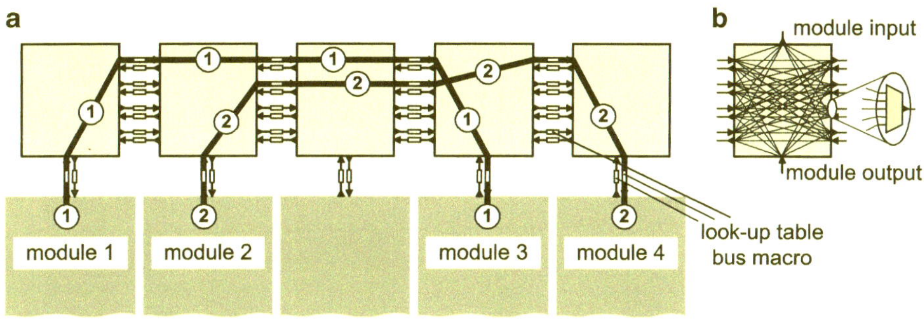

Fig. 2.8 (**a**) Reconfigurable Modular Bus (RMB) example. The first established link is channel 1 from module 1 to module 3. After this, channel 2 was established from module 2 to module 4. During link construction, the state machine in each crosspoint (starting from source to destination) will select the most top wire segment that is currently available. (**b**) The crosspoints feature a full crossbar that is implemented with multiplexers

switches (called crosspoints) with respect to the currently used routing resources, as illustrated in Fig. 2.8. This process is completely carried out in hardware and does not require additional software assistance. Setting up a channel starts from the source towards one destination. In each channel, the currently topmost available wire segment is chosen, until the destination crosspoint is reached. With this policy, a valid route is always being found if there are sufficient wire segments between the crosspoints available. The approach requires fully equipped crossbars, hence, it is unlikely scalable with the number of wire segments (quadratic), the bit width (about linear, as shown in Table 2.2), and the number of crosspoints (linear). However, in [FvdVMT06, FvdVA$^+$07] ILP based solutions are presented for finding module positions to minimize the wire segments between the crosspoints, as well as shortening the longest path and to reduce the total amount of required wire segments in a system. With respect to the example in Fig. 2.8, all of these objectives will be enhanced by swapping module 2 with module 3.

First Channel Straight to Destination (FCSD) Routing Instead of using the entire first available wire segment in each slot like in the RMB routing, it is possible to route the path using the same wire segment in all slots from the start module to the destination. This approach is called first channel strait to destination routing (FSCD routing) [KT11a]. As shown in Fig. 2.9, FSCD routing can fulfill the same routing demand as RMB routing. However, for FSCD routing, smaller multiplexers can be used. Consequently FSCD routing will result in better performance at lower implementation cost. Furthermore, FSCD routing can be implemented directly inside the switch matrices of most Xilinx FPGAs, as shown in Fig. 2.10. Consequently, FSCD routing saves a substantial amount of logic and latency for the signal switching.

Virtual Wires Originally, the *virtual wire* [BT93, BTD$^+$97] concept was proposed for systems containing a large design that is partitioned into multiple FPGAs. This is

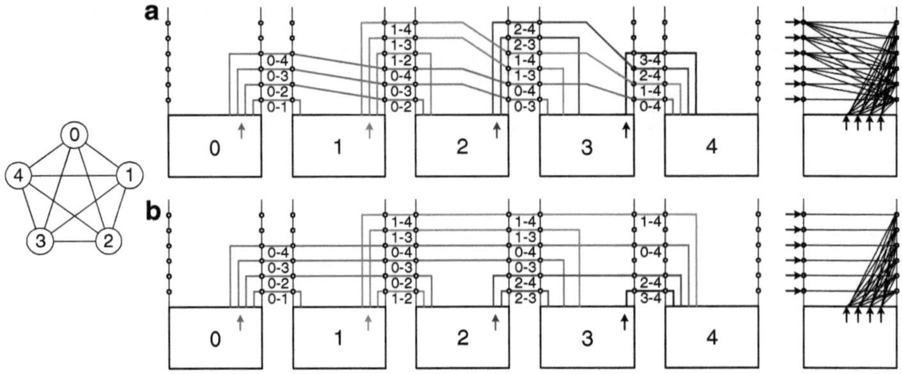

Fig. 2.9 Mapping a complete graph to 1d arrays, (1) RMB routing, (2) FCSD routing. Only the left-to-right routing is shown

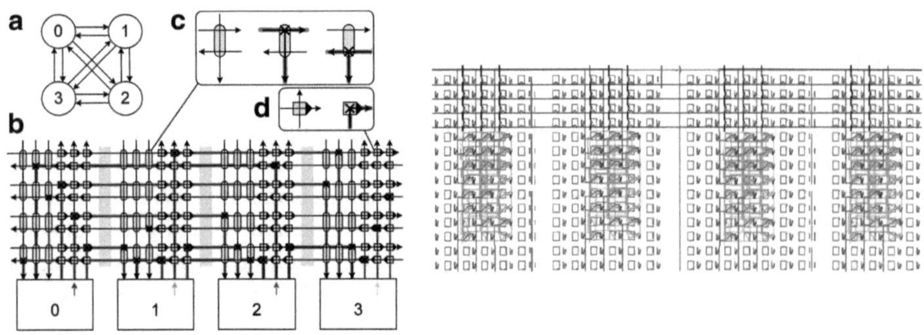

Fig. 2.10 FCSD case study; (a) dataflow graph, (b) communication architecture for direct FPGA routing fabric implementation, (c) module input connection, (d) module output connection. FPGA editor screenshot of a Spartan-6 implementation using serial links

highly relevant for building FPGA-based logic emulators, as the amount of logic on one FPGA scales quadratically with the length of the die (assuming a quadratic chip) and also quadratically with enhancing the technology (transistor shrinking), while the number of I/O pins scales only linear with the die length and even less than linear with the process technology. Consequently, there is a much lower I/O bandwidth per amount of logic resources available on larger FPGAs. When the amount of I/O pins between the FPGAs is insufficient for linking all signals, some or all signals are serialized at the source FPGA and transmitted to the destination FPGA where the signals are deserialized again. Therefore, a much lower amount of I/O pins is required to transfer some signals between the partitions located on different FPGAs.

Beside logic emulation, the authors propose to also use virtual wires for integrating reconfigurable modules into a system at runtime. This is an interesting option to be combined with circuit switching as the crossbars can be implemented

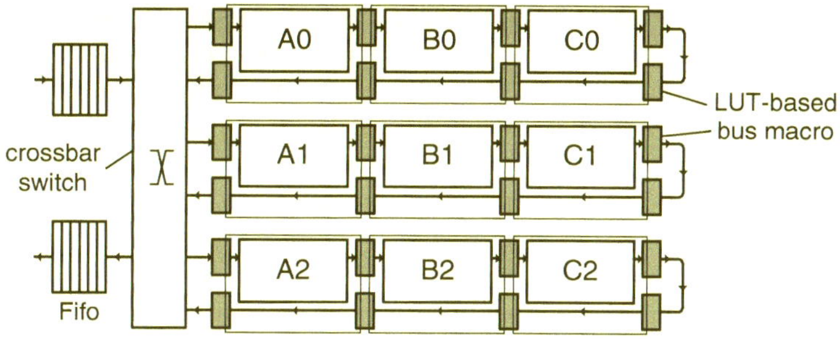

Fig. 2.11 Circuit switching network proposed in [BHB+07]. Multiple modules can be integrated in a daisy chained fashion within horizontal slots. The system includes multiple one-dimensional slots that can be linked in any order with the help of a crossbar switch

with much less logic, as the amount of logic resources for the multiplexers scales linear with the bit width of the switched data channels. However, this is only applicable in situations, where the throughput allows a serialization of the data.

Two-dimensional Circuit Switching In [HSKB06], a two-dimensional circuit switching system that is based on templates for connecting, turning, forking, crossing, or routing through a set of signals is presented. By attaching these templates together through partial reconfiguration, it is possible to set a fixed routing path between some modules at runtime. Consequently, communication with modules placed in a two dimensional fashion can be achieved in a very flexible manner. However, this approach requires a significant amount of area on the FPGA around the modules in order to build the routing, because of the templates for implementing the routing tracks cannot be shared with the module resources. In other words, modules are obstacles for the routing tracks. In addition, this circuit switching technique requires complex online routing algorithms and also an online timing verification.

Another approach for a simple two-dimensional circuit switching network was presented in [BHB+07]. In that work, multiple one-dimensional resource slots can be shared by multiple modules in a daisy chained fashion within a particular slot, as depicted in Fig. 2.11. The proposed system does not facilitate direct links between modules that are located in different resource slots, or joining of different slots. Any signal crossing the boundary of a chained module has to pass a look-up table based connection macro. This is regardless to the question if a module is accessing a particular signal or not.

As an improvement, it is possible to separate the resources for the circuit switching network from the module implementation, as proposed in [KBT10]. This means that a routing wire of the FPGA fabric can be either allocated to the network or a module. By this constraining, there will be no resource conflicts and modules are no longer obstacles for the circuit switching. Routing is updated by manipulating the switch matrix configuration. This is possible without side effects on all recent Xilinx FPGAs. By simply counting the used wire segments, the latency of a switched path can be determined at runtime.

Wires on Demand Athanas et al. proposed to freely route signals between some reconfigurable modules. They call this concept *wires on demand* [ABD+07] and it is based on reserving routing channels exclusively for performing fine grained online routing. In [SAP08] the authors demonstrate that this routing can be performed with low memory requirements (96 kB for the routing resource database) within an embedded system. The approach cannot guarantee to fully route all signals for any placement scenario at runtime. Furthermore, modules are typically integrated incrementally into a system, hence, any rerouting of an existing routing path will lead to an interruption. The routing channels are not shared with the modules but constitute obstacles, and are thus restricting the module placement.

2.4.2 *Networks on a Chip*

The motivation for packet switching for on-chip communication comes from the ASIC domain where more and more functional units are integrated. In order to deal with multiple clock domains, modularity for IP-reuse, and to support parallelism in communication and computation, *Networks on a Chip* (NoC) [BM02] have been proposed. In [DT01], Dally and Towles suggested *packet routing* for ASIC design. This is justified by the authors because in ASICs some key portions of a chip are often limited by the wire density and another observation that an average wire is used (toggling) less than 10% of the time. By spending about 6.6% more in chip area in an example implementation, the wire utilization could be materially enhanced without increasing latency because critical communication was pre-scheduled.

NoCs in FPGA-based Reconfigurable Systems An early work on implementing a NoC on an FPGA was introduced in [MBV+02]. The network implements a two-dimensional torus topology and uses wormhole routing. The authors report a throughput of 77.6 MB/s on 16-bit wide data channels between adjacent routers when the system is clocked at 40 MHz. The reported synthesis result for a 2×2 torus is 3,227 slices which is 9.5% of an Xilinx Virtex-II-6000 FPGA. The system was implemented by folding the two-dimensional torus in a one-dimensional slot based structure with the help of tristate-based bus macros (see Page 68 for details on bus macros).

Another FPGA-based NoC implementation [SBKV05] consisting of a 3×3 mesh, requires 3,934 slices on a Xilinx Virtex-2Pro 30 FPGA which is 28.7% of the total logic resources. The achieved clock frequency was 32 MHz and the width of the channels was chosen to be 8 bit. The routers employ an XY-routing strategy with store and forward routing. The target router is accessed by its coordinate. With the proposed 8 bit wide address field, the network is limited to a maximum of 4×4 routers. A router has four ports, connecting adjacent routers, plus a fifth port for connecting a module to the network. All ports can simultaneously receive and send data.

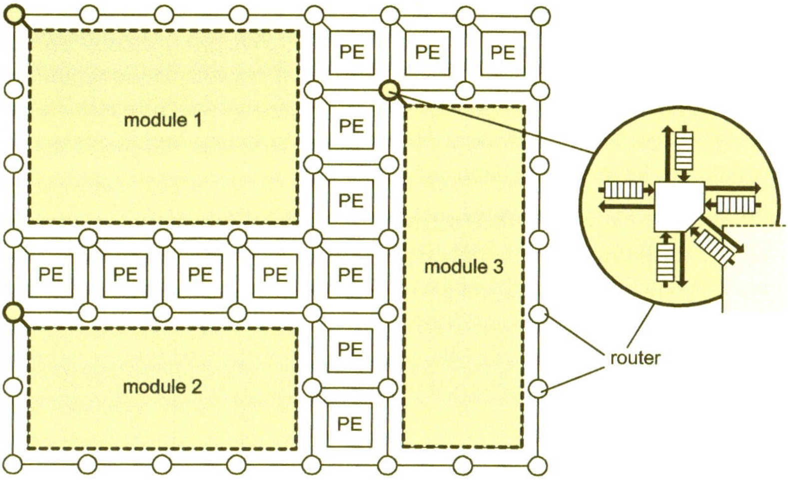

Fig. 2.12 Example of a DyNoC. Modules can be implemented in one or more consecutive processing elements (PE). With the help of partial runtime reconfiguration, routers inside a module can be detached from the network and their logic resources will be used for implementing the module. After terminating a module, the routers can be included again. In a DyNoC, modules must be placed such that it is always surrounded with routers at all four edges

A very similar NoC for FPGAs was previously presented in [ZKS04]. In that work, instead of a complete NoC, a single router that occupies 570–1,830 LUTs on an Altera Flex 10K FPGA, depending on the data channel width (8–32 bit), has been implemented.

To achieve larger networks and to scale down the grid without a dramatic logic overhead, Bobda et al. [BMK+04, ABM+05] propose to detach routers from the network for alternatively implementing the module logic in the resources of the detached routers. As depicted in Fig. 2.12, the routers are arranged in a mesh structure and larger modules may replace routers by reconfiguring the FPGA. By removing routers from the network and replacing them with module logic, modules become obstacles. The XY routing protocol is capable to deal with obstacles by routing packets around them. As this will increase the distance to the destination, the packets are temporarily stamped for forcing the routers not to send the packet back (which would reduce the distance to the destination) but to route the packet around the obstacle. However, this XY obstacle-routing requires that modules are always surrounded by routers at all four edges, which implies that there is a ring of routers around the reconfigurable area.

The largest implemented DyNoC contains a 4 × 4 mesh [BMK+04]. A single router requires 819 (1,229/2,150) look-up tables for implementing routers with 8 (16/32) bit data path width and the maximum clock frequency is reported with 75 MHz [ABM+05].

The listed NoC implementations point out that these network structures are not appropriate for FPGA implementations. The logic overhead is enormous while the achieved data rates are relatively low. Furthermore, the implementations are very restricted and the here mentioned NoCs cannot synchronize between multiple clock domains. In a nutshell, packet-switching is not suitable for integrating partial modules into a reconfigurable system. However, packet-switching may be directly included into the architecture of future FPGAs for providing backbone connectivity between multiple larger logic islands.

Packet-Switching versus Time-Multiplexing on FPGAs Both, packet switching and time multiplexing aim at enhancing the utilization of interconnect resources. While packet switching allocates a bandwidth to modules with respect to the current demands of the modules, time multiplexing is based on schedules for virtualizing interconnect links. In [KMD$^+$06], the authors examined tradeoffs (area, throughput and latency) between time-multiplexed and packet-switched networks. For time-multiplexed networks, schedules have been computed offline with two different algorithms: a greedy heuristic and Pathfinder [ME95]. The greedy heuristic allocates the shortest paths and the earliest possible time slots for a piece of data first, while avoiding congestion. Pathfinder is an iterative algorithm where in each iteration the shortest path is computed for all routes independently. For example, by using Dijkstras algorithm [Dij59] or an A* variant [Tes98]. After each iteration, Pathfinder adapts the cost function with respect to the usage of a particular wire segment and to the available slack of the nets passing a particular wire segment. Consequently, Pathfinder minimizes congestion and latency.

The packet switching network contains simple routers that use a round robin policy for selecting between the input queues of a router. The time-multiplexing is implemented with a simple state machine, but requires some extra storage for the scheduling table. For providing a fair comparison, the authors of [KMD$^+$06] optimized design parameters, such as buffer size or channel widths according to a synthetic benchmark. It was found that packet-switching requires less logic than time-multiplexing with increasing traffic, because of the scheduling tables will become unlikely large.

2.4.3 Buses

Buses are the native way to link communicating modules together within a system-on-a-chip (SoC). All major FPGA vendors offer tools that allow easily integrating a set of user-defined modules or IP cores into complete systems by the use of on-chip buses. For example, Xilinx offers the Embedded Development Kit (EDK) that allows to compose systems based on the CoreConnect bus [Int99] and Altera respectively provides the System on a Programmable Chip Builder (SOPC) for designing systems based on the AVALON bus [Alt07]. Apart from these two commercial bus specifications, the WISHBONE specification [Sil02] is public domain, issued by Silicore. For this bus standard, SoC design tools have been developed by academia [KKH$^+$06] and as open source projects (for example the

WISHBONE Builder [ope]).Because of the prevalence of buses in SoC design, buses are good candidates for also integrating partially reconfigurable modules into a system at runtime.

The following paragraphs will first of all compile an overview about on-chip buses for runtime reconfigurable systems that are physically implemented with tristate drivers. After this, related work on buses that physically link reconfigurable modules with logic resources will be introduced.

On-chip Tristate Buses for FPGAs Plenty of papers [WP04, KJdlTR05, KPR04b, BMG06, PdMM$^+$02] that were published in this field are based on older Xilinx Virtex FPGA architectures that provide wires, spanning over the complete horizontal width of the devices and that can be used to build buses with tristate drivers (see also the paragraph about Xilinx bus macros on Page 68). However, newer FPGA architectures have no support for tristate wires and none of these approaches deal efficiently with dedicated signals, e.g., interrupt lines and chip select signals.

In [WP04], [KD06b], and [KD06a], systems are proposed where fixed resource areas with also fixed connection points to a bus interface are used for the integration of partially reconfigurable modules into a runtime system. With these solutions, all modules have to fit into a resource area of fixed size and one resource area cannot be directly shared by multiple modules, even if the logic of multiple modules would fit into the same resource area. This will typically result in an enormous overhead due to internal fragmentation, as demonstrated in the introduction to this chapter. A suitable bus infrastructure for integrating reconfigurable modules should be able to connect modules of different sizes efficiently with the rest of the system.

A more flexible approach for integrating partially reconfigurable modules into a system at runtime has been published before by Wirthlin and Hutchings. In the *dynamic instruction set computer* (DISC) [WH95, WH96], modules spanning over the full width of the FPGA (a CLAy31 device from National Semiconductor) can be integrated at runtime with the help of a vertical aligned on-FPGA tristate backplane bus. Different resource requirements of the modules are fulfilled by adjusting their height. The module positions can be varied in a one-dimensional fashion.

Hagemeyer et al. [HKP06] propose a similar tristate driver based bus architecture for Xilinx Virtex-II FPGAs. In that system, module sizes may vary in terms of one-dimensional aligned and adjacent resource slots. Another contribution presented in [HKP06] are four different approaches for distributing dedicated write signals, such as enable signals to reconfigurable modules that are connected to the bus. The approaches are illustrated in Fig. 2.13 and include the following techniques:

(a) The *position-based* dedicated signal distribution routes the enable signals in a non-uniformed manner within the resource slots. Consequently, this approach prevents module relocation and does not allow multiple instantiations of the same module.
(b) The *module-based* dedicated signal distribution uses a uniform horizontal routing for the enable signals within the resource slots. The entire vertical

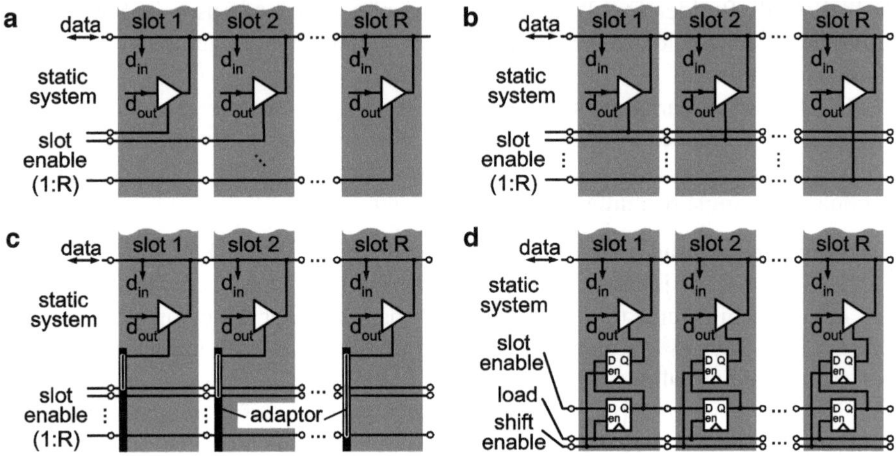

Fig. 2.13 Different techniques for distributing a select signal to reconfigurable modules (from [HKP06]). (**a**) Position-based dedicated enable signals, (**b**) Module-based dedicated signals (**c**) Slot-dedicated enable signals with adaptor, and (**d**) Enabling / Disabling via shift registers

routing to the enable terminal of the tristate driver is part of the module routing. This allows module relocation but still prevents multiple instances of the same module.

(c) The *slot-based* dedicated signal distribution uses a so-called *adaptor* for connecting the tristate enable terminal in vertical direction. The adaptor bridges the different lengths in vertical direction between a fixed connection to the tristate driver and one of the uniform routed internal enable signals. By adjusting the adaptor, multiple instances of the same module can be integrated into the system. In [HKP06], the authors present an approach to use a dedicated logic column (a CLB column in Xilinx FPGAs) for implementing the adaptor which results in an enormous resource overhead for this variant.

(d) The last approach in [HKP06] uses a serial *shift register* with parallel load that is distributed over all resource slots with 1 bit in each slot. The enable signal to a specific slot is distributed by shifting in a one hot encoded enable stream combined with a following load enable for the output flip-flops which store the tristate driver control signals. This technique provides the same flexibility as variant (c) and entails a reasonable implementation cost but requires $R+1$ clock cycles for setting select signals in R resource slots. Furthermore, the approach requires some additional logic within the static part of the system for connecting the shift register interface.

With the above listed approaches, a tradeoff between placement flexibility or overhead in terms of logic or latency is provided. The same group published an improved dedicated write signal distribution scheme in [HKKP07b] with reasonable overhead and low latency that allows free module relocation, as well as multiple instances of the same module. As depicted in Fig. 2.14a, comparators that are

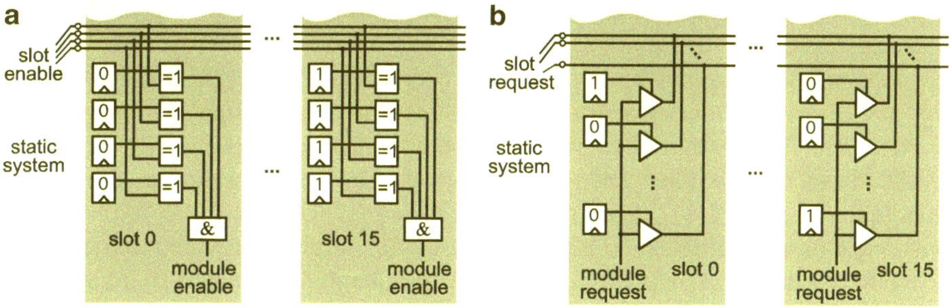

Fig. 2.14 Improved dedicated signal linking from [HKKP07b]. (**a**) Module enabling generation by a comparator that is decoding a binary encoded slot address. (**b**) Module request signal linking (e.g., a module interrupt) by decoding a one-hot encoded slot address

identically placed and routed within the resource slots are used for decoding a slot-dependent address that is stored in some extra flip-flops. This approach exploits a feature of Xilinx Virtex FPGAs that allows to set the initial values of the flip-flops during the initial configuration of the device. Further partial reconfiguration steps will not touch this initial flip-flop state that represents the particular slot addresses. Beside enable signal distribution, [HKKP07b] reveals how individual slot addresses can be used to send request signals via dedicated signal wires from a module to the static part of the system. This is achieved while maintaining relocatability and the capability of multiple module instantiation, as depicted in Fig. 2.14b. The disadvantage of these techniques is that the logic overhead scales with the amount of resource slots which is reasonable for the proposed system with 16 resource slots but unlikely if slots sizes are reduced in order to minimize internal fragmentation. Following the considerations on internal fragmentation in Sect. 2.2, the resource slot size of the proposed system should be about an order of magnitude smaller. Consequently, the amount of resource slots will rise by an order of magnitude, and therefore, prevents the depicted approaches in Fig. 2.14 that are based on fixed slot addresses.

For multimedia processing systems, Sedcole et al. designed the *Sonic-on-a-Chip* communication architecture [SCCL04] that consists of a global bus similar to the one proposed by Hagemeyer [HKP06]. In addition, [SCCL04] reveals a chained bus for connecting adjacent modules. This is also implemented with tristate wires and is possible as these wires are segmented on Xilinx Virtex FPGAs.

On-chip Logic-based Buses for FPGAs Tristate wires require multiple drivers to a single wire, hence, leading to a poor chip utilization if such wires are used for static routing. Static means that the drivers to a tristate wire are not changed. The utilization is poor, because each driver requires some area on the die, but only one driver can be active at a certain point of time. Using unidirectional wires instead would allow multiple independent active wires at the same cost [LLTY04]. Another problem of tristate wires spanning long on-chip distances is their propagation delay which enormously reduces throughput. With respect to the chip area and latency, it is more advantageous to split a wire in segments with small drivers for each segment, instead of using a single large driver [MSCL08]. Finally, tristate wires possess multiple drivers which may produce short circuits that potentially damage the device or may introduce further side effects including supply voltage drops, loose of signal integrity and an increased power consumption.

All these reasons have led to avoiding on-chip tristate wires in ASICs and consequently in FPGAs. Accordingly, all common SoC buses are implemented with logic based multiplexers. Such buses are often implemented hierarchically for reducing the size of the multiplexers and consequently enhancing speed for high performance systems. For example, in a typical system that is communicating via an AMBA bus [Ltd07], a high speed AHB bus (Advanced High-Performance Bus) may directly connect a CPU with the memory while communicating through a bridge with some peripherals connected via a slower APB bus (Advanced Peripheral Bus).

When designing buses for reconfigurable systems, hierarchical buses are unsuitable for connecting multiple modules within one reconfigurable area when these modules should be freely placeable. For such systems, the bus interface has to be uniformed and the challenge in designing on-FPGA buses for reconfigurable systems is to build buses providing low latency at low implementation cost that can integrate modules placed at many different positions.

A straight forward method of connecting modules with the help of logic resources to a bus is to simply bridge the signals from a configuration island into the static part of the system that contains all bus logic. This is implemented by Koh and Diessel with the *COMMA* architecture [KD06b, KD06a] that provides multiple reconfigurable islands, located in a two-dimensional fashion. By using look-up table bus macros, simple point-to-point bridges are implemented for linking partial modules with the bus logic. This logic is completely located within the static part of the system. The system was implemented on a Xilinx Virtex-4 FPGA that can be configured sub-column wise instead of configuring the device over the full height. By choosing the height of an island to exactly one sub-column, modules can be exchanged without influencing other parts of the system. The bus macro bridges deploy two look-up tables per signal wire that include no additional functionality, thus, resulting in pure logic and latency overhead.

A more advanced communication architecture was developed by Hagemeyer et al. [HKKP07b] as an alternative for tristate buses. In that approach, parts of the bus logic are joined together with the look-up tables used for connecting the bus with the modules. More details on this work are depicted in Fig. 2.41.

Serial Buses for FPGAs Some systems do not require high speed bus communication. For example, a FIR filter module may require some register file accesses for initializing and parameterizing the filter, while the entire data for computation may be transferred via dedicated resources for the complete execution time of the module. Then a serial bus may be appropriate to connect the control CPU with the filter module.

Such a serial bus for reconfigurable systems is presented in [KKT06] for a cryptography application. In that system, a master sends an address followed by a command and optionally the data to the slaves that are daisy chained. In other words, the address, the control, and the data signals of a master are serialized and transmitted to the slaves. In each slave module, the address value is decremented and sent further to the succeeding modules. If the incoming slot address is equal to zero, the module is selected and the command is executed. Commands can include reading or writing a particular register as well as some simple module control signals, such as a selective module reset. Except for the physical voltage levels, the serial interfaces in [KKT06] follow the RS-232 specification. Including a register file, a UART, a control FSM and the address computation, one module interface is reported to cost 830 look-up tables.

It must be mentioned that this interface features some fault tolerance against temporal interruptions of the chain as well as against bit errors, because the proposed serial bus was used to link partial modules together that may be spread over multiple FPGA boards. A test implementation demonstrated, that the interface logic can be scaled down to less than 300 look-up tables when omitting fault tolerance. Each module provides one RX-TX (receive and transmit) pair for the upstream from a master to the slave modules as well as a second pair for the downstream link. Therefore, only four wires are sufficient for connecting partial modules via this serial bus with the rest of the system. In [KKT06], modified look-up table-based macros - located at the left and right border of a module – have been used for building a system featuring a chained reconfigurable bus along adjacent modules.

2.4.4 Physical Implementation Techniques

The communication architecture mainly restricts the placement style (as explained in Sect. 1.6). While an island style reconfigurable system may be implemented by locating all communication logic within the static part of the system [WP04, KV06, KD06a, LBM+06], more complex systems require a corresponding communication architecture, which may include large portions of the logic transparently within the reconfigurable modules [HKKP07b, KPR04b]. So far, suitable architectures for efficiently supporting a grid style module placement have been rarely proposed (e.g., [OM07, BHB+07]). The challenge in designing runtime reconfigurable systems featuring a two-dimensional grid style is to include variably placeable modules that can be freely scaled in height and width.

Partial reconfiguration implies that modules are integrated into a system at runtime. This means that the module interface signals – for example all signals of a VHDL entity description of a module – have to be connected to the rest of the system. Hence, partial runtime reconfiguration demands a strict separation of the module computation from module communication. This allows handling the module functionality as a black box while only requiring to adjust the communication for a particular reconfigurable module at runtime after or during reconfiguration. The modules themselves as well as the static system are physically stored as completely routed components in bitstream repositories, as it is virtually impossible to perform

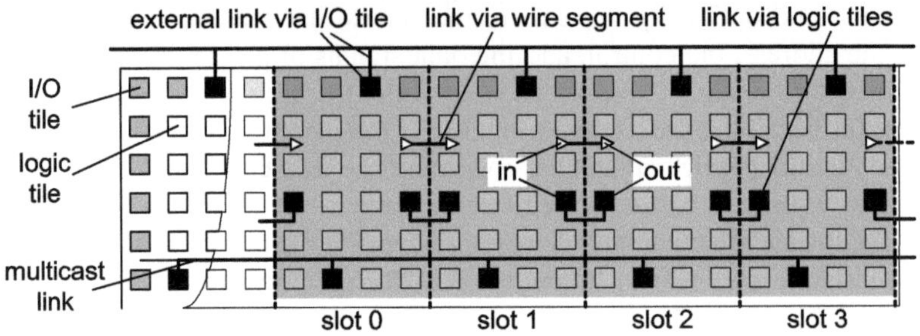

Fig. 2.15 Four resource slots assigned logically on top of the logic and routing fabric of an FPGA. Communication is either established via external I/O tiles or via dedicated wire segments of the FPGA fabric. Depending on the tools, the latter one may be implemented directly or with the help of additional logic tiles acting as connection points

full place and route in an embedded system at runtime. As a consequence, *physical implementation techniques* for providing low level connections are necessary.

For providing links among the interfaces of multiple reconfigurable modules or between the static part of this system, modules must be designed with corresponding connection points acting as fixed transfer points for the module interface signals. Basically, this could be achieved by simply binding specific signals that have to cross the module boundary to dedicated wire segments of the FPGA fabric. Depending on the design tools, such wire segments are typical not directly usable at higher abstraction levels. To overcome this issue, links via I/O pins [AS93,LNTT01, MTAB07] for connecting reconfigurable modules have been proposed, as well as so-called *hard macros* [LBM+06] that are build upon additional FPGA functional units. The connection points must be bound to predefined resources on the FPGA, that have an equivalent placement in the static system as well as inside the partial modules. Equivalent means that specific signals are located at the same fixed relative locations inside all resource slot, as illustrated in the examples of Fig. 2.15.

2.4.4.1 Partial Module Linking via I/O Pins

Using I/O pins is a simple, straight forward method for providing communication to and from reconfigurable modules. This was used in pioneering approaches for exchanging full FPGA configurations. In such systems, a resource slot engages a complete device and partial reconfiguration is achieved on system level by exchanging some full FPGA configurations while keeping other devices operating. However, the progress in silicon industry results in the ability to implement full systems on a reconfigurable chip (SoRC). Hence, partial reconfiguration will be defined as exchanging fractions of the configuration of a single FPGA throughout the rest of this book.

An early work using runtime reconfiguration was proposed by Athanas and Silverman in the *PRISM* project [AS93]. PRISM is an acronym for "processor reconfiguration through instruction set metamorphosis". The proposed system comprises some Xilinx 3090 FPGAs for extending the instruction set of a M68010 processor. In this system, the FPGAs are fully reconfigured for adaptations which was called *instruction set metamorphosis*. This concept was retained in a further development of *PRISM-II* [WAL$^+$93], which features Xilinx 4010 FPGAs. In these systems, the coupling with the static part (here the CPU) was completely carried out via the I/O pins of the FPGAs that are individually but only fully reconfigured. The FPGA I/O pins are directly connected to the PCB backplane bus.

Lockwood et al. proposed a system called *field programmable port extender* (FPX) [LNTT01] for network packet processing with partially exchangeable accelerator modules. In this system, all interface signals of a partial module are routed to I/O pins, located inside the module bounding box. And most of these pins are connected to external memory buses. These buses are designed for high throughput with a width of 64 (SDRAM) and 36 (SRAM) bits. Consequently, a single partial module requires a considerable number of external pins, thus, making this approach very costly and unlikely for scaling the number of simultaneous partial modules inside the system. The approach follows an inflexible island style placement of the modules, hence, the external I/O pins are exclusively allocated to one particular module.

Linking partial modules via I/O pins has also been proposed in a project called the *Erlangen Slot Machine* (ESM) [MTAB07], as just one possibility of providing physical links to reconfigurable modules. Beside other techniques (listed later throughout this section), the ESM can directly connect modules to external peripherals (for example, a video DAC or an AC97 audio chip) with the help of an external crossbar chip. This allows a fine-grained, and therefore flexible placement grid, while still providing access to the external peripherals. These advantages are achieved by the cost of an extra crossbar device that causes additional latency and is consequently limiting throughput. A further restriction is that the module size is related to the maximum of the resource requirements for implementing the logic and for providing the external I/O pin interface. The ESM follows a one-dimensional placement style with columns spanning the complete height of the device. Consequently, the size of a reconfigurable module is directly related to its width; and the used main FPGA of the ESM (XC2V6000) provides only three I/O pins at the bottom side of the device per logic column. Whereas one column contains 768 4-bit look-up tables on this device. These three bottom I/O pins are linked to the crossbar chip. In other words, the I/O-pin density is 3 pins per 768 LUTs that is 12 pins for an average 32 bit softcore CPU with about 3,000 LUTs.

In the ESM, I/O pins are not only used to link peripherals to the reconfigurable modules but also for directly connecting dedicated external SRAM memory banks. The ESM platform provides six memory banks that are connected to the top I/O cells located above a logic column. A distinctive feature of this scheme is that modules may be relocated to different positions (start columns) while still being capable to access a private external memory bank.

2.4.4.2 Partial Module Linking via Tristate Buffers

While modern FPGA architectures are no longer supporting internal tristate drivers,[4] this was common in many older FPGA architectures. Such tristate signals have been mainly used for implementing wide input functions, such as wired-OR or wired-AND gates, as well as for global communication purposes.

A pioneering project using internal tristate buffers for integrating partially reconfigurable modules in a single FPGA is the *dynamic instruction set computer* (DISC) [WH95, WH96] proposed by Wirthlin and Huthings. Based on two National Semiconductor CLAy31 FPGAs – one for control and one for actually accelerating computations – the system is able to partially exchange simple instructions with special configuration commands (as part of the instruction set). The accelerators span the full width of the device for connecting tristate wires distend over the full height of the device. The modules height is adjustable to the entire logic demands. As the tristate drivers are located in a fine vertical grid, modules can be placed freely into the system following a vertical one-dimensional style, while still being able to communicate with the rest of the system. Consequently, the system features module relocation. Furthermore, the device could accommodate multiple reconfigurable modules at the same time. In this approach, the tristate drivers are actually used as connection points for linking the signals of the hardware accelerators to specific long lines.

Xilinx Bus Macros Most projects using tristate drivers for implementing a module interface to partially reconfigurable systems on Xilinx FPGAs are based on so-called *Bus Macros* [Inc02]. These macros define the positions of a set of tristate drivers acting as connection points to predefined horizontal long lines, hence, following the same idea of the vertically aligned DISC approach. The reason for using horizontal long lines stems from the fact that Xilinx FPGAs follow a column based reconfiguration style. This means that configurations cannot be written randomly to the device but in larger portions of so-called vertically frames that span the complete height of the device. Ergo, partial modules for Xilinx FPGAs should have thin and tall shape.

For Xilinx FPGAs that provide internal tristate drivers and according wires, the propagation delay on these resources is higher as compared to other routing resources. On the other hand, these routing resources can be used independently from the look-up tables that implement the logic of a module. As a result, linking partially reconfigurable modules with tristate drivers on Xilinx FPGAs can be implemented with low logic resource overhead.

---

[4]The last FPGAs with internal tristate drivers to internal wires are the Xilinx Virtex-II and Virtex-II Pro FPGAs that have been released in 2000 and 2002 respectively for the II Pro series. Tristate signals for partially reconfigurable systems come along with additional place and route restrictions, as typical FPGAs provide only very few of such routing resources, and require timing parameters that must be met to turn buffers on-and-off. This leads typically to lower clock speeds as compared to logic based approaches.

Walder and Platzner proposed a *reconfigurable runtime environment* [WP04] using tristate driver based macros for Xilinx FPGAs that allows including multiple modules of different size in a one-dimensional slot based fashion into the system. The reconfigurable modules access the bus via a *bus access controller* for encapsulating the module interface to the bus arbiter. Consequently, the data transfer protocol is transparent to the modules and permits concurrent bus transactions. In this approach, modules are not allowed to have direct connections to some I/O pins or to other on chip modules inside the FPGA.

In [KPR04b] Kalte et al., a further approach of a one-dimensional tristate driver based bus is presented for the *RAPTOR2000* [KPR02] prototyping platform. Here, focus is put on tightening the resource slot width in order to reduce the effects of internal fragmentation. In addition, the authors present how the tristate wires can be used for implementing segmented buses. Therefore, multiple bus transactions on the same horizontal wire can be performed. This is possible because the used Xilinx FPGAs provide configurable switches to link tristate buffer lines to longer wires. A major drawback of this approach is that so-called *bridge modules* must be placed into the system just for activating the switches between two consecutive wire segments.

In this system, the authors propose linking partial modules with the help of tristate drivers not only for bus signals but also for I/O pins. Consequently, partial modules can be relocated to different positions inside the FPGA while capable of directly accessing external peripheral components.

Further Approaches Because of a lack of available and easily usable tools, multiple other systems have been developed that are basically variants following the techniques of the three above mentioned papers. For example, Bieser and Müller-Glaser [BMG06] presented an FPGA-based rapid prototyping system containing a simple lightweight tristate driver based system bus for directly linking modules in a slot based manner. Krasteva et al. [KJdlTR05] reintroduced the virtual wire concept and implemented a tool called *BITPOS* [KdlTRJ06] for merging partial module configurations into the static configuration bitsteam. In [SGS05], Sunwoo et al. propose a reconfigurable system based on the Atmel AT94K FPGA architecture for providing RAM connections for built-in self-testing (BIST).

2.4.4.3 Partial Module Linking via Logic Resources

As an alternative to the tristate driver approach, it is possible to build hard macros from the look-up tables (*look-up table bus macro*) that can also be instantiated within some high level design tools (e.g., the Xilinx design tools). By instantiating two LUTs, one at the beginning and one at the end of a wire resource, it is possible to bind a particular signal to a specific wire. The macros are designed and placed such that one LUT is placed inside the reconfigurable module bounding box, and the other one inside the static part of the system or another module, as shown in Fig. 2.16. As a consequence, the LUTs act as connection points to the wire inside the macro that passes the boundary of a partial module in a predefined manner.

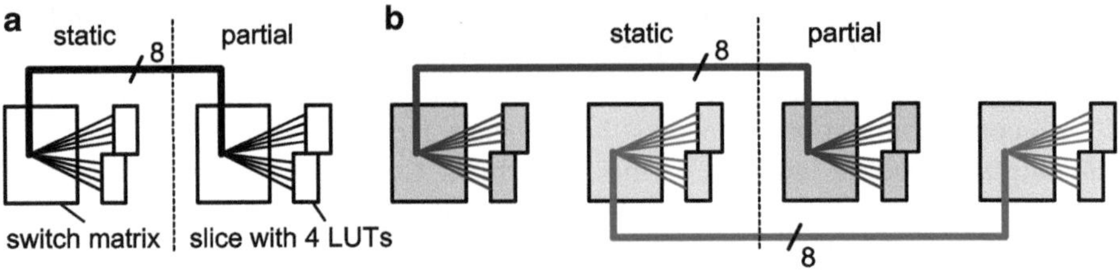

Fig. 2.16 Look-up table bus macro. (**a**) Narrow macro, (**b**) two nested wide bus macros for linking up to 16 signal wires in a single CLB row

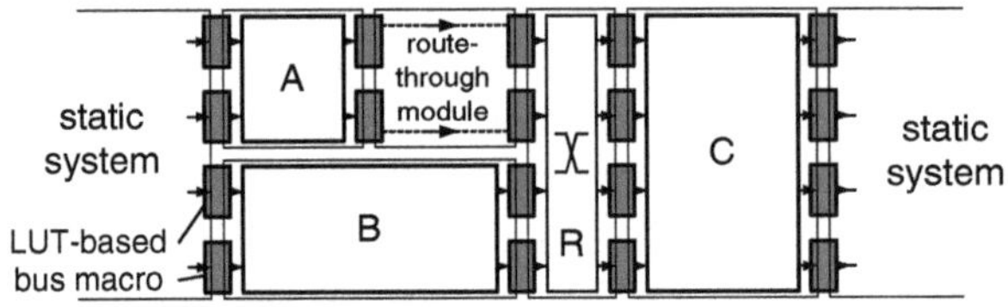

Fig. 2.17 Row-based two-dimensional module linking as proposed in [SF08]. Modules may be of one or more rows height and communicating via LUT-based bus macros. Route-through modules bridge horizontal gaps between modules and dedicated routing modules carry out vertical routing between the rows (see Module R)

As compared to the tristate driver approach, the look-up table bus macros allow a higher density of signals and a lower latency by the cost of two look-up tables just for implementing the communication between the static system and a module. Look-up table bus macros are obligatory for all modern Xilinx FPGA families including Spartan-3, Virtex-4, and Virtex-5 that are not equipped with internal tristate drivers. A design flow based on LUT-based bus macros was presented in [LBM^{+}06].

In [KD06b, KD06a] LUT-based bus macros have been used to link partial modules in a vertically aligned resource slot style into a system that is based on Xilinx Virtex-4 FPGAs. The proposed system provides multiple different reconfigurable islands and a tool assists in instantiating and placing the communication macros at the module sides in a regular manner. This allows variable module sizes and relocating modules to different slot positions. The proposed system incorporates that Virtex-4 FPGAs can be reconfigured sub-column-wise in multiples of 128 4-bit look-up-tables. This allows adjusting the module height in multiples of a sub-column. The logic overhead for this communication architecture is enormous with two LUTs for each signal that is passing a resource slot boundary. The approach does not allow direct connections between partial modules and the complete communication is carried out via the static part of the system with multiple wide multiplexers.

By tiling the reconfigurable area in a few rows and in restricting modules into bounding boxes that are allowed to be one or more rows height, Silva and Ferreira designed a reconfigurable system allowing to place modules in a two dimensional fashion [SF08]. As illustrated in Fig. 2.17, modules can only receive data from the left side and send data to the right side via LUT-based bus macros. Vertical routing

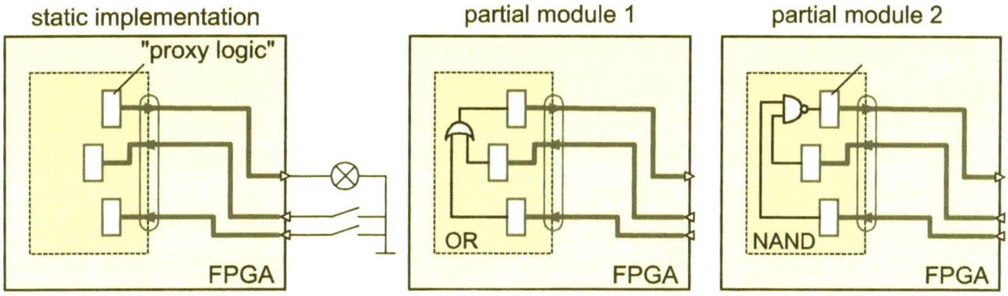

Fig. 2.18 Partial module integration using *proxy logic*. After an initial static implementation, the routing is used and preserved for incrementally building all partial modules

is provided by special switching macros that can route an input stream from one row to another one that may be located in a different plane. To adjust modules of unequal width, route through modules can be instantiated that expand the interface of a module, so that multiple modules end at the same switching macro or to bridge modules with the static system.

LUT-based bus macros can be implemented in all horizontal and vertical directions. This is used in [OM07] for implementing reconfigurable systems with prerouted modules possessing macro connections at the edges for linking adjacent modules or for connecting the static system. By providing implementation alternatives offering varying shapes and different signal directions at the borders of the modules, the runtime system is able to plug modules together very flexible in a two dimensional grid style. This approach requires that all data streams are routed through a chain of adjacent modules making this approach unlikely for larger systems containing, many modules and many different data streams.

Dynamic Hardware Plugins (DHP) In [HLTP02], Horta et al. published an early concept for integrating partial modules with look-up tables that connect input and output antennas into a so-called "gasket area". The gasket was used to overcome limitations of early FPGA design tools in strictly following area constraints for routing. Here, the idea is to leave a logic corridor between the reconfigurable islands and the static part of the system unused. This leads to an oversupply in routing resources inside the gasket for supporting the place and route tool not to use routing resources within the static system when implementing a reconfigurable module or not to use resources inside the reconfigurable islands for the static part of the system, respectively. Only the antennas area allowed to completely cross the gasket for linking partial modules with the static system. Obviously, this approach is very costly. Furthermore, with improved tools, the gasket technique has become obsolete.

2.4.5 Proxy Logic

The bus macro approach costs two LUTs per signal wire and additional latency. This was improved in the resent Xilinx vendor tools using an incremental partial design flow [Xil11]. As shown in Fig. 2.18b, anchor LUTs in route-through mode (called *proxy logic*) are placed in the partial region for each signal crossing the partial module border. During the implementation of the static system, the partial interface signals are routed to the anchors. The partial modules are implemented incrementally from this static system without modifying any of the already implemented static routing. Consequently, the partial module interface wires are constrained by preserving the initial static routing.

The proxy logic approach costs still one LUT per signal wire. Its main drawback is that the routing is different for each reconfigurable island. This prevents module relocation even if the islands provide an identical logic or memory layout. Moreover, changes in the static system will in general result in a different routing to the proxy logic. Consequently, all permutations of module type and placement position have to be rerouted on each modification of the static system. Hence, this approach is only suitable for systems of low complexity.

The Xilinx bus macro approach has dominated the FPGA runtime reconfiguration community for more than a decade. However, Xilinx moved completely over to their new fourth generation PR flow based on proxy logic while removing the support for bus macros in their latest tools. For this flow, Xilinx included some extra features in their flooplanning and constraint generation program PlanAhead. This includes resource budgeting, defining reconfigurable islands, and an automatic placement of proxy logic. While especially the last feature has substantially simplified the design flow, a lack of routing constraints still prevents this flow to implement multi-island or slot and grid style reconfiguration as described in Sect. 1.6.

In addition to Xilinx, Altera (as another major FPGA vendor) has announced to support partial reconfiguration in their devices and design tools. According to [Mar11], designing partial systems will be based on an incremental design flow very similar to the recent flow proposed by Xilinx. Again, logic primitives will be placed into the partial region and connected via a static implementation phase and further incremental module implementation steps. Consequently, these systems will include the same limitations.

2.4.6 Zero Logic Overhead Integration

As an alternative to the Xilinx busmacro and proxy logic approach, it is possible to fully remove the logic overhead for the communication to or from reconfigurable modules. The *zero logic overhead integration* combines the Xilinx bus macro technique with proxy logic. As shown in Fig. 2.19a, modules can be plugged directly into a reconfigurable region without additional logic resources for the communication and only the wires act as plugs, called *PR link*. When implementing the static system (Fig. 2.19b), connecting primitives are placed *into* the partial region that act as a placeholder module for the non existing partial modules. So far, this follows basically the proxy logic flow from the last section. However, by

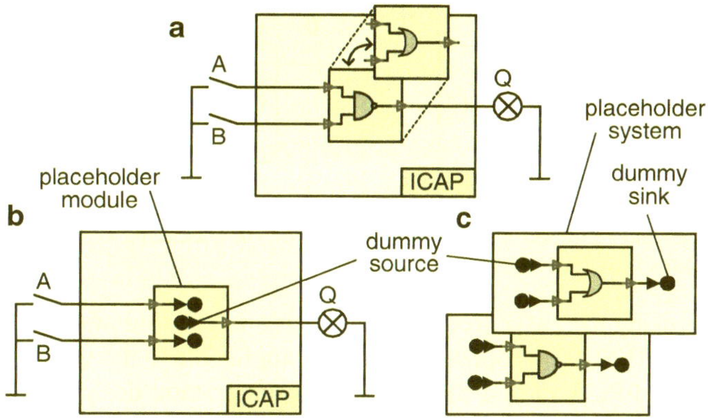

Fig. 2.19 (**a**) Partially reconfigurable system using zero logic overhead integration. The system is composed of an implementation of (**b**) a static system and (**c**) independently generated modules

constraining the routing *before* implementing the static routing, we can define which wire of the FPGA fabric will route a signal to or from the reconfigurable region. Note that no extra logic will be added in the static part of the system.

For implementing the reconfigurable modules (Fig. 2.19c), we will place connecting primitives *outside* the partial region that act as a placeholder for the non existing static system. Again, we constrain the routing in such a way that it is compatible to the routing used for the static system. This means that the same wires will be used for the same interface signals in the static system and all partial modules. For this implementation, no extra logic is added into the reconfigurable region. Consequently, when placing the partial modules into the static system at runtime (Fig. 2.19c), no placeholder primitive will remain in the system.

The difficulty of constraining the routing was solved with blocker macros. Such a macro is placed into the partial region when implementing the static system. There it will occupy all routing resources in order to force the Xilinx vendor router not to use resources inside the reconfigurable region. However, by leaving *tunnels* to the placeholder modules inside the blocker, the routing will be constrained. A tunnel is a path of wires that are excluded from the blocking. The partial modules are implemented similar. Here, the blocker will be located around the partial region and tunnels will guide the routing to the connection primitives that replace the static system. Note that the static and partial module implementations can be carried out independently to each other.

2.4.6.1 Alternative Methods for Physically Linking Reconfigurable Modules

For the sake of completeness, alternative approaches for integrating reconfigurable modules into a system at runtime will be presented.

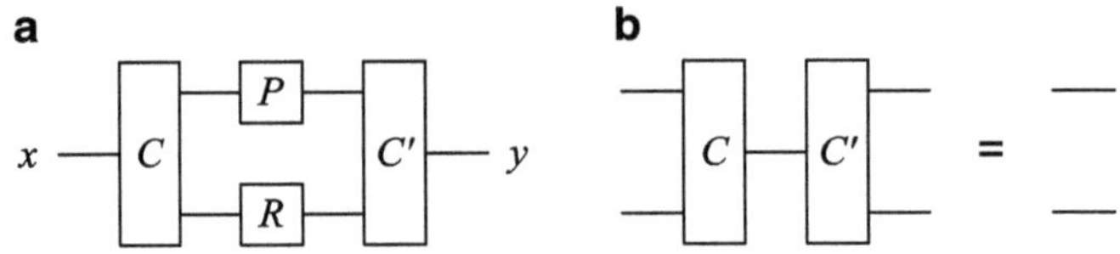

Fig. 2.20 (**a**) Static modeling of a system that can behave either as P or as R depending on the control blocks C and C' (control signals are not shown). (**b**) Cascading C' and C should behave like a pair of wires [LSC96]

Instead of using the gasket technique to overcome the limitations of early design tools, Luk at al. [LSC97] proposed a solution, applicable to the vendor tools for the Xilinx XC6200 FPGA at that time. Starting from a static design, all modules are built one after another, using an incremental design flow. With their tool *ConfigDiff*, the authors have been able to generate differential bitstreams that contain just the differences between two full configurations. Obviously, this approach is not scalable, because it requires $|\mathcal{M}| \cdot (|\mathcal{M}|-1)$ differential bitstreams for a system with $|\mathcal{M}|$ partial modules that can be exclusively loaded to the FPGA, if the modules can be configured in any order.

A similar approach was presented in [KT04] to partially configure any kind of FPGA. In this work, an incremental design flow was used to build a full static configuration bitstream and multiple complete bitstreams containing the static part and one or more exchangeable modules. By correlating the complete module bitstreams with the full static bitstreams a resulting bitstream was produced that mostly contains the fraction of a particular exchangeable module. This correlated bitstream was compressed for the runtime bitstream repository. At runtime, a dedicated module bitstream can be decompressed and merged into the full static bitstream. The goal of this approach is to reduce the bitstream repository size by preventing to store multiple full bitstreams. This technique and the ConfigDiff approach omit physical connections that are specially designed for partial runtime reconfigurable systems. Consequently, connections to the reconfigurable modules and some routing nets of the static system may be interrupted when swapping a module. This is because of some nets may be routed on different paths from each place and route run for implementing a module.

A straight forward approach for modeling reconfiguration with connection points on an FPGA was presented by Luk et al. [LSC96] based on the *Ruby language* [JS90]. Opposed to VHDL or Verilog, Ruby facilitates the encoding of layout information to a design. For example, ruby allows to express that function block A is arranged beside B: $(A \leftrightarrow B)$ and that C is grouped beside D: $(C \leftrightarrow D)$ while (A and B) are grouped below (C and D): $(A \leftrightarrow B) \updownarrow (C \leftrightarrow D)$. These capabilities have been used in [LSC96] to guide the physical implementation of a dynamically reconfigurable design. The idea is to include control blocks into the design for switching between multiple function blocks as illustrated in Fig. 2.20a. In the model, the control blocks behave like demultiplexers and multiplexers (like introduced in Sect. 1.4). Within the physical implementation, the control blocks represent plugs at fixed positions that act as connection points behaving like simple wires (see Fig. 2.20a).

2.5 Bus-Based Communication for Reconfigurable Systems

Buses provide the most common and cheap communication technique between multiple modules and memory. Therefore, buses are candidates for the communication with runtime reconfigurable modules. Comparable with plugging an extension card (e.g., a network adaptor) into a PCI backplane bus on a PC mainboard, a reconfigurable module should be as easily integrated on a reconfigurable device.

In this section, new concepts of a bus-based communication architecture for implementing reconfigurable SoCs will be discussed and analyzed. The communication architecture includes physical connection techniques and is not restricting the overlaying bus protocol. Consequently, the communication architecture is compatible with most commonly used bus standards found in SoCs, including AMBA [Ltd07], CoreConnect [Int99], AVALON [Alt07], and Wishbone [Sil02]. The aim of the bus architecture is to allow changes to the system, for example, by loading a new module configuration to the FPGA, without interfering other running modules connected to the same bus. Instead of using tristate bus lines (see Fig. 2.21a) that are often used for buses at board level or for backplanes, this work considers buses that are based on unidirectional signal lines that are typically linked together by multiplexers (see Fig. 2.21b) or logic gates (see Fig. 2.21c) (see also the footnote on Page 68).

A bus is a set of signals where different signals or groups of different signals have special purposes, as shown in Fig. 2.22. A bus provides a shared medium for the communication between various modules connected to the bus. These modules can be divided into two classes, namely (1) master modules and (2) slave modules. Only masters are allowed to control the bus, e.g., by driving an address to the bus, whereas slaves are only allowed to respond to requests from a master. However, by the use of interrupt lines, slaves can notify a master to initialize a communication with a master. Table 2.1 presents an overview of signal classes found in typical on-chip buses. The table distinguishes between signals that are driven by masters (*master write*) or slaves (*master read*). Furthermore, the table differentiates if a signal is *shared* among different slaves, or if there is a *dedicated* (non-shared) connection between a slave or a master module connected to the bus.

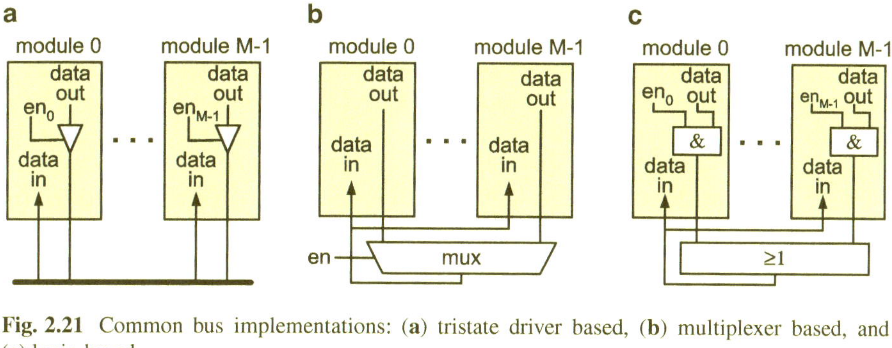

Fig. 2.21 Common bus implementations: (**a**) tristate driver based, (**b**) multiplexer based, and (**c**) logic-based

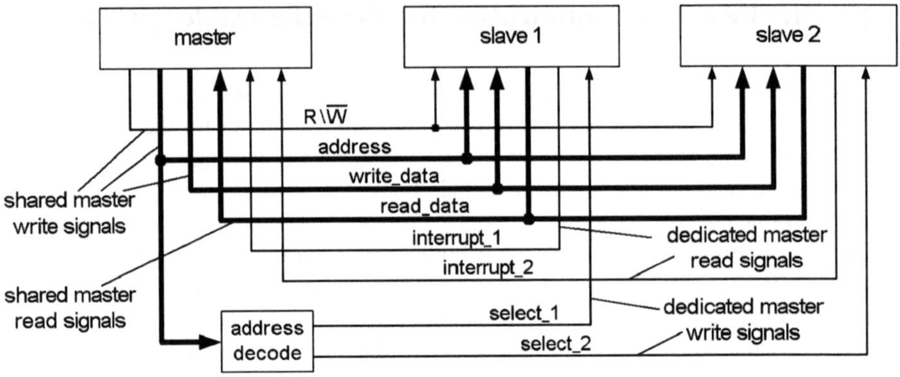

Fig. 2.22 Example of a system consisting of a master and two slave modules communicating through a bus. The shown bus signals give examples for all problem classes classified in Table 2.1

Table 2.1 Classification of bus signals. All bus protocols can be implemented by combining signals from the four problem classes listed in the table. Examples for each problem class can be found in Fig. 2.22

	Shared	Dedicated (non-shared)
Master	`address`	`module_enable`
	`data_out`	`bus_request`[a]
write	`write_enable`	
	`read_enable`	
	`byte_select`	
Master	`data_in`	`interrupt`
read	`address_in`	`wait_request`
		`bus_grant`[a]

[a] Signal for master-arbiter communication

If multiple masters are connected to the same bus, an arbitration scheme is required to resolve conflicts. For example, if two masters want to access a slave at the same time, besides *masters* and *slaves*, there must be a third class of

bus subscriber namely an *arbiter*, connected to the bus. The arbiter communication takes only place with the master modules by using a dedicated bus signal for each master (e.g., `bus_request` or `bus_grant`). However, with respect to reconfigurable on-chip buses, these bus signals form no extra problem class. For instance, the signal distribution of an interrupt request of a slave to a master is basically the same problem as the distribution of a bus request signal to an arbiter. Therefore, the communication with an arbiter will not be examined any further. All approaches presented in the following sections discuss the communication between a master (located in the static part of the system) and one or more slaves (implemented as partially reconfigurable modules). However, the techniques can be easily extended for multi-master operations on a reconfigurable bus architecture.

For the following, it is assumed that partially reconfigurable modules are separable from the overall system. This can be achieved by applying some logic and routing constraints to the interchangeable modules. For example, by applying area constraints in forms of bounding boxes, synthesis tools can be forced to fit modules exclusively within a specific region of the FPGA while further preventing the usage of logic resources located inside the bounding box by other parts of the system. In addition, all bus interface signals have to be routed via dedicated wires, such that a static part of a system can communicate with a dynamic reconfigurable part of the system over these wires only. The wires will be used for linking together the static and the dynamic part of the system among different modules placed into an area allocated for dynamic partial reconfiguration. This is comparable to connecting some cards via plugs in a traditional backplane based bus. Here, interface signals are connected via dedicated pins inside the plugs and the pin layout must be equivalent if the same card should operate in any plug.

There exists a lack of adequate constraints on wires inside commonly used synthesis tools that prevent the tools to directly map a signal to a specific (user-defined) wire or a set of wires. However, this lack can be avoided by utilizing special macros built of additional resources connected to both sides of the desired wire. With respect to the Xilinx design tools, it is possible to use any look-up-table or the attached flip-flops behind the LUTs for this purpose. These resources provide connection points for such a user-defined wire. Instead of assigning a wire directly within the tools, an associated macro is instantiated and placed to a user-defined position with the help of location constraints. Such macros act as plugs for the module interface signals (the top-level interface of the partial modules) that have to cross the module border. The here proposed ReCoBus and connection bar communication architecture will be physically implemented as monolithic hard macros that can be directly instantiated by the Xilinx design tools. The here used term "macros" should not be mixed up with the term "bus macros" used in an alternative design flow provided by Xilinx (see also the paragraph on Xilinx bus macros on Page 68).

Each quadrant in Table 2.1 represents a different problem class for a specific set of bus signals that has to be addressed individually when dynamic reconfigurable on-chip buses should be implemented efficiently. Based on an example with a master located in the static part of the system and some partially reconfigurable slaves located in a set of horizontally aligned resource slots, multiple methodologies for implementing the different bus signal classes will be proposed and analyzed separately in the following four sections.

2.5.1 Shared Write Signals

Examples for write signals that are driven by a master and shared among multiple slaves are address lines, write data lines, or read/write control signals. The implementation of shared write bus signals is trivial for static only systems, where these signals can be freely routed on the FPGA. For dynamic reconfigurable buses, most related approaches are based on a *macro approach* [LBM$^+$06] that requires

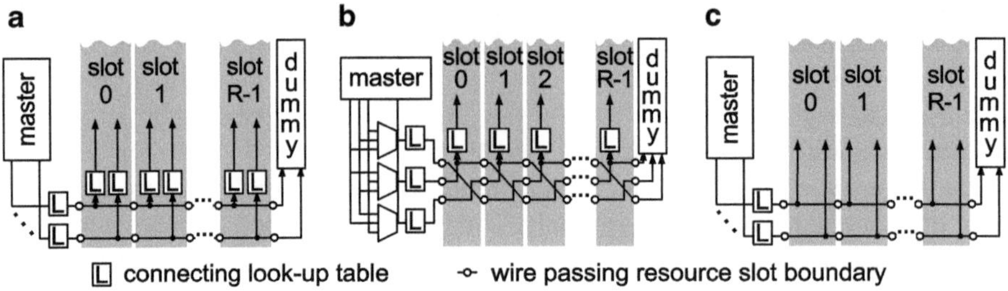

Fig. 2.23 Shared write distribution. (**a**) Common macro approach requiring a LUT per shared write signal in each resource slot. (**b**) Interleaved connection scheme exemplary for $N = 3$ interleaved chains for reducing the amount of connecting LUTs within the resource slots by N. (**c**) Implicit connection scheme without any connecting LUTs inside the resource slots. A dummy ensures a homogeneous routing also for the last resource slot such that it is compatible to all other resource slots

some resources in addition to the wires inside every module connection point for providing shared signals. Consequently, the amount S_{SW} of shared[5] write signals would occupy the same number of look-up tables (or some other logic resources), multiplied by the amount of resource slots R when the module connection points are *explicitly* provided within each resource slot. In addition, S_{SW} LUTs are used for the connection to the static system as illustrated in Fig. 2.23. Therefore, the resource overhead L^c_{SW} (in terms of look-up tables) for implementing S_{SW} shared write signals using the common macro approach is:

$$L^c_{SW} = (1 + R) \cdot S_{SW} \quad (2.4)$$

The examples in Fig. 2.23 comprise a dummy sink that is required to arrange the routing of the last resource slot (slot R-1) just as homogeneous as the other slots. It is possible to avoid this dummy by simply extending the wire into the static area located right beside the reconfigurable area while leaving the end unconnected as an antenna. Alternatively, the dummy can be implemented by instantiating some look-up tables for terminating the shared write signal with the LUT inputs.[6] This requires $\lceil \frac{S_{SW}}{k} \rceil$ k-input LUTs. As a further alternative, the last resource slot will be left with another internal routing as the other slots. As revealed in Sect. 3.2.6 this will not necessary restrict the module placement as the routing can be adjusted at runtime.

For reducing the overhead of the common macro approach, a so-called *interleaved connection scheme* has been proposed [KHT08]. Here, a shared write wire is not tabbed within each resource slot but once every N-th slot, as shown in Fig. 2.23b. The idea behind interleaving shared write signals is that in typical systems smaller

[5]The letters that are also used as indices will be underlined throughout this section.

[6]Connecting shared write signals to a LUT after the last resource slot will work more stable with the Xilinx design tools as compared to the antenna approach.

modules (requiring less resource slots) also require smaller interfaces. Analogously, larger modules require a respectively wider interface. Consequently, interleaving allows reducing the resource slot width, and therefore, reducing internal fragmentation, without additional look-up tables inside the resource slots. In order to allow a free module placement to the resource slots, additional *alignment multiplexers* in the static part of the system will ensure that a particular shared write signal is driven to an interleaved chain throughout the resource slots depending on the placement position of the currently selected module. Note that this approach only works for one selected module, hence it is unlikely to be used for control signals. However, the majority of write signals like write data and address lines will take advantage of this technique, and it may be combined with the common macro approach for distributing the remaining control signals. A more detailed discussion on interleaving bus signals can be found in Sect. 2.5.3. Let N be the number of interleaved chains and $MUX(N)$ be the number of LUTs required to implement one N-input multiplexer, then the resource overhead (in look-up tables) for implementing S_{SW} shared write signals using the interleaving approach is:

$$L_{SW}^{in} = S_{SW} \cdot MUX(N) + S_{SW} + \left\lceil \frac{S_{SW}}{N} \right\rceil \cdot R \tag{2.5}$$

The logic overhead can be reduced further by connecting the reconfigurable modules directly to the *internal bus wires* that implement the shared write signals, as shown in Fig. 2.23c. For allowing a relocation of modules connected to the bus, the routing has to be constrained, such that, the internal bus signals always occupy the same equivalent routing resources within each resource slot. In the *common macro based* approach proposed by Xilinx [LBM+06], the routing is not required to be arranged in a uniform manner, because the flow follows an island reconfiguration style, where modules share a single fixed area over time. In slot or grid style placement, multiple modules can be located freely to different positions, hence, requiring a uniform communication architecture throughout all possible placement positions. With respect to the proposed communication architecture, a routing is called uniform if it is identical inside all resource slots and also at the slot borders. Using the implicit connection scheme is not directly supported by the Xilinx router. However, in Sect. 3.2 it is revealed how systems can be easily constrained for allowing this technique together with the standard Xilinx design tools. The logic overhead for the implicit connection scheme is one look-up table per shared write signal:

$$L_{SWS}^{im} = S_{SW} \tag{2.6}$$

Note that the logic overhead of the implicit connection scheme is independent of the amount of resource slots. Consequently, this technique allows a very fine resource slot grid without additional logic overhead.

2.5.2 Dedicated Write Signals

Besides shared write signals that are connected to all modules at the bus, it requires dedicated write signals for linking a signal to a specific module. These signals are typically control signals. An example for such a dedicated control signal is a `module_select` signal that is not shared among multiple modules. In the following, this example will be used to describe how a master can selectively address a slave that is implemented as a reconfigurable module.

As a straight forward approach, it is possible to use non-uniform routing within the bus as illustrated in Fig. 2.24a. In order to allow module relocation this requires implementing the dedicated write signals outside the resource slots. If a module configuration is loaded to the FPGA at runtime, it may be further required to merge the module bitstream into the configuration of the static system. This will connect the module with the heterogeneously routed dedicated write signals located in the static part of the system. Such a bitstream merging is obligatory when using Xilinx Spartan-3 or Virtex-II devices. Newer Xilinx devices such as Spartan-6 and Virtex-6 enable sub-column-wise reconfiguration of the device, thus, supporting configuration merging by the FPGA architecture. However, this requires accurately following the device restrictions during floorplanning, where the FPGA is partitioned into a static region and multiple resource slots. If these restrictions do not correspond to the system requirements, this results in a lower logic utilization of the FPGA. However, the main limitation of a non-uniform routing is that it is not usable for a two-dimensional grid style alignment of the resource slots (see also Fig. 2.24a).

To overcome these limitations for module select signals, three novel techniques have been proposed and will be examined in the following: (1) a *slot-wise rotated module select distribution scheme* where a set of dedicated wires is homogeneously routed through the resource slots, (2) a *counter-based module select scheme* requiring half the amount of routing resources, and (3) techniques that are based on *reconfigurable select generators* where the number of routing resources scales logarithmically with the number of modules connected to the bus.

2.5.2.1 Slot-wise Rotated Distribution Schemes

This technique routes a set of wires, one for each resource slot and dedicated write signal, slot-wise from resource slot to resource slot. In each resource slot, the routing is rotated by one wire position and then passed to the next resource slot. As presented in Fig. 2.24b, this scheme allows modules to connect to an exclusive dedicated wire in each resource slot at exactly the same relative position while maintaining a homogeneous routing within each resource slot.

This signal distribution scheme has the predicate of a uniform interconnect structure with respect to the module boundary while still being able to supply an

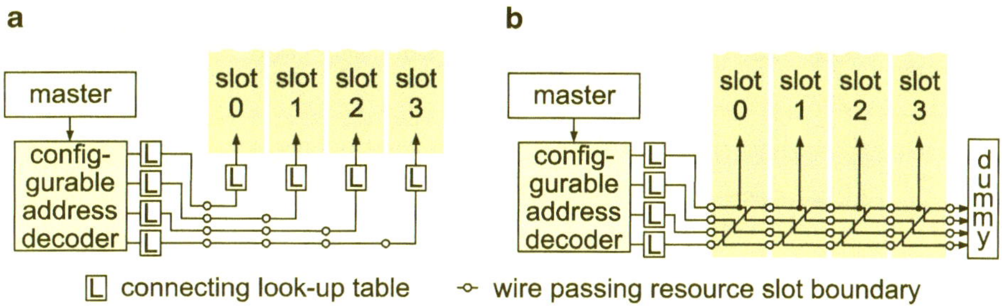

Fig. 2.24 Distribution of module enable signals (**a**) using inhomogeneous routing within the static part of the system and (**b**) homogeneous routing formed by a rotated module select signal distribution scheme

individual signal to each slave by rotating the complete set of dedicated signals inside the module boundary. The technique can be implemented with wires spanning over multiple basic elements to perform the slot-wise rotation scheme, because of such wires, spanning over S consecutive BEs, will automatically rotate dedicated signals for up to S resource slots (see also Fig. 2.4 that introduces, among others, the phrases wire span S and basic element BE).

If the number of resource slots (R), multiplied by the number of basic elements per slot (W), is larger than the span (S), this scheme can be cascaded to any arbitrary width by driving the incoming signal from the configurable address decoder to a new unused channel as shown in Fig. 2.25.

Figure 2.25a gives an example where $R = 12$ module slots of width $W = 1$ are connected with an individual module select signal for every slot by the use of local wires that span over $S = 6$ basic building blocks. The fat drawn module select signal points out how a dedicated signal is cascaded after S resource slots to another channel that spans again over S resource slots.

Figure 2.25b reveals a second example where the resource slot width granularity is larger than one basic element ($W = 2$). In this case, the module select signal has to be cascaded after $\lfloor \frac{S}{W} \rfloor$ consecutive BEs, as shown by the fat drawn module select signal. The figure points out that this cascading can be applied within the same channel. Note that the routing from module slot to module slot is equivalent for each possible resource slot, hence making this approach suitable for a free module placement to the bus.

By using implicit connections (see also Sect. 2.5.2), the slot-wise rotated distribution scheme will not require any logic resources inside the resource slots. However, there are plenty of look-up tables required for the address decoder logic in the static part of the system. Furthermore, the number of routing wires inside the bus scales unlikely with $\lceil \frac{R}{S} \rceil \cdot S$, thus, preventing this approach for a larger amount of resource slots.

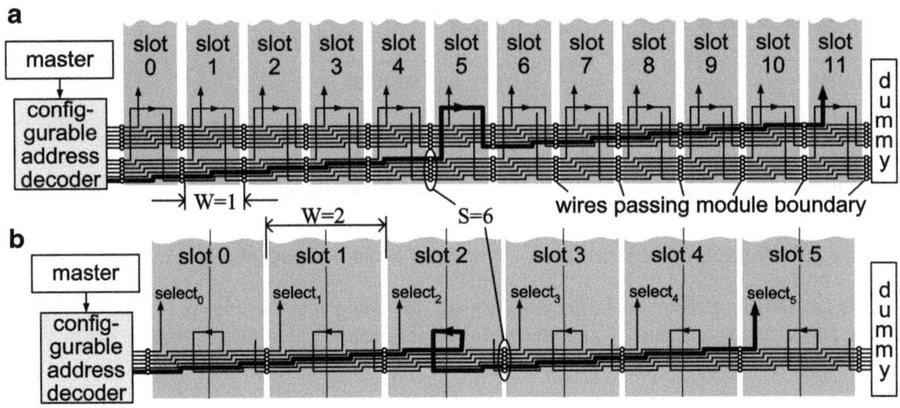

Fig. 2.25 Module select distribution by exploiting the regular structure of an FPGA architecture. Wires spanning over multiple basic elements (here $S = 6$) automatically perform the slot-wise rotated distribution scheme, as shown in Fig. 2.24b. (**a**) Cascading the slot-wise rotating scheme when the number of resource slots R is greater than the span of a channel (i.e., $R > C_i^S$). See also Sect. 2.3 that introduces the here used notation. Whenever a channel has connected S dedicated signals to S resource slots, we will use a new channel to cascade the scheme. For instance, slot 0 to slot 5 use channel C_0^6 to route dedicated signals from the configurable decoder while the last slots are cascaded via channel C_1^6. (**b**) Cascading the slot-wise rotating scheme within a channel. If a resource slot is wider than one basic element (BE), i.e., $W \geq 2$, then the routing scheme can be cascaded within the same channel after $\lfloor \frac{S}{W} \rfloor$ BEs. Note that both cascading schemes have exactly the same routing structure within each resource slot

2.5.2.2 Counter-Based Module Select Distribution

If a master shall be able to access any arbitrary combination of slaves by dedicated signals and if a module can be placed freely to each resource slot, the slot-wise rotated distribution scheme requires an exceptionally high amount of routing resources within the bus and a lot of logic inside the static part of the system. Normally, there is only one or no slave being enabled at the bus. Thus, $\lceil ld(M+1) \rceil$ internal bus wires determine the theoretical lower bound to generate dedicated module select signals for a bus, connecting up to M modules at the same time. Such a system would typically demand a noticeable amount of logic to decode the module select signals with the help of a comparator in each module resource slot of the bus (see also Fig. 2.14). In contrast, an alternative scheme was developed that is based on Möbius-coded counters in order to generate individual module select signals by the use of $\lceil R/2 \rceil$ internal bus wires. Here, the input is compared with the same constant in each resource slot for generating a module select and decremented for the following slots.

A Möbius-code is obtained by a ring-shifter, where the input shift value is the inverted output of the shifter. The longest non-repeating sequence of different encodings of such a ring-shift register will be used for generating the module select signals. An example for a 3-bit Möbius-code is the sequence {`000`, `001`, `011`,

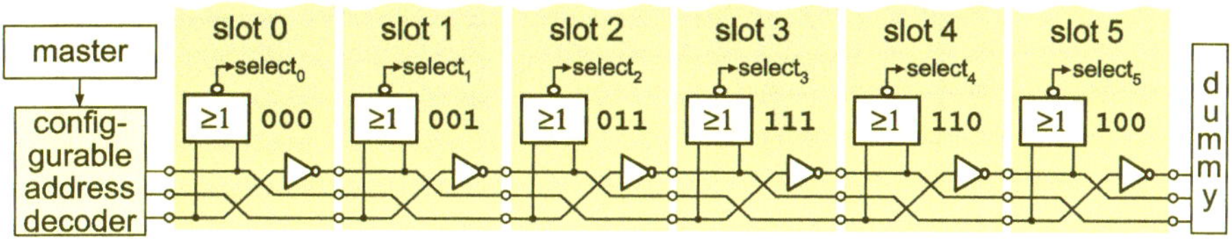

Fig. 2.26 Chipselect generation. In each module slot, the input value is passed through a combinatory Möbius-counter stage. Upon an input value of `000` at a slots' input, the according module select signal is activated

`111, 110, 100`}. All Möbius-code words can be decoded with a single two-input gate. For example, the code word containing only `'0'`-bits can be decoded with a NOR-gate evaluating the most left and the most right shift register bit of a particular Möbius-code. Note, that this property holds for any width of a Möbius-code.

A module select generation, using a 3-bit Möbius-Code, is depicted in Fig. 2.26 for six resource slots ($R = 6$). In each slot the incoming data is ring-shifted and the value of the wrap around wire is inverted. The result is supplied slot-wise to the next stage performing the same operation, thus forming a combinatorial Möbius-counter stage in each slot. When the input value of a slot is `000`, the module select signal is activated. For this purpose, a single NOR-gate allows decoding the code at the input of a slot. This NOR-gate is further used as a macro resource such that high-level tools can easily apply this technique. If this approach is used in Xilinx Virtex FPGAs, the inverter will not demand an exclusive lookup table, because these devices provide dedicated inverters usable inside the logic resources that can be concurrently used with the LUTs. It is not required to decode a `000` inside the resource slots, it is only important that the same element of the Möbius-code sequence is decoded in every single slot. When the system wants to activate a particular resource slot select signal, it has to apply a Möbius-code value to the bus that has been counted backwards in such a way that the count value reaches `000` at the slot intended to be activated. For instance, in Fig. 2.26 the configurable address decoder has to apply a `001` at its output to activate slot 1. Note that all bus interfaces are equal. Thus, the modules using this technique are fully relocatable. In this example, there is always one slot selected. If there must be the situation that no module is selected, the last encoding (here `100`) is used for this purpose. A system may be designed in such a way that a master drives addresses directly to the combinatory Möbius counter.

2.5.2.3 Reconfigurable Select Generators

A completely different methodology to generate individual module select signals is based on reconfigurable module select generators that decode an address signal within each resource slot. These select generators are configured at two levels. After loading a module onto the FPGA device through partial reconfiguration

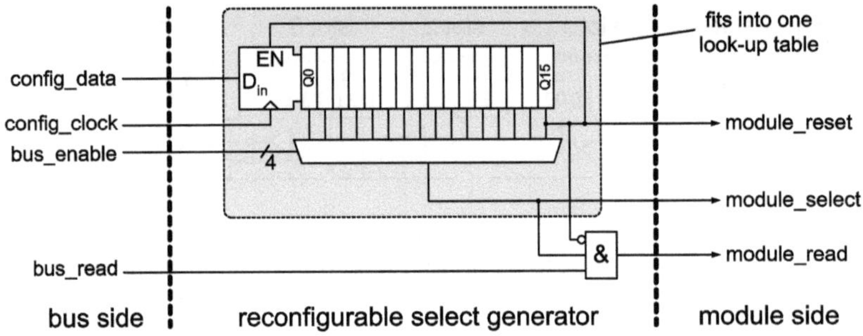

Fig. 2.27 Reconfigurable select generator based on a shift register primitive with a self-locking feedback loop. At FPGA reconfiguration time, the shift register is enabled by configuring a `1` to all LUT values (equal to initializing the shift register with `1` values). In the following system reconfiguration step, a configuration word starting with a `0` followed by a one-hot encoded sequence is shifted into the LUT (via `config_clock` and `config_data`). Therefore, the shift register will lock itself when a complete new table has been shifted into the LUT. After finishing this process, `module_reset` is automatically released, and the register is used in LUT mode, and thus, generating the `module_select` signal by decoding the `bus_enable` signal vector. A `module_read` is prevented until accomplishing both configuration steps

(*device level*), it will be in a state where the module is deactivated (e.g., by applying a module wide reset). In this state, the module select generator can be configured at the second level, the *system level*, in order to write some specific data for the module individually. This data may include information for a reconfigurable module select generator that decodes bus signals in order to generate dedicated module control signals.

Figure 2.27 presents an approach tailored to distributed memory FPGA architectures, such as all Xilinx Virtex architectures, where look-up tables can be configured as memory and shift registers. Whenever the FPGA is configured with the logic of a reconfigurable select generator, one look-up table inside this logic is configured to use the shift register mode and this shift register is initialized with the same bit value at all positions. The cascading output of the shift register – normally used to build larger shift registers – is connected to the enable input pin of the same shift register. This signal is further used as the `module_reset` signal supplied to the logic of the module. The preload bit value has to be chosen to enable the shift register. By setting the shift registers enable input to be active high or low, it is possible to select whenever the reset signal to the module should be active high or low, too. After configuring the FPGA with the complete module, the system can configure the reconfigurable module select generator in a second step.

This second step is performed by shifting in a new look-up table value to the shift register by the `config_data` input and the `config_clock` input of this register. The first bit shifted into the register must be set inverted to the shift register input value in order to lock the shift register after finishing the second level reconfiguration process. This first shifted value will be stored at the cascade output of the shift register that is connected to the shift enable input for locking the shift

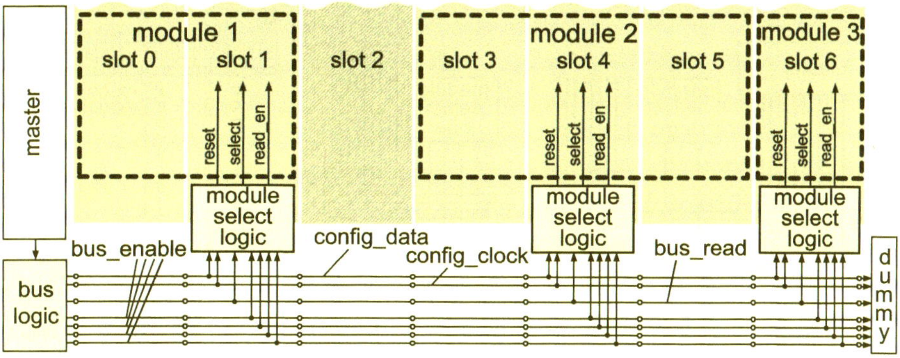

Fig. 2.28 System composed of a master and some modules of different widths. Each module requires one module select block that contains the reconfigurable select generator shown in Fig. 2.27

register after logic level configuration. After this, a function table for decoding the `bus_enable` signal must follow. Typically, the function table will contain a one-hot encoded value. Assuming the example in Fig. 2.27 with a `module_select` signal to be activated only when the `bus_enable` signal is equal to `0011`, then the value `0001 0000 0000 0000` has to be shifted into the look-up table. Note that in the case when `bus_enable` is equal to the maximum value (e.g., `1111` in the example in Fig. 2.27), the `module_select` signal will take the signal value of the cascading output locking the shift register. Thus, it is possible to generate $2^k - 1$ individual module select signals when using a k-input look-up table as a shift register inside the reconfigurable module select generator. As a consequence, this approach requires for M modules the minimal theoretical amount of $\lceil ld(M+1) \rceil$ internal bus wires in order to generate individual module select signals for each module. Note that this alternative scales with the maximum number of modules M and that it is independent of the number of resource slots R provided by the reconfigurable bus.

This module select generation methodology has the predicate that it is completely independent with respect to the resource slot position of a module at the bus. Figure 2.28 presents an example of a system using the module select logic introduced in Fig. 2.27. Due to the homogeneity of the bus interconnect resources, it is possible to relocate modules on the bus. Each time, after a new module is configured at the FPGA device level, this module is the only one with an activated shift register in its configurable module select generator logic. This holds true under the assumption that the system also performs the second reconfiguration step at the system level before writing the next module configuration to the FPGA. Note that a selective reconfiguration of the shift register look-up table values can be used anytime to apply `module_reset` and to change the value of the `bus_enable` signal that will actually select the connected module.

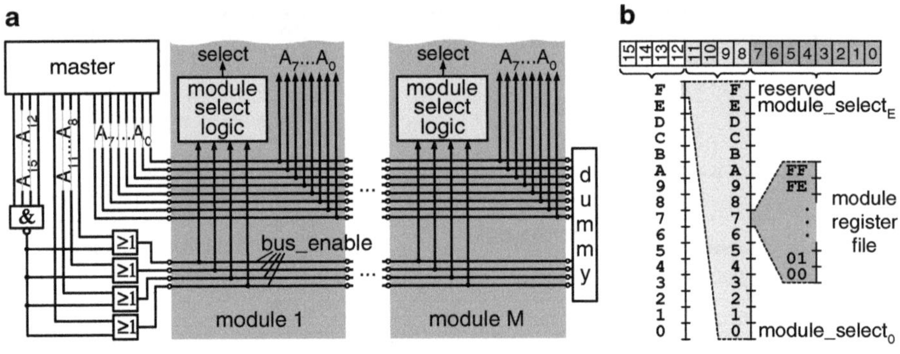

Fig. 2.29 (**a**) Example of a reconfigurable system with a 16-bit address bus, driven by a master. The eight least significant address bits are linked directly to the modules via shared write signal wires. The module select logic blocks contain reconfigurable select generators, (see also Fig. 2.27) interconnected by an internal `bus_enable` signal. This signal contains a `1` at all positions if all four most significant address bits do not contain the base address prefix `F---`. In the case that $A_{15}, \ldots, A_{12}$ match the prefix, the reconfigurable select generators will decode the address bits $A_{11}, \ldots, A_8$. (**b**) The corresponding memory map for the 16-bit address space

If the address ranges of the different reconfigurable modules can be chosen freely and within one consecutive address block, then the bus logic may be substantially simplified. In this situation, the current master can directly drive all low address lines on the bus via shared write signals for accessing a register file inside each individual module. The `bus_enable` signal is then connected straight to the next higher set of address lines of the master. Finally, all other higher address lines are used to decode the address range of the modules connected to the reconfigurable bus. Assuming the system, illustrated in Fig. 2.29, as an example: the address range for a module's individual register file is chosen to be 8-bit (256 registers) and a maximum of up to $M = 15$ modules may be connected to the bus at the same time, while mapping the complete address space of the reconfigurable modules at the upper limit of a 16-bit address range. Then the address lines $A_7 \ldots A_0$ of the master will be connected directly to the bus via shared write signal wires. The bus internal `bus_enable` lines will contain the next address lines $(A_{11}, \ldots, A_8)$ if all remaining higher address lines $(A_{12} \ldots A_{15})$ contain a `1`. In the case that the remaining higher address lines $(A_{11}, \ldots, A_8)$ are not in the upper limit address space (`F---`) that is evaluated by the shown NAND-gate, the `bus_enable` lines will contain a `1` for every line, thus preventing the selection of any module. This is because of the highest look-up table entry is reserved to encode that no modules are accessed.

The reconfigurable select generator approach, depicted in Fig. 2.27, can also be used for multicast bus transfers, where more than one reconfigurable module can be accessed at the same time. In this case, the shift registers of the reconfigurable select generators of two or more modules will contain a `1` at the same position. For instance, assuming that we have a module 1, mapped to the `bus_enable` address `0001`, and a module 2, mapped to `0010`, and that both modules should further be accessible at the multicast address `0100`. In this case, we have to shift the value `0100 1000 0000 0000` into the look-up table of the select generator in module 1 and the value `0010 1000 0000 0000` into the LUT in module 2. Note, that both modules can be accessed individually or simultaneously. This multicast operation can be used for write and read modes. In the latter case, the data outputs of all selected modules are bitwise linked by OR-gates (see also Sect. 2.30).

Online Bitstream Manipulation It is possible to omit the second configuration step at the system level for setting the reconfigurable select generator (see Fig. 2.27) by manipulating the configuration data at runtime. If multiple instances of the same module are intended to be connected to the bus at the same time, different look-up table values are required for each module. Consequently, the LUT cannot be hard coded directly in the partial module bitstream, if multiple instances of the same module have to be loaded into the same reconfigurable system. This can be performed during the reconfiguration phase by replacing the fraction of the reconfiguration data that is carrying the look-up table values of the module select comparator. The required bitstreams can be generated with the Xilinx tools by performing the LUT table manipulations in the netlist (using the proprietary XDL language [BKT11]). From the resulting netlists, we can generate differential bitstreams where only the LUT entries of the reconfigurable select generator will differ.

As compared to the original reconfigurable select generator approach, the online bitstream manipulation avoids the logic overhead required in the static part of the system for shifting in the comparator values into the select generator look-up-tables. Furthermore, the bus internal wires for distributing the configuration clock and the configuration data signals are now useable for implementing the module itself. Moreover, manipulating the bitstream removes the need to have a logic primitive that provides the shift register feature. This is important as on Spartan-6 FPGAs, only every other logic column provides distributed memory.

Regardless of whether the LUT-values are set by shifting in a new table or by manipulating the configuration bitstream at runtime, the logic overhead L_{DW} for providing S_{DW} dedicated write signals in a system with up to M simultaneously running modules is:

$$L_{DW} = (R) \cdot S_{DW} \cdot \left\lceil \frac{M}{2^k - 1} \right\rceil + S_{DW} \tag{2.7}$$

This summarizes the logic overhead in the resource slots and the connecting LUTs that are required to connect S_{DW} dedicated signals in each resource slot. One k-bit LUT that evaluates an internal bus enable signal to a module select can decode $2^k - 1$ different modules. One codeword is required for encoding that no module is selected. If more individual modules have to be accessed ($M \geq 2^k$) it is possible to combine multiple LUTs to a larger one on all Xilinx FPGAs. However, the amount of 15 individually addressable modules in the case of $k = 4$ bit look-up tables will be sufficient for most practical systems. It is also possible to share the same encoding

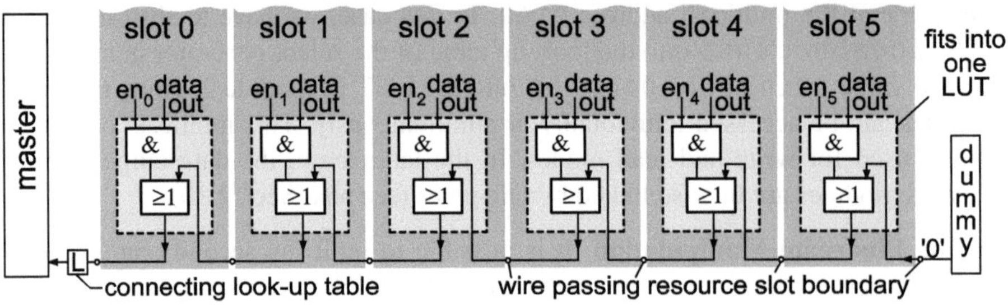

Fig. 2.30 Implementation of a shared read signal via a distributed logic-based bus. The uniform structure of the *distributed read multiplexer chain* allows a flexible connection of modules to the bus

of a dedicated write signal among multiple modules. For example, by mapping the register files of two modules into disjoint address ranges. Consequently, the logic overhead will be $(1 + R) \cdot S_{DW}$ LUTs in practical systems.

2.5.3 *Shared Read Signals*

As a counterpart to shared write signals, a bus also has to provide signals in the opposite direction. Therefore, selected modules must be able to drive signals to a set of shared wires of the bus. An example for shared read signals is the read data bus from several slaves to a master. As already mentioned in the introduction, typical on-chip buses use unidirectional wires in order to avoid on-chip tristate buses as depicted in Fig. 2.21a. Figure 2.21b,c present examples of common on-chip bus implementations.

Reconfigurable buses have to be designed in a uniform and modular manner for allowing to include the bus structure directly within the modules connected to the bus. This was proposed before for read signals in [HKKP07b]. There, the large OR-gate in Fig. 2.21c is split into small chunks of two-input OR-gates for arranging the signal routing between the resource slots in a uniform and modular manner, as illustrated in Fig. 2.30. The output of the last OR-gate represents the shared signal supplied to the destination in the static part of the system (here a master). Such a regular construct will be called a *distributed read multiplexer chain* in the following.

Under all circumstances, the bus logic has to prevent modules to drive a logic value to a shared signal if it is not selected. As shown in Fig. 2.30, this can be achieved by an additional AND-gate that fits into one look-up table together with an OR-gate. For this purpose, the enable signals (`en_i`) shown in Fig. 2.30 are connected to the `module_read` output signal of a reconfigurable select generator (see Fig. 2.27) where each resource slot contains its own select generator. As a consequence, the output data of an individual module can only be applied to the bus if (1) the corresponding module select generator has been properly configured at the system level, (2) the module is selected by an according `bus_enable` assignment from the static system, and (3) a read operation is requested by activating the `bus_read` signal (see also Fig. 2.27). Note that all approaches presented in Sect. 2.5.2 are suitable to distribute a dedicated write signal to each $\texttt{en}_\texttt{i}$ signal, but the reconfigurable select generator is the most adequate one. If not only a single shared read signal is connected to the master but many, it requires one look-up table for each shared read data line in each resource slot. In this case, the $\texttt{en}_\texttt{i}$ select signal inputs of the LUTs, used to implement the read multiplexer chains within one resource slot, are connected to the same `bus_read` signal of the same reconfigurable select generator. Then, the logic overhead L_{SR}^{SC} (in terms of LUTs) for implementing S_{SR} shared read signals by simple read multiplexer chains is:

$$L_{SR}^{SC} = (1 + R)S_{SR}. \tag{2.8}$$

2.5.3.1 Performance Considerations

Besides the logic overhead, shared read multiplexers increase the combinatorial path by one look-up-table for each resource slot. This may considerably limit the maximum amount of allowed resource slots in a system when implementing this structure. Before discussing further details of the ReCoBus structure, the relevance of this effect will be quantified. The impact of increasing the amount of resource slots was determined for a Xilinx Virtex-II device (XC2V6000-6) for one horizontal distributed read multiplexer chain of different sizes. In order to obtain realistic timing values, the propagation delay between two I/O pins was measured. In the test circuits, a signal is routed directly to the look-up table which represents a connection to the resource slot with the most critical propagation delay. On the way back, this signal is propagated through R LUTs in every W-th logic column. The results, presented in Fig. 2.31, quote the propagation delay between the two I/O-pins.

The width W of a resource slot was varied between one and eight basic elements called CLBs for Xilinx FPGAs. Figure 2.31 indicates how the pin-to-pin propagation delay increases with the amount of resource slots R provided by the bus. This delay provides a realistic estimate of the propagation delay for a master accessing a module. The figure also allows determining possible parameters of the reconfigurable bus if a certain clock frequency is required. A 100 MHz system clock, for example, is equivalent to 10 ns of propagation delay. This is then the time allowed for the combinatorial paths, including routing delays and the flip-flops (setup-hold times and clock to output delays). In this situation, a reconfigurable bus that is based on a distributed read multiplexer chain is limited to a size of $R = 8$ resource slots when the width of the slots is chosen to $W = 2$ CLBs, or in the case of $W = 6$ CLBs, the bus is limited to just $R = 4$ resource slots. Note that complete bus is $R_h^{BE} = R \cdot W$ CLBs wide. Consequently, the first example provides a bus that is $R_h^{BE} = 8 \cdot 2 = 16$ CLBs wide, while in the second case, the bus spans over $R_h^{BE} = 4 \cdot 6 = 24$ CLBs.

The long combinatory path of the read-multiplexer chains is inapplicable for pipelining, as this would lead to variable latencies depending on the module

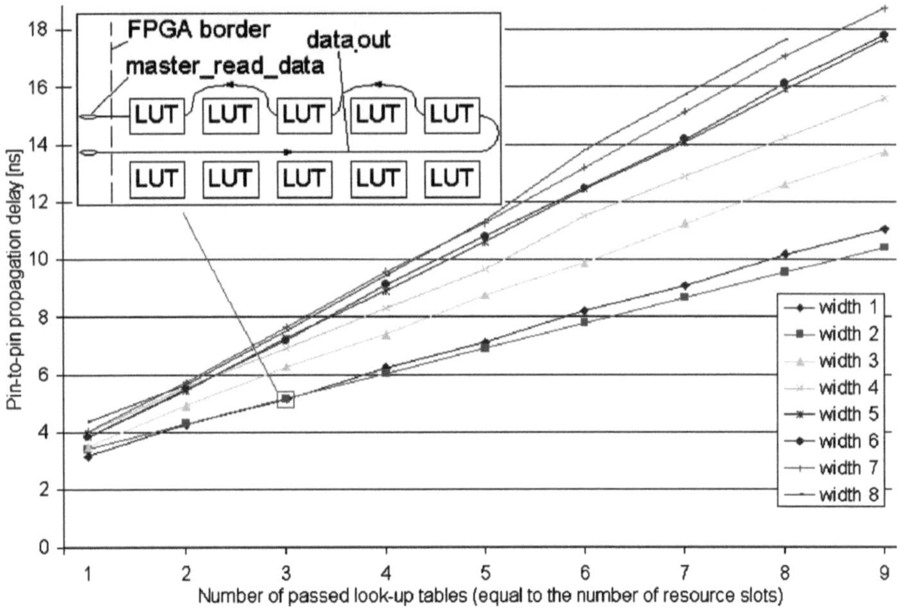

Fig. 2.31 The impact of the bus size on the propagation delay. The graphs show how the propagation delay increases with the number of resource slots R and by the width W of the resource slots. The values are determined for Xilinx Virtex-II XC2V6000-6 FPGA

placement position at the bus. However, by duplicating the amount of shared read signals while using them as multi-cycle paths, the throughput can be enhanced to a wide range of speed and elongation requirements. This is illustrated in Fig. 2.32b for enhancing throughput and in Fig. 2.32c for extending the length of a read multiplexer chain.

In order to increase flexibility and reduce the internal fragmentation in case that a slot is not completely filled with the logic of a module, it is advantageous to keep the width of the resource slots small to reduce internal fragmentation (see Sect. 2.1). Nonetheless, it is possible to apply the read multiplexer methodology to large scale systems with small slot widths, while still keeping the bus operation frequency high. Instead of using one distributed read multiplexer chain that links one slot after the other, multiple chains can be used in an interleaved fashion, as illustrated in Fig. 2.33. As a consequence, buses implemented with a wire span of S basic elements and a module width of W BEs can save up to S/W times the propagation delay of a single look-up table. Furthermore, the usage of wires, spanning over a larger amount of elementary building blocks, leads to a lower overall signal propagation delay. This is because less switch matrices have to be passed for implementing the read multiplexer chain. Note that in case, the FPGA architecture does not offer the desired wire span S, it is easily possible to cascade multiple wires to route a signal in a uniform manner over longer distances by using the approaches presented in Fig. 2.25. For connecting a module in each resource

a

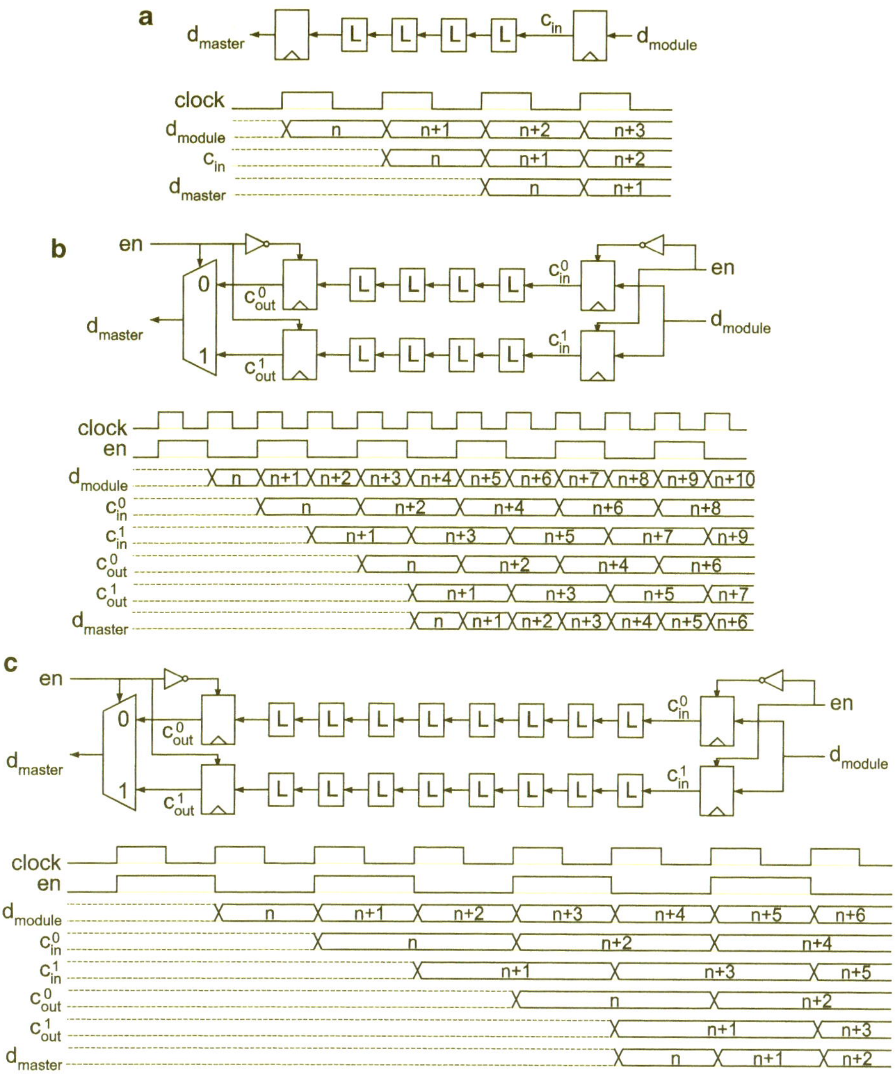

Fig. 2.32 (**a**) Critical path of a read multiplexer chain, passing four look-up tables, and the according wave form. (**b**) By duplicating the read multiplexer chain, the throughput is doubled, while each chain holds the same propagation delay as in (**a**), by setting a multi-cycle path to the chain. (**c**) Instead of increasing the clock frequency, the additional read multiplexer chain could alternatively be used for extending the length of the chain while allowing the same clock frequency as in (**a**)

slot with this technique, the module must provide access to all bus signals within each resource slot. However, this flexibility for the dynamically reconfigurable bus comes along with some unused logic resources in the case when modules span over multiple slots, as usual in a fine-grained resource slot layout. The examples

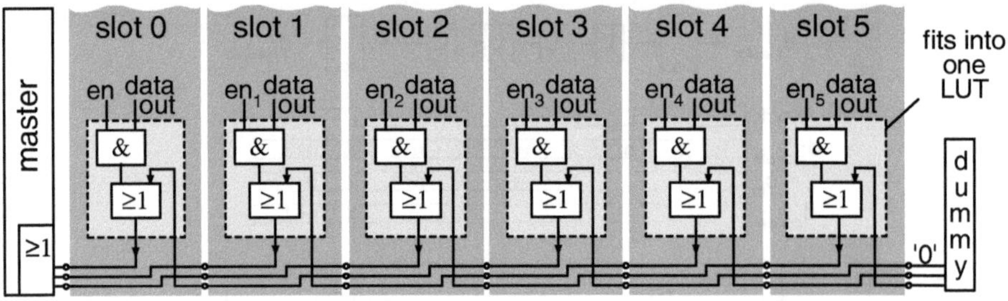

Fig. 2.33 Interleaved read multiplexers. Shared read signals from different slots (`data out`) are linked by interleaving multiple distributed logic-based bus chains

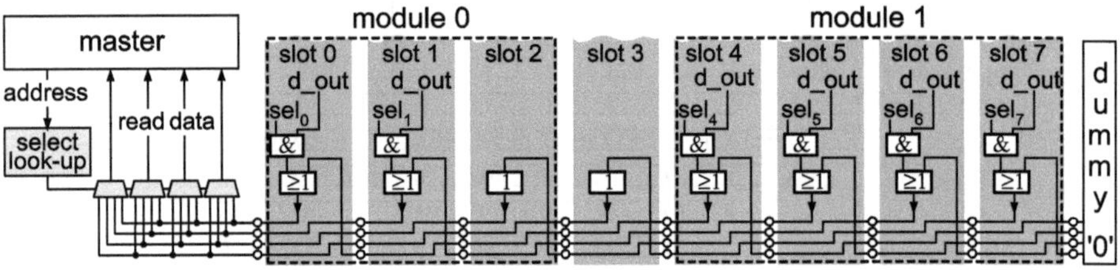

Fig. 2.34 Multi-slot read technique. The wider the required module bus interface is, the more slots are used to connect shared read signals by independent chains of interleaved distributed logic-based bus chains. A set of multiplexers that are inserted in front of the master ensure that the data signals are aligned in the correct order regardless of the respective start position of the currently selected module at the bus

in Figs. 2.30 and 2.33 show how a single shared read signal is connected through chains of look-up tables. In order to build buses of any arbitrary bit-width, multiple chains, one upon the other, have to be instantiated. Alternatively, slots may be more than one elementary basic block (BE) wide, thus allowing to cascade multiple chains side by side or in any mixed combination.

2.5.3.2 Interleaved Bus Signal Distribution

Different modules may have individual interface widths in terms of data bus widths. In most cases, modules of higher complexity typically demand a wider interface and vice versa. For example, although a simple UART is typically connected via an 8-bit bus interface, a complex Ethernet core will be linked in most cases with a 32-bit wide interface. In order to benefit from this observation, a *multi-slot read technique* is proposed, as revealed in Fig. 2.34. In this signal distribution scheme, not all of the S_{SR} shared read signals are connected in each resource slot, but only a subset of $\lceil \frac{S_{SR}}{N} \rceil$ signals. The signal subsets form N individual read multiplexer chains that are connected once per N resource slots. Furthermore, the N chains are interleaved along the bus by displacing the logic of the chains among each other in N adjacent slots. This leads to a repeating connection scheme with an identical

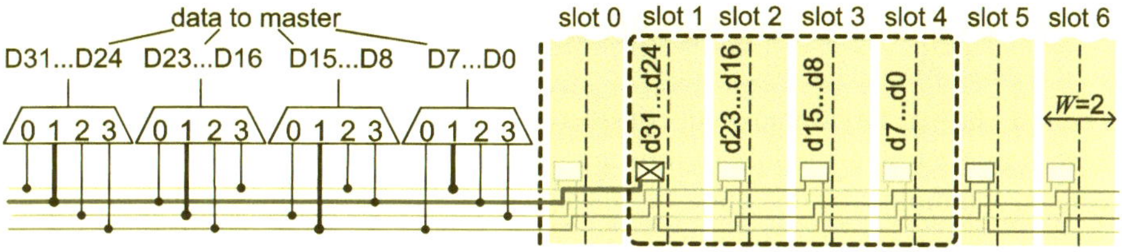

Fig. 2.35 Alignment multiplexing for the multi-slot read technique. Whenever a module is accessed at the bus, the alignment multiplexer select signal must be set to the resource slot index modulo the interleaving factor N. In this example, the start resource slot index of the currently selected module is 1 and the interleaving factor is 4. By setting the alignment multiplexers to input 1, the module data (d31,... d0) is send correctly adjusted to the destination port (D31,... D0) within the static system

logic and routing footprint throughout all resource slots. Interleaving reduces the logic cost per slot to $\lceil \frac{S_{SR}}{N} \rceil$ as compared to S_{SR} for the original bus implementation depicted in Fig. 2.30.

In order to establish a better comparison with existing systems, it is assumed that each signal shown in Fig. 2.34 represents 1 byte and that the data bus is up to 32 bits wide. Then `module 0`, possesses a 16-bit interface while `module 1` uses the complete interface width of the bus. The bus is arranged by four interleaved logic-based bus chains for the shared read signals. In this example, modules that are 1, 2, 3, or 4 consecutive resource slots wide can connect a module containing an interface widths of up to 8, 16, 24, or respectively 32 bits. These slots can be as narrow as the width of one basic element with respect to assumed FPGA model (see Sect. 2.3).

When a master reads data from a module, multiplexers inserted between the different independent multiplexer chains and the master will switch the different bus chains in such an order that the master can read a correctly aligned 32-bit word. This is then independent of the position of the start resource slot of the currently selected module. The select signals for these *alignment multiplexers* can be generated by a tiny look-up table. This table is written by the operating system whenever a new module is included into the system and it delivers the alignment select signal according to the current module select address. Consequently, this look-up table requires as much registers as the maximum amount of simultaneously integrated modules M in the system. The interleaving factor N affects the width of the registers. The total amount of flip-flops in the table is then: $M \cdot \lceil log_2 N \rceil$. Note that all Xilinx FPGAs allow using some or all look-up tables alternatively as small registers, thus allowing to implement the alignment select generation efficiently without dedicated flip-flops. If the module is placed at another slot position, the multiplexer select signals have to be adjusted by the start resource slot index, modulo the amount of interleaved distributed logic-based bus chains N, as it is illustrated in Fig. 2.35. The register file is directly addressed by the same address lines that are decoded within the bus to produce the `bus_enable` signal (see Fig. 2.28). Note that the overhead for the alignment multiplexers and the select look-up is not related to the amount of resource slots provided by the bus.

2.5.4 *Dedicated Read Signals*

The last class of bus signals, not considered so far, are dedicated (non-shared) read signals that allow a module to transmit state information to the static part of the system. Examples for this class of bus signals are interrupt lines and wait request signals.[7] These two examples have one main difference: as interrupt lines can be activated in any combination, all used interrupt lines form a signal that is binary encoded. In contrast to this, a wait request signal may be activated only by a single selected module, thus leading to a one-hot encoding over all wait request signals of the modules. Therefore, wait request signals can be connected to a master using the distributed logic-based bus implementation as presented in Fig. 2.30. By enabling a specific module by a master, it is clear that only this activated module may enable the wait request signal.

For dedicated read signals, such as interrupt lines, the approaches presented in Figs. 2.24 and 2.25 for dedicated write signals could be applied with the change of the signal direction. These techniques, however, are not applicable to large scale systems that are combined with a small slot width W. This is because of the required routing and logic resources to implement these techniques will scale unlikely scale the number of module resource slots. The same holds for related approaches that use a demultiplexer for connecting a dedicated read signal per resource slot to the static part of the system (see Sect. 2.4.3). In the following, alternatives will be examined where the amount of routing wires that are required for connecting dedicated read signals is constant or scaling with the number of modules connected to the bus.

2.5.4.1 Time-Multiplexing

In many systems, it may be acceptable to handle dedicated read signal events (e.g., an interrupt request) with a delay of a few clock cycles. Then, the approach presented in Fig. 2.36 can be applied, where such signals are captured using time-multiplexing via a reconfigurable bus. Whenever a module has a dedicated read signal output, this signal has to be connected to a distributed logic-based read multiplexer chain (see Fig. 2.30). The particular dedicated module signal is chosen by a select generator, analogous to the one presented in Fig. 2.27. The select generator can be configured to send state information from a specific slot to a particular flip-flop and, consequently, to a particular dedicated read signal line, connected within the static part of the system. If required, modules may interconnect multiple dedicated read signals via several resource slots towards the static system.

Implementing time multiplexed shared read signals requires one LUT per resource slot for the distributed read chain and an amount of LUTs related to the

[7] Wait request signals are used by an accessed module (e.g., a slave) to request extra idle cycles for a bus transfer. The wait request concept allows including a variable amount of wait states for the bus access.

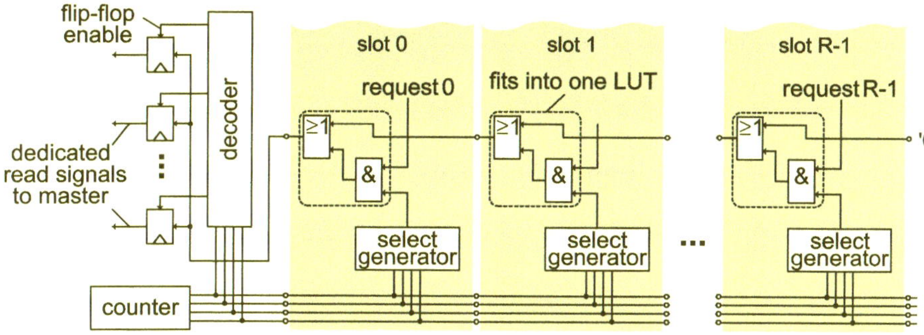

Fig. 2.36 Time-multiplexed connection scheme for dedicated read signals. A counter enables cyclically a set of flip-flops and simultaneously an according select generator within a definable resource slot for writing the request state (e.g., an interrupt or a bus request) of a module to a flip-flop in the static part of the system

total number of modules M (assuming that each module requires one request signal to be transferred to the static part of the system). Beside the logic in the resource slots, it further requires some logic in the static part of the system for the flip-flop select decoder and the state counter. The logic overhead for implementing shared read signals by time-multiplexing dedicated shared read signals is summarized by:

$$L_{DR}^{TM} = \underbrace{R}_{read\ chain} + \underbrace{R \cdot \left\lceil \frac{M}{2^k} \right\rceil}_{select\ generator} + \underbrace{M \cdot \left\lceil \frac{M}{2^k} \right\rceil}_{decoder\ in\ static\ system} + \underbrace{log_2 M}_{counter} . \tag{2.9}$$

2.5.4.2 Logic-Based Demultiplexing

While it might be applicable to distribute an interrupt signal in a time-multiplexed manner, this kind of distribution is disadvantageous for bus request signals. For example, this may be disadvantageous if multiple masters at the reconfigurable bus can request control over the bus only by an elongated time required for arbitration. As a consequence, each master should have an individual connection to the bus arbiter. One solution to this problem is to use a single distributed logic-based bus signal (see Fig. 2.33) for each possible bus request instead of one for all in a time-multiplexed fashion. A solution using this technique is proposed in Fig. 2.37. In order to gain flexibility for module placement at the reconfigurable bus, the approach allows configuring a `bus_request` signal to a connection point within a specific resource slot at the system level (by the operating system). The logic in each slot provides a demultiplexer and a configuration register for the select signal of this demultiplexer. Figure 2.37 points out that such an approach comes along with a considerable logic overhead that is at least $M \cdot R$ look-up tables for M modules

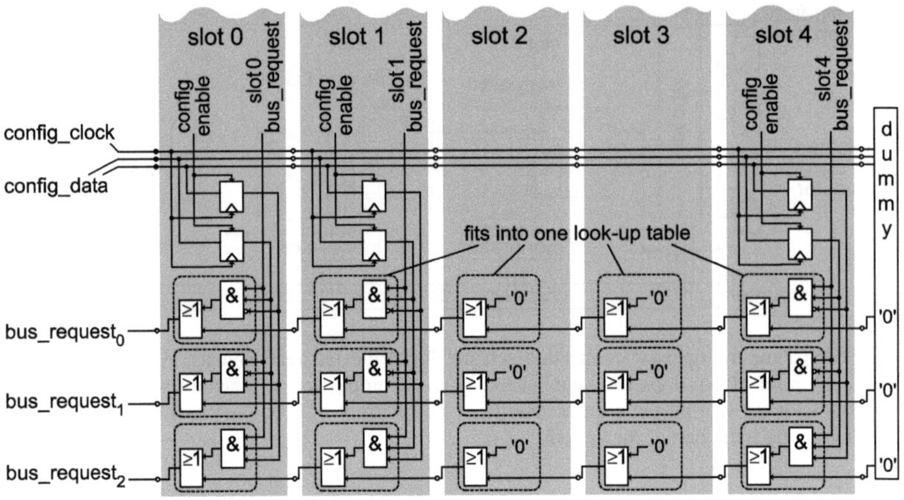

Fig. 2.37 Configurable bus request demultiplexer. In each resource slot, with a `slot bus_request` signal output, this signal is connected to multiple distributed logic-based read signals (like three in this example). Each read signal is selected on a different select value that is stored in a set of flip-flops for each resource slot driving a bus request signal. Such a set of flip-flops can be written individually on the system level

connecting a bus request signal in each resource slot. A related approach [HKP06] uses one chain of look-up tables per resource slot R to link one dedicated signal per slot to the static part of the system (see also Sect. 2.4.3). Consequently, this approach scales worse and requires R^2 LUTs only for the logic within the reconfigurable area.

As already mentioned for the multi-slot read technique, modules typically span over multiple elementary resource slots. This holds true especially for masters connected to the bus, as they consume a considerable amount of logic just to implement the master interface. Based on this observation, a *multi-slot dedicated read technique*, as presented in Fig. 2.38, was proposed, where N demultiplexers are built upon N interleaved, horizontally aligned bus request chains. Consequently, the resource overhead can be reduced by N. After device reconfiguration, the multiplexer in each resource slot is configured to pass through the preceding incoming signal. In a second system level configuration step, the bus request signal of the module may be connected in one particular slot to one of the horizontal bus request chains.

The interleaved horizontal chains can be directly implemented with routing wires, spanning N resource slots (like four as shown in the example). However, the routing may be cascaded to any desired span by using the approaches presented in Fig. 2.25. The chosen span determines the number of interleaved horizontal chains and consequently the number of dedicated read signals for all modules together. If this span does not accommodate all required dedicated read signals, multiple interleaved channels can be instantiated, one above the other.

The configurable multiplexers shown in Figs. 2.38 and 2.39a are used for a second reconfiguration step at the system level. This means that after writing the

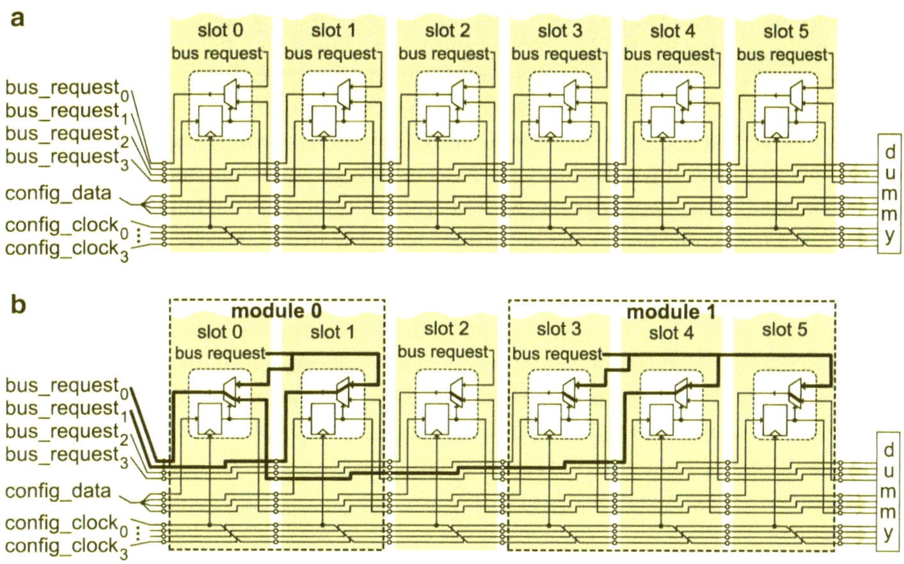

Fig. 2.38 Multi-slot dedicated read technique, exemplarily for connecting bus request signals. (**a**) General structure. Multiple distributed logic-based bus request signals are interleaved according to the routing infrastructure offered by the FPGA architecture (in this example, quad lines spanning over four resource slots). In each resource slot, a reconfigurable multiplexer allows selecting between the local bus request signal or the incoming preceding request signal. Each distributed logic chain can be configured individually. (**b**) Configuration example for two modules connected to the reconfigurable bus. Note that all local bus request signals within a module are connected to the same bus request source of the module

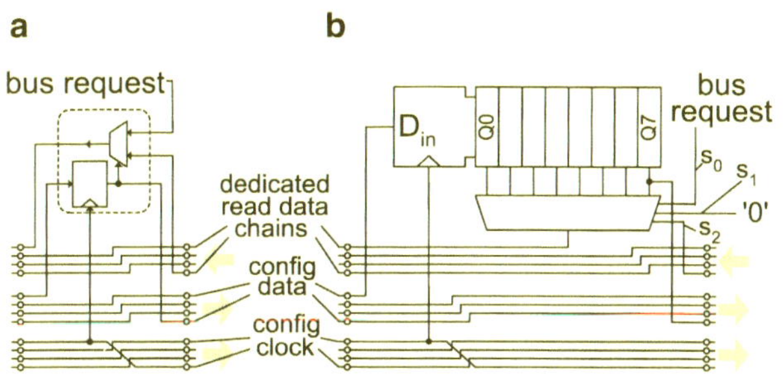

LUT Entry $s_2\ s_1\ s_0$	Connect to bus requ.	Connect to route through
0 0 0	0	0
0 0 1	1	0
0 1 0	0	0
0 1 1	1	0
1 0 0	0	1
1 0 1	1	1
1 1 0	0	1
1 1 1	1	1

Fig. 2.39 (**a**) Switching element for a configurable multi-slot demultiplexer chain that is suitable of implementing dedicated read signals. (**b**) Implementation alternative tailored to distributed memory architectures, when look-up tables can be used as a shift registers (like in all Xilinx Virtex FPGA families). The table in the *right* gives the values to be shifted into the look-up table in order to set the switching element to connect to the `bus_request` signal or to route through bus requests from preceding resource slots

configuration to the FPGA, the values of the shown LUT can be adjusted by a driver without using the FPGA configuration port again. Here, the same approach, as presented in Fig. 2.27 for the reconfigurable select generators, is used. This approach can be efficiently implemented on all current Virtex or Spartan FPGA devices, shipped by Xilinx Inc., within a single look-up table per configurable multiplexer. As shown in Fig. 2.39b exemplary for a 3-bit look-up table, this look-up table also has to be used in the shift register mode. In the following, it is exemplarily assumed that $N = 4$ horizontal traces are interleaved with a span of 4 resource slots. Then the configuration data output `config_data` from the configuration unit is connected to the first N configurable multiplexers in the first N resource slots (see Fig. 2.38). The configuration logic is located in the static part of the system. Each cascading output of the first N shift registers is connected to a further configurable multiplexer, located N resource slots further. This is repeated over all resource slots, and thus, each interleaved multiplexer channel forms one shift register over the look-up tables that are all instantiated in shift register mode. Therefore, there exist N interleaved shift register chains. Each shift register chain comes along with its own clock signal. This allows to shift data from the `config_data` terminal into a selectable shift register chain, without interfering the other chains and consequently, the present signal state in all other chains. The look-up table in Fig. 2.39b is configured either to pass through the incoming request signal from a preceding look-up table or to connect the chain to the local bus request signal. The according entries for the example in Fig. 2.39b are given in the adjoining table. In order to configure the bus request signals as shown in Fig. 2.38b, the values from the table have to be shifted into the corresponding shift register chain. For example, to connect the signal $\texttt{bus\_request}_0$ in resource slot 4 to the local bus request signal of module 1, the sequence `0101 0101` has to be shifted with $\texttt{config\_clock}_0$ followed by the sequence `0000 1111` for passing through the signal in resource slot 0. Analogous, $\texttt{bus\_request}_1$ is connected to module 0 via resource slot 1 by shifting in the sequence `0000 1111 0101 0101` into chain 1. Note that this will also set the reconfigurable multiplexer in slot 5 to route through potential bus requests from preceding slots. Whenever the resources of a module at the bus are removed, the system can shift in the sequence `0000 1111` into all look-up tables of the utilized bus request chain for disconnecting the local bus request signal from the bus. Instead of using individual clocks for each interleaved chain, a single clock can be used in combination with enable logic to individually activate a specific shift register chain. Note that the configuration clock signals are distributed by using local wires instead of using global clock trees.

2.5.4.3 Demultiplexing Through FPGA Configuration Manipulation

All presented approaches for connecting a dedicated read signal from a particular reconfigurable module to a destination in the static part of the system have in common that they omit a modification of the configuration data written to the FPGA. The only exception, where the configuration data is required to be modified during

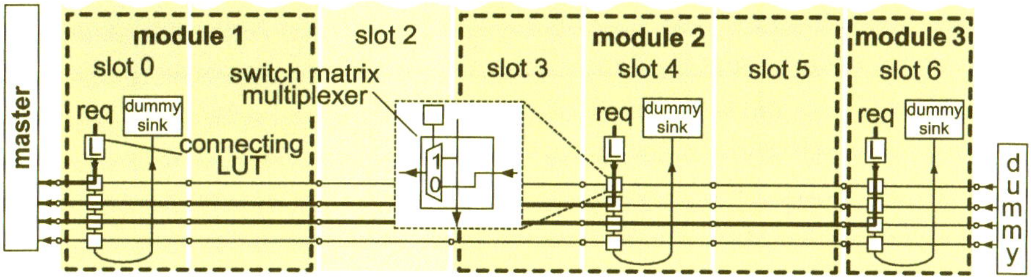

Fig. 2.40 Dedicated read signal distribution by FPGA configuration data manipulation. A request output from a module is driven by a connecting LUT vertically over a set of horizontally routed wires. The routing resources are chosen such that a single multiplexer within the switch matrices can be configured for either routing through the horizontal wire or connecting vertically to the request signal

the reconfiguration process, is the address information that determines the position of the reconfigurable module on the device. For all Xilinx devices, the required documentation for changing address information inside the configuration data can be found in the respective user guides (e.g., [Inc07]).

Instead of implementing demultiplexers that connect a request output signal to a horizontal dedicated routing resource towards the static system, it is alternatively possible to use the switch matrix resources of the FPGA fabric. An example for this approach is illustrated in Fig. 2.40. Here, the amount of horizontal routing tracks for implementing dedicated read signals scales with the amount of modules located at the bus. This is an important improvement against all other presented approaches that require at least one LUT per resource slot and typically many more in most practical systems. Figure 2.40 reveals that one connecting look-up table is sufficient per module and dedicated read signal within the resource slots. This connection may be carried out by sharing resources that are used to implement other signal classes, such as dedicated read signals. The maximum number of concurrent, horizontally aligned, dedicated read lines has to be specified at design-time and has to be selected according to the maximum number of simultaneously integrated modules.

Each module is constrained to drive its outgoing dedicated read signal (typically an interrupt or bus request signal) straight vertically across a set of horizontal wires that are routed in a homogeneous manner from resource slot to resource slot towards the static part of the system. Within the FPGA fabric, there exist switch matrices at the crosspoints. Horizontal and vertical wires for implementing dedicated read signals have to be selected such that the horizontal output wire towards the static system can either be connected with its preceding horizontal input wire from the preceding resource slot or with the vertical request signal output of a particular module. The crosspoints are part of the FPGA routing resources and require no logic in terms of look-up tables. The required configuration data to modify the setting of a crosspoint can be generated on-the-fly by exchanging fractions of the original module configuration data at runtime The necessary bitstream information is not published but could be easily figured out by reverse engineering. The configuration data of Xilinx FPGAs was identified to be highly regular and very suitable for

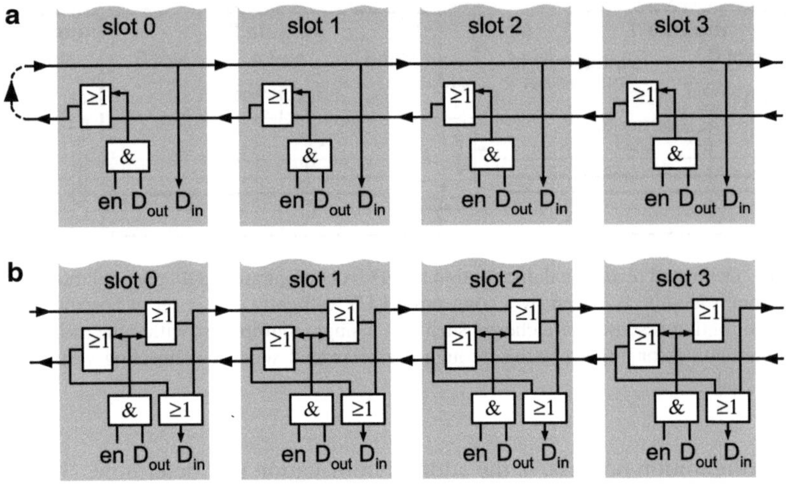

Fig. 2.41 (**a**) Unidirectional read multiplexer chain versus (**b**) bidirectional multiplexer chain [HKKP07b]. The unidirectional multiplexer chain (**a**) requires one look-up table per resource slot and signal bit, while the bidirectional variant (**b**) requires three times more LUTs. Beside looping back the output of the unidirectional chain to the reconfigurable modules (e.g., for transferring data from one partial module to another one) it is possible to use the read and write signals concurrently, hence, allowing the static part of the system reading one partial module while sending data to another one at the same time. This is not possible in case of the bidirectional multiplexer chain. Interleaving is only applicable to the unidirectional chain and allows faster operation despite the routing in backward direction. Note that the signal in backward direction can be pipelined in the static system

manipulations at runtime. Alternatively, a set of configuration data templates, one for each dedicated read signal crosspoint, can be built offline and stored in a repository for the runtime system.

2.5.5 *Module-to-Module ReCoBus Communication*

For all proposed bus-based interconnect techniques, communication with a reconfigurable module takes only place always with the static part of the system and two reconfigurable modules cannot directly communicate with each other. Some related approaches [LCK07, HKKP07b] propose buses, allowing direct links between two reconfigurable modules with the help of a bidirectional read multiplexer chain (see Fig. 2.41). However, these approaches require three times more logic as compared to simple read multiplexer chains and they do not allow to use the write and read signals concurrently for simultaneous read and write operations by the static part of the system. Furthermore, communication structures for direct bus transfers within the reconfigurable area are unsuitable to be used for interleaved buses, because this would demand an extra alignment multiplexer in each module.

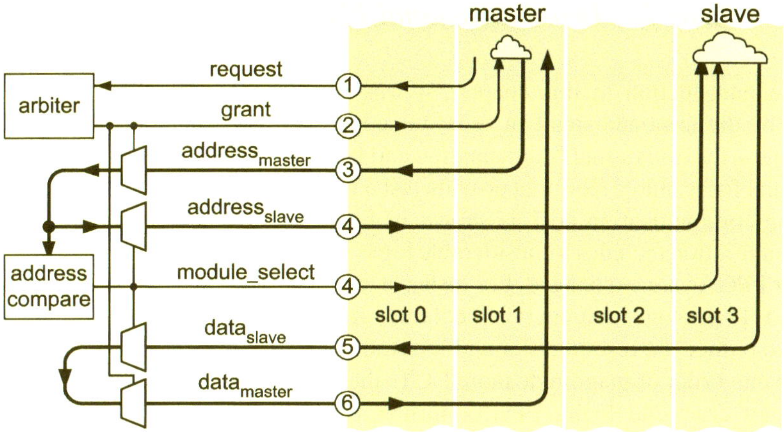

Fig. 2.42 Example of a reconfigurable master module reading a register from a reconfigurable slave. In the first step, the master requests bus access from the arbiter. Next, the arbiter grants bus control to the master (step 2). Upon this grant, the master drives the slave address to the static part in step 3. In the case of an interleaved bus, this address has to be firstly aligned according to the start position of the master that is currently selected by the grant signal. The resulting address is then evaluated inside an address comparator that will activate a module select for the slave in step 4. Within the same step, the address data will be aligned by a demultiplexer according to the start position of the selected slave. With respect to these two signals, the selected slave will drive its data onto the bus in step 5. Finally, in step 6, the data is transferred to the master. If the data signals are interleaved, it requires an alignment multiplexer for adjusting the slave data as well as an alignment demultiplexer for ordering the data to the master. The figure does not show the alignment (de-)multiplexer look-up tables

If a ReCoBus is implemented on Xilinx-FPGAs, it is possible to share the same look-up tables for read and write data signals. Furthermore, these signals can be used concurrently, hence data can be sent from one reconfigurable module to the static part of the system and directly back to another reconfigurable module connected to the bus. With this procedure, it is possible to communicate between two reconfigurable modules connected to the same bus. This is illustrated in Fig. 2.42 for a reconfigurable master module that accesses a slave module, which is also located in the reconfigurable area. In this example, the signal path is propagated three times towards the static part of the system: (1) for the arbitration (step 1 and step 2), (2) for driving the address from the master to the slave (step 3 and step 4), and (3) for sending the data from the slave to the master (step 5 and step 6). By interleaving the address and data bus signals, it is possible to reduce the combinatorial depth (in terms of look-up tables), and hence, the overall propagation delay. The throughput can be further enhanced by pipelining the signal path within the master, the slave, and/or in the static part of the system. An experimental evaluation of the propagation delays and on the impact of pipelining will be presented in Sect. 2.7.1.

2.5.6 *Buses for Two-dimensional Placement*

In the introduction to this chapter, it was demonstrated that the resource slots shall be dimensioned small in order to reduce internal fragmentation. The waste of logic – due to internal fragmentation and because of the communication cost – is minimal for resource slots that provide just a few hundred look-up tables (depending on the communication cost as shown in Fig. 2.2). Consequently, partial runtime reconfiguration includes a considerable logic overhead if complete logic columns on larger FPGAs are exchanged. For instance, a Xilinx Virtex-II 6000 device provides 768 LUTs per logic column (CLB column) and the best reported systems have been able to reduce the resource slot width to four columns [MTAB07], hence providing about one order of magnitude more LUTs than the optimal value.

Apart from the logic, FPGAs further provide dedicated resources, such as multipliers and dedicated memory. On FPGAs from Xilinx or Altera, these special resources are located in columns, spanning the complete height of the FPGA conjoined with the logic columns. The dedicated resources are very valuable and should therefore be highly utilized by the system. For example, an 18×18 bit multiplier would require nearly 400 LUTs for an implementation with the help of logic tiles. And a single block RAM can substitute more than 2,000 LUTs (also in the case that the LUTs are used as distributed memory) on Xilinx Virtex-II or Virtex-IV FPGAs.[8]

The only way to tackle these issues is to tile the reconfigurable area into a two-dimensional resource slot grid. This holds many advantages including that multiple modules can share the same multiplier column or memory column and that bounding boxes can be better adjusted to fulfill different requirements with respect to the logic or some dedicated resources as exemplary shown in Fig. 2.43. A further benefit is that the FPGA I/O pins can be used more flexible as it is possible to assign logic and I/O pins above or below the reconfigurable area. A two-dimensional module placement is natively supported by all newer Xilinx FPGAs (starting with the Virtex-4 family). For instance, the configuration frames of a Spartan-6 FPGA span the hight of 16 CLBs that provide 128 6-input LUTs (or alternatively 256 5-input LUTs). This is in the optimal slot size region, as examined in Sect. 2.2. The main penalty of a two dimensional placement is a rather complicated resource and configuration management as revealed in Sect. 3.4.

[8]In [BLC07] it is proposed to pay this price for enhancing module relocatability. The idea is omitting the usage of multipliers or BRAMs such that the same module (equal to the same configuration bitstream) can be placed to different resource slots that provide once a BRAM column or once a multiplier column at the same relative position (proposed for Virtex-4 devices). The goal of this approach is to reduce the number of configuration bitstreams. However, this is bought by a poor device utilization and it must be mentioned that in some cases two individual modules – one for the BRAM resource slot and one for the multiplier slot – require the same amount of memory in the bitstream repository as the relocatable one. In other words, modules should also achieve high device utilization in the case of specialized resources provided by the FPGA fabric.

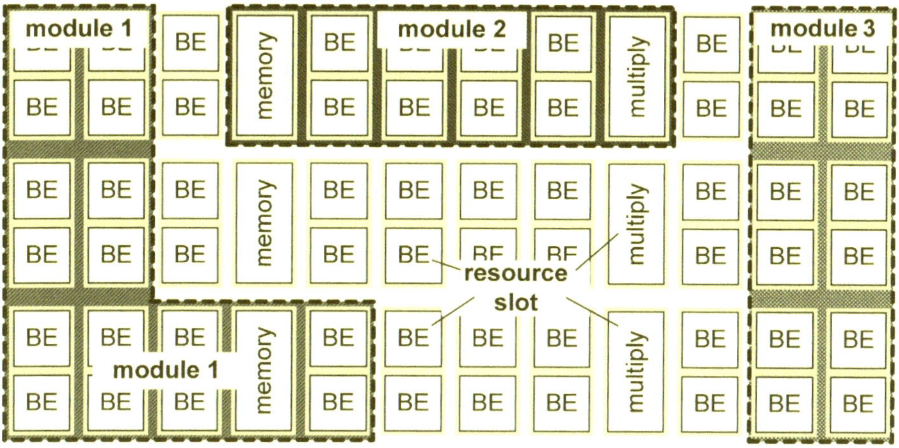

Fig. 2.43 Example of three modules placed on a two-dimensional grid of resource slots. The resource slots are one BEs wide and two BEs height. All modules have different resource requirements with respect to the amount of logic BEs, multipliers or memory blocks. The module shape can be arbitrary as shown for module 1 that has an 'L'-shape

The ReCoBus interleaving scheme, where the module interface can grow in relation to the module size, can be extended for two-dimensional systems as illustrated in Fig. 2.44 for shared read signals as well. Each row of resource slots is interleaved, as revealed in Sect. 2.5 for the one-dimensional case. The interleaving scheme in each new row is horizontally shifted one position further as compared to the preceding row, hence interleaving the bus also in vertical direction. Note that all rows are exactly the same and that the vertical interleaving is achieved by cyclically rotating the connections in the static part of the system. The bus does not require any vertical connections between resource slots, and the logic overhead of the resource slots is the same as in the one-dimensional case. Also, the alignment multiplexer is the same as in the one-dimensional case. The only additional logic overhead is a set of OR-gates for implementing shared read signals. In the case of shared write signals, another set of unidirectional wires will be used in opposite direction as the one shown in Fig. 2.44, hence the alignment multiplexer will become a demultiplexer and the OR-gates will be removed. Consequently, the resource overhead for shared write signals is the same for the one-dimensional and for the two-dimensional case. The same holds true for dedicated write signals, because all internal enable signals for the reconfigurable select generators within one row are connected to the same source within the static part of the system without any further logic.

The last signal class, required for implementing buses suitable for a two-dimensional placement of reconfigurable modules, are dedicated read signals. As two-dimensional reconfigurable systems require advanced configuration bitstream manipulations when implemented on Xilinx Virtex-II FPGAs (see also Sect. 3.4), the dedicated read signal distribution that is based on the manipulation of the

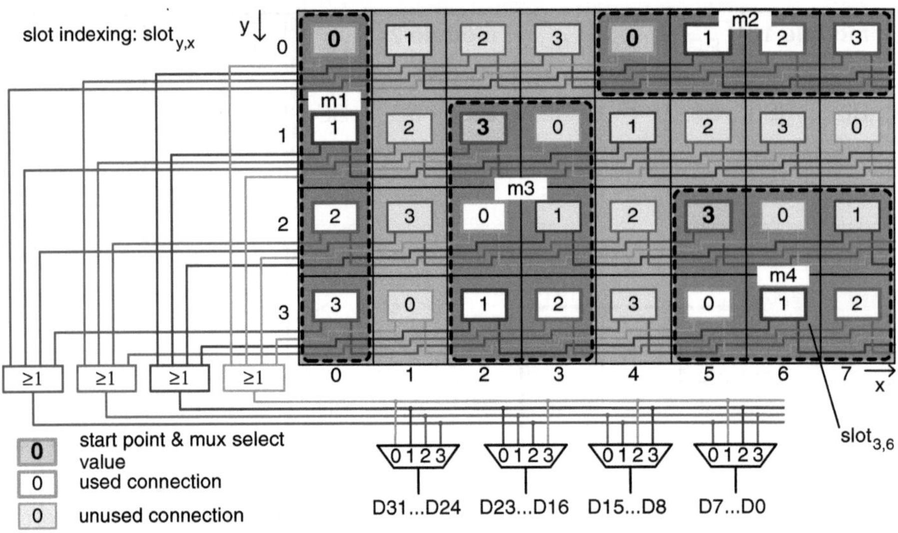

Fig. 2.44 Interleaved ReCoBus (only shared read data) implementation for two-dimensional module placement. All modules provide 32-bit access via four resource slots that are vertically (m1), horizontally (m2), or in an 'L'-shape (m3 and m4) aligned. The *highlighted top left* resource slot is the start resource slot, connecting the most significant byte within a reconfigurable module. The numbers in the resource slots denote the multiplexer setting for the alignment multiplexer if the slot is the start point of a reconfigurable module. The value is the sum of the x and y start point coordinates modulo the interleaving factor. For example, if module m4 is selected, the alignment multiplexers must be set to position $(5 + 2)\, mod\ 4 = 3$

configuration bitstream is most suitable. In this case, the techniques proposed for one-dimensional slot based systems, depicted in Fig. 2.40, can be duplicated for each row of resource slots. As it is possible to share the connecting resources with other bus logic, this approach does not include additional logic cost within the resource slots.

It can be summarized that the interleaving technique can also be efficiently applied to systems that allow placing modules in a two-dimensional resource slot grid. Using a grid style configuration instead of a one-dimensional slot arrangement will not increase the logic overhead per resource slot. This allows reducing the size of a resource slot in general and enhancing the utilization of the dedicated resources, like, for example, dedicated memory blocks.

2.6 Point-to-Point Communication

Besides buses for directly accessing registers inside a reconfigurable module (slave mode) or for addressing system memory by a reconfigurable module (master mode), a reconfigurable module may require some further signals to carry out the

Table 2.2 Relationship between the look-up table count and the input width of a multiplexer implementation

Mux inputs	2	3	4	5	6	7	8	9	10	11	12
LUT-4 (Virtex-II)	1	2	2	3	4	4	4	5	5	6	6
LUT-6 (Spartan-6)	1	1	1	2	2	2	2	3	3	3	4

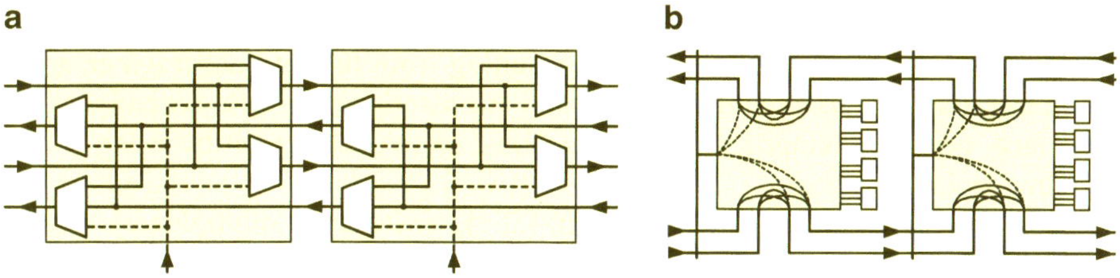

Fig. 2.45 (**a**) Two simple circuit switching elements implemented with multiplexers (LUTs) and (**b**) alternative implementation directly with the help of the FPGA switch matrices. For the latter case, only the important possible switch settings are shown that are selected by a multiplexer within the FPGA routing fabric. Direct implementation of circuit switching inside the switch matrices allows using the LUT resources by the reconfigurable modules. The module output is marked by *dashed lines*

communication with the environment. These could be other reconfigurable modules, I/O pins, or some dedicated modules in the static part of the system. The module environment communication should preserve the flexibility provided by a ReCoBus, such as a fine resource slot grid or module relocation.

If we analyze related work on point-to-point communication or circuit switching for runtime reconfigurable systems, we will observe that in virtually all systems the channels are established directly after exchanging or loading a reconfigurable module. These channels will then be retained until the module terminates or if it is replaced. Therefore, there is only switching activity in the point-to-point terminals upon a module swap with the help of partial reconfiguration. As reconfiguration is typically performed rarely, the multiplexers in the point-to-point terminals should be directly implemented within the switch matrix resources of the FPGA fabric, rather than in the look-up-table resources. Despite the fact that the FPGA look-up tables are basically multiplexers (with SRAM cells at the inputs and the select inputs being connected to the routing fabric of the FPGA), look-up tables are not very efficient for implementing wide input multiplexers as listed in Table 2.2.

For example, a simple one-dimensional RMB-like (see also Page 54) circuit switching architecture with two horizontal channels requires four multiplexers for the left and right channel outputs that have three inputs each. This allows selecting between one of the two preceding channels or the module output, as illustrated in Fig. 2.45. For this example, a switching element requires 4 times two 4-input look-up tables, which fits exactly into one CLB on a Xilinx Virtex-II FPGA (or one Spartan-6 slice). This is the logic overhead per bit within a channel. The total logic overhead in CLBs within the reconfigurable area is therefore the overhead per switched bit, multiplied by the channel width, multiplied by the amount of switching elements (equal to the number of resource slots). For instance, the simple example with two times two horizontal channels requires for 50 slots and 16 bit wide channels 800 CLBs on a Virtex-II FPGA. This is 9.5% of a Virtex-II-6000 device if implemented within the LUTs of the FPGA fabric or two to three times the amount of logic required for implementing a 32 bit softcore CPU. Note that this is pure overhead that originates only from using runtime reconfiguration. And this overhead contributes to more overhead resulting from the internal fragmentation, overhead for the bus communication, and the overhead for the reconfiguration interface, hence, making it difficult to achieve an overall benefit by using partial runtime reconfiguration.

As shown in the introduction (Sect. 1.2.1), switchable multiplexers can be implemented by look-up tables that are configured in route through mode (see Fig. 1.3 on Page 7 for examples). This approach requires about half of the amount of LUTs as compared to the values listed in Table 2.2. In route through mode, the LUT truth table function is the identity for the selected LUT input. The multiplexer setting is exchanged by modifying the LUT function and not by changing the state of some multiplexer select signals. In this approach, the multiplexer state – determining which particular input is connected to the output – is stored in the internal look-up table SRAM cells (the LUT function table) and not within some user flip-flops. This allows reducing the overhead against classical multiplexer implementations to roughly one half, which will still be substantial in many systems with a fine-grained resource slot grid. For example, a 4-input LUT can directly implement a 4:1 multiplexer when changing the LUT configuration according to the selected input. When implementing the same 4:1 multiplexer in the common technique, two 4-input LUTs are required (see also Table 2.2).

Alternatively, switching can be carried out directly within the switch matrices of the FPGA fabric that are basically reconfigurable multiplexers. This is depicted in Fig. 2.45b for the simple two channel one-dimensional circuit example discussed above. The figure only shows the FPGA routing resources relevant for implementing the switching elements. However, there are plenty of additional wiring resources available on typical FPGA architectures, thus only a relatively small amount of the available routing resources are required for implementing the switching elements. According to the Xilinx Virtex-II architecture, there exist 20 outgoing wires (called hex lines and double lines) towards each horizontal direction as plotted in Fig. 2.5. Consequently, a switching element of the assumed example would occupy only 10% of the horizontal routing resources while leaving all look-up tables and the remaining routing resources for implementing the reconfigurable modules. On the latest Xilinx FPGAs (e.g., Virtex-6/7, Spartan-6), there exist roughly the same number of wires. However these wires are spread into more directions and the upper (useful) limit for a regular circuit switching communication architecture is four wires per CLB.

The following two sections will discuss how point-to-point links can be implemented between a reconfigurable module and the environment by directly using an FPGA routing fabric. In Sect. 2.6.1, a one-dimensional circuit switching architecture will be proposed that is suitable for a fine resource slot grid and wide data channels. An architecture for switching dedicated signals in a two-dimensional fashion for reconfigurable systems on an FPGA will be presented in Sect. 2.6.2.

2.6.1 One-dimensional Online Circuit Switching

In Sect. 2.6 it was assumed that all channels between the circuit switching elements are equal and that each channel may be able to transfer any data from or to a reconfigurable module. In related approaches (e.g., [ABD$^+$05]) this model was used and it requires full crossbar switches for relatively wide data paths. Regardless of whether the multiplexers of the switching elements are implemented with the help of look-up tables or directly within the FPGA switch matrices, wide input multiplexers are inefficient to implement in both cases. Consequently, a suitable circuit switching architecture for runtime reconfigurable systems should minimize the amount and the size of the multiplexers inside the circuit switching elements.

Circuit switching is used for switching a channel from one source to one (point-to-point) or more (multicast) destinations. The channel must match the entire application and different channels with different requirements may be used within one system. For example, a multimedia system may require a video channel and an independent audio channel. While a video channel typically requires a throughput of some tens to a few hundreds of megabytes per second, an audio channel demands just 1.4 Mbit in CD quality. Consequently, diversely adjusted channels for the independent data streams are in most cases much more efficient to implement than using a homogeneous communication architecture with identical channels for any type of stream. Another reason for using independent channels is that the switching elements require a lower degree of freedom, hence, simplifying online routing algorithms. Finally, a system should always provide feasible switching for all allowed placement positions of the reconfigurable modules. In most applications, it cannot be accepted to reject a module or a service because of insufficient online communication resources for integrating a partial module into the system.

These observations have been taken into account for the design of so-called *connection bars* that feature an adjustable set of wires, which create switched connections from resource slot to resource slot in a homogeneous manner. In order to allow a module to connect to some resources outside its boundary, multiple adjusted connection bars are allocated at design time along all resource slots. Within each resource slot, a bar may be tapped to receive a down stream or ripped for driving an upstream towards the bar. Note that in the latter case, the module may still receive a downstream, hence, allowing to receive data, then process it, and finally send it to some other modules or the static part of the system further along the bar in a read-manipulate-write manner.

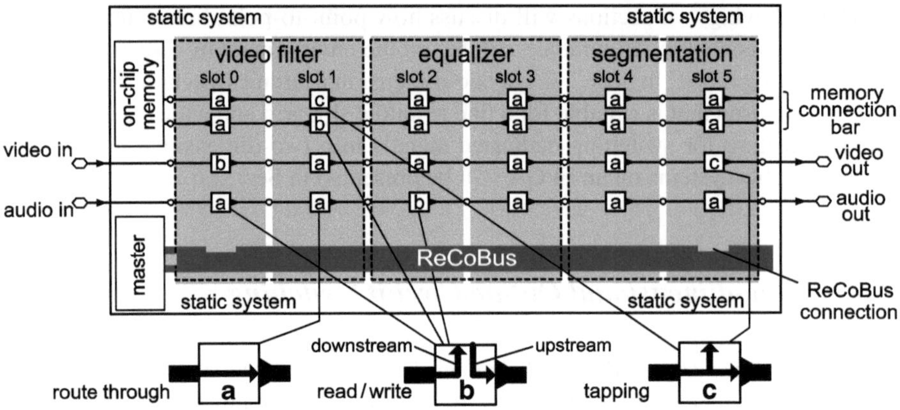

Fig. 2.46 Example of a system with an audio module and two video modules linked together by a ReCoBus and three connection bars. The top bar allows the video filter module to connect to on-chip memory located outside the reconfigurable area, while the other connection bars are used to connect the audio and video streams to the I/O pins. A module can pass through the signals of a particular connection bar (*box a*), access these signals in a read-manipulate-write fashion (*box b*), or tap the stream without manipulation (*box c*). The settings of the switching elements inside the resource slots are part of the FPGA routing fabric and controlled by configuration of SRAM cells

Figure 2.46 presents an example of such a system, where three modules communicate with the environment using the connection bar approach. A bar can utilize any arbitrary amount of input or output wires. All resource slots that do not write a bar have to pass through all internal bar wires. Regardless of whether a bar is accessed, the input and output wires have to be equivalent in all resource slots and for all possible settings of the connection bar switching elements. Consequently, modules can be placed freely to the resource slots while still being able to communicate with dedicated resources or modules outside the module bounding box. The communication bar approach does not require any online routing algorithm and the switching elements are part of the module and can be automatically set during module configuration. This is possible by selectively instantiating connection bars that pass through a set of wires or provide access to these. The entire bar instance is related to a particular module. For example, a video module will access the video connection bar with an associated connection bar containing connection primitives, while passing through the bar for audio data. As an alternative for setting the connection bar switching primitives directly during module reconfiguration, the switching element setting can be pushed before or after the entire module configuration. For example, if a video filter is exchanged by another, the associated video bar may be firstly configured in feed-through mode for masking disturbances related to the reconfiguration process from the video stream. Later, after writing the configuration of the new module to the FPGA, the connection bar switch may be changed from feed through to read-manipulate-write operation.

The connection bar approach is ideal for many applications processing data streams. The same set of horizontal wires of a connection bar can be segmented into multiple point-to-point links or for streaming data from one source to multiple destinations or for a mixture of both. This matches a wide range of networking or multimedia processing systems. A video processing system, for example, may send the video stream to a reconfigurable filter module first, from where the preprocessed data can be distributed to multiple parallel working pattern recognition modules. In this example, the connection bar is segmented in the preprocessing module, allowing a point-to-point link from the video source and, at the same time, a multicast link to the recognition modules, using the same horizontal connection bar. The only restriction in such a system is that the preprocessing module must be placed before the recognition modules in the data stream. Note that a connection bar can be alternatively used for providing connections between both sides of a static system that is located at the left and right neighborhood of the reconfigurable area. This allows the static system to route signals via a bar through a reconfigurable region.

The latency of a connection bar becomes large in systems with many resource slots. For reducing the combinatory path of a connection bar, the internal wires can be interleaved, as discussed for the bus architecture in Sect. 2.5.1. By interleaving a connection bar, the S_C signals of a particular bar are distributed via N wire subsets, containing $\lceil \frac{S_C}{N} \rceil$ wires each. The wire subsets provide a subset of switching elements after N resource slots and each subset is displaced against the preceding subset by one resource slot position. Hence, each resource slot is connected to a subset of $\lceil \frac{S_C}{N} \rceil$ internal bar wires. As the wire subsets skip $N - 1$ resource slots, routing may be carried out faster. For example, when implementing a connection bar on a Xilinx Virtex-II XC2V6000-6 FPGA with double lines (a wire spanning a distance of two basic elements), the propagation delay for three lines in a row is $3 \cdot 0.282\,\text{ns} = 0.846\,\text{ns}$. Interconnecting the same distance with interleaved hexlines allows the same resource slot grid, while reducing the propagation delay on the same routing distance to 0.356 ns or 42%. Note that interleaving allows to use routing wires, spanning over a wider distance than the width W of a particular resource slot. For full connection bar access, a module must be at least N consecutive resource slots wide.

Free module placement in systems using interleaved connection bars requires a set S_C alignment multiplexers (each possessing N inputs) once per module, if the data is streamed to the module (and not bypassing it). in case the module sends data to the connection bar, S_C alignment demultiplexers are necessary. There is no further alignment required in the static part of the system. On Xilinx Virtex-II or Spartan-3 FPGAs, the alignment multiplexer can be directly implemented within a CLB switch matrix. For example, a system with one CLB column wide resource slots ($W = 1$) that is twofold interleaved ($N = 2$), will use double lines for implementing connection bars. Such double lines span a distance of two CLBs and can be tapped in the middle, after one CLB. The same holds respectively true for hex lines that connect to CLBs, located three and six columns away from the source CLB. The end points and the middle connections of these routing resources are connected to the same configurable input multiplexer within a CLB switch matrix. This allows switching between the first and the second interleaved wire subset by changing the associated switch matrix multiplexers between end point or the middle connection. More details on online bitstream manipulation are revealed in Sect. 3.4.

2.6.2 Two-dimensional Online Circuit Switching

Connection bars are very efficient for distributing wide circuit switching channels in a one-dimensional fashion at low latency. Multiple connection bars located in different rows of resource slots aligned in a grid style can also be used for a two-dimensional placement. As it will be shown in Sect. 2.7.3, many hundreds of individual horizontal wires of the FPGA fabric can be allocated for implementing a communication architecture capable to integrate reconfigurable modules before dropping the achievable clock frequency or getting unroutable designs. For slot based reconfigurable systems (tailored to Xilinx FPGAs), this only holds for horizontal routing resources, because partially reconfigurable modules possess typically a narrow but high form factor. Therefore, many horizontal wires are not usable for implementing the reconfigurable modules, as these resources can be driven within the module boundary but only sensed outside the bounding box.[9] In other words, the required modular design style of strictly forcing modules into bounding boxes will disallow using some interconnect resources at the module boundaries. Employing these resources in order to provide a communication architecture for integrating partial modules into the runtime system comes basically for free. However, any vertical routing between the different multiple horizontal communication bars must then be performed within the static part of the system.

As an alternative to two-dimensional online circuit switching with multiple connection bars, a mesh topology for free two dimensional online routing was examined. Such an approach is only suitable for narrow channels with very few wires, because almost all vertical wires are used for implementing the interconnection of a partial module, hence, conflicting with the online routing architecture. On current FPGAs, this is possible with a reasonable performance. A Xilinx Virtex-II 6000-6 FPGA, for example, allows to establish a signal path composed of up to 35 double lines (or 26 hex lines) with a 10 ns latency budget. This equates to a Manhattan distance of 70 (156) CLBs or 73% (163%) of the device height or width. A single wire channel at this speed is sufficient for Fast Ethernet, 70 CD audio streams or 41 uncompressed CIF (Common Intermediate Format) video frames per second (24 bit RGB).

---

[9]With some restrictions, some if the wires can still be used as demonstrated in [KT10]. This is for example possible, if there exist paths such that a wire that leaves a reconfigurable region can route strait back into it without any further hop. Then that path cannot be in conflict with other adjacent modules or the static system.

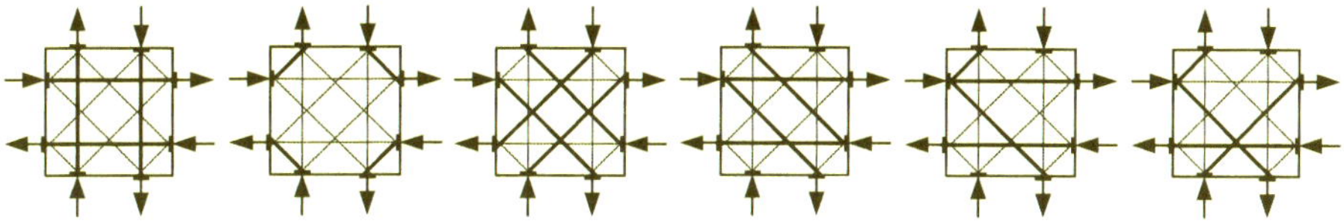

Fig. 2.47 Examples of some switch positions of a routing cell that is obtained by reserving one routing wire in each direction (north, east, south, west) of the device. When each output can be connected with three different inputs by partial reconfiguration, there exist $3^4 = 81$ different switch permutations of such a routing cell

For free two-dimensional online circuit switching, an interconnect network is allocated logically on top of the FPGA fabric, whereas the switching elements are allocated within the switch matrices. This can be implemented by reserving allocating a set of wires, for example, one for each direction (north, east, south, west) within each basic element (see also Sect. 3.2.4) of the reconfigurable area that will be used for performing the circuit switching. These wires are permitted for implementing the reconfigurable modules or any other hardware of the static system. The blocked wires will represent a separate interconnect layer for module-to-module and module-to-environment communication. By using disjoint interconnect resources for the module implementation and the online circuit switching network, routing can be performed regardless of the module placement positions. Therefore, the modules are no obstacles for the circuit switching. The blocked wires are directly linked together by the routing resources supplied by the BE switch matrices of the FPGA, thus, not requiring any extra logic. Only at the source and the destination(s) of a connection bar there may be some logic required for connecting the path with the modules. The reserved wires should be chosen in such a way that they form a regular two-dimensional routing grid. In each cell of the grid, it should be possible to route strait through, or to turn left, or to turn right. In addition, an incoming signal may be connected to multiple outputs of a cell for allowing multicast operations on the same path.

Figure 2.47 depicts an example of how two-dimensional circuit switching could be implemented directly within the switch matrices of an FPGA. For supporting this approach efficiently, a set of wires must be chosen that allows a replicated routing by directly configurable connections from the input wires of a cell to the allocated output wires, without using additional resources. A replicated routing scheme allows a free rectangular circuit switching on the device by using the same equivalent routing resources in each routing cell. For example, when reserving one dedicated wire in all four directions for providing a two-dimensional online circuit switching network on an FPGA, then it should be possible, to establish any path within the network. This implies that it must be possible to set a zigzag path (like towards east then towards north and in east direction again), such that only the reserved wires are used without additional wiring resources. Most common, this is not possible in a trivial manner, because the approach requires that connections can be arbitrarily set. In general, this is not possible, because FPGA switch matrices cannot provide full crossbar switches. Consequently, online circuit switching is possible, if the allocated wires fulfill the following three properties:

1. Incoming wires must be extendable to the same corresponding channel index for allowing straight forward routing,
2. There must exist a cycle with four hops in clockwise direction for turning to the right, and
3. A cycle with four hops in anticlockwise direction for turning to the left.

The switch matrix settings of the circuit switching network can be computed with standard algorithms such as Dijkstra [Dij59], Belman-Ford, or A* [Tes98]. Congestion can be prevented with iterative techniques such as proposed for VPR [BR97] or Pathfinder [ME95] where cost functions are adapted over the iterations such that congested paths become more expensive. This will force the search algorithms to use alternative resources in following iterations.

2.6.2.1 FPGA Limitations

Implementing two dimensional online circuit switching on existing FPGAs may restrict the communication architecture because of an insufficient degree of routing freedom or because of some inhomogeneous resources found in the otherwise regular FPGA fabric. With respect to the amount of wires connected to an FPGA basic element, there can only exist a relatively small subset of all possible connections within a configurable switch matrix, as fully connected crossbars would be too expensive. Therefore, the approach presented in Fig. 2.47 may not be applicable to all available FPGA architectures. In Fig. 2.48, the connectivity of a switch matrix is exemplary illustrated by an adjacency matrix in form of a bitmap for a Xilinx Virtex-II and Virtex-5 FPGA. A black dot in the figure denotes a possible connection between a switch matrix input and an output. While a fully connected crossbar would lead to one black plane in the figure, the FPGA adjacency matrices are sparsely populated and the routing multiplexers inside the switch matrix are connected to just 6.3% (Virtex-II) and 7.5% (Virtex-5) of the available inputs, in average.

By examining the Xilinx Virtex-II switch matrices in more detail, no set of routing wires can be found, allowing to directly implement the approach in Fig. 2.47, because of not all of the above mentioned properties can be fulfilled at the same time. While property (1) for straight routing tracks can always be met on Virtex-II FPGAs, only one further property can be directly implemented for making a turn in the routing path. A replicated routing scheme, however, can also be obtained, if some connections between the set of blocked wires is not available. For example, if there is no possibility to turn left at a routing cell, it is feasible to implement a loop by routing three times in turning right direction. This leads to a *clockwise* or in turn to an *anticlockwise* routing scheme. Examples for these routing schemes tailored to Xilinx Virtex-II FPGAs (also suitable for Virtex-IV, or Spartan-3 FPGAs) are illustrated in Fig. 2.49. Here, each output wire of the network can be directly

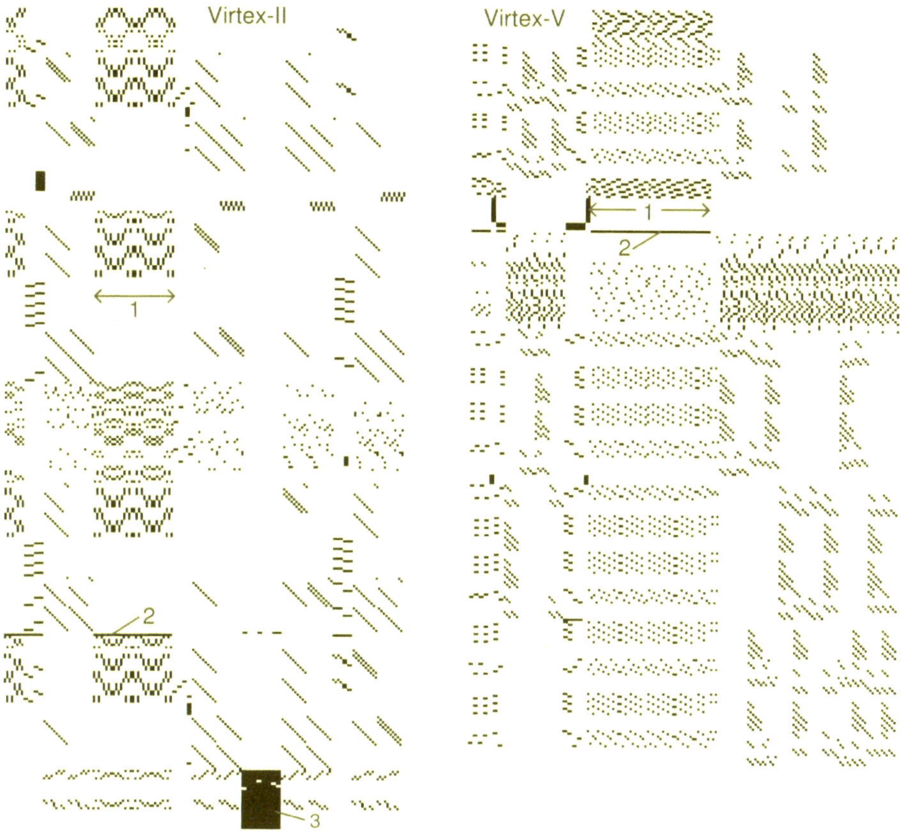

Fig. 2.48 Adjacency matrix of a Xilinx Virtex-II and a Virtex-5 CLB switch matrix, illustrated by a bitmap graphic. Each *black dot* represents a possible configuration from one source (the rows →, e.g., a LUT output or wire end port) to one particular sink (the columns ↓, e.g., a LUT input or wire begin port). In a Virtex-II (Virtex-5) switch matrix, there exist 332 (305) shared inputs for 160 (172) multiplexers. There exist 3,328 (3,936) possible settings, which is 6.3% (7.5%) the amount a fully interconnected crossbar would provide (that would be a black plane). The marked details in the bitmaps are: (1) the LUT input multiplexers, (2) constant input value, and (3) LUT and flip-flop outputs to OMUX wires that also act as hops in Virtex-II. A row is a reconfigurable multiplexer and the number of dots within a row indicates the number of multiplexer inputs. Note that Virtex-5 CLBs provide eight 6-input LUTs which can roughly implement twice the logic of a Virtex-II CLB with eight 4-input LUTs

connected with a wire, starting in a straight preceding switch matrix or with a wire starting in one bended cell. Therefore, each wire has two switch positions and a complete cell that is providing one signal per direction can be described with just 4 bits. The two switch positions are set by configuring the multiplexer for a particular wire with one out of the two corresponding configuration encodings.

While the adjacency of Virtex-5 switch matrices is not well suited for implementing two-dimensional circuit switching networks, this is nicely supported by Virtex-6

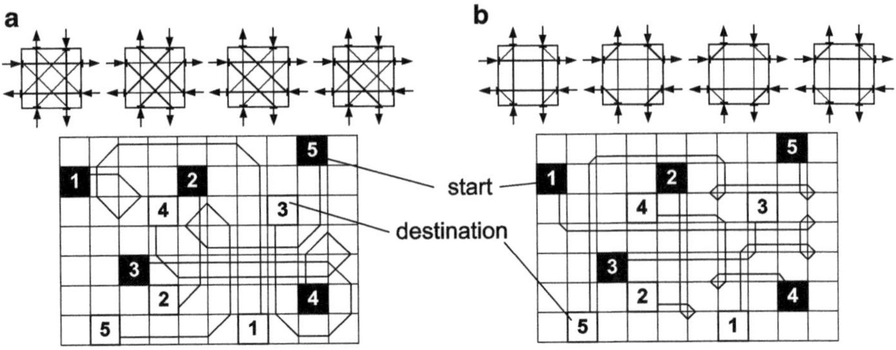

Fig. 2.49 Two-dimensional on-line circuit switching scheme tailored to Xilinx Virtex-II, Virtex-IV, or Spartan-3 FPGA. On these devices it is possible to arbitrarily route within a switch matrix straight ahead and either turning (**a**) clockwise or (**b**) anticlockwise directly by allocating one wire per direction (north, east, south, and west) but without any additional routing resources. (**a**) and (**b**) are exclusive as there exists no set of wires that allows directly implementing a zig-zag path, like turning right, then left, and right again

and Spartan-6 FPGAs. For the latter devices, we can arbitrary allocate double lines or quad lines that fulfill the properties to extend and to turn left or right within each CLB, BRAM, or DSP switch matrix.

Apart from the routing freedom, the online circuit switching architecture should further be capable to respect inhomogeneous FPGA resources, such as dedicated memories or multipliers. Furthermore, the architecture and the resource slots should be selected with respect to the available routing resources of the FPGA fabric. As depicted in Fig. 2.50, the resource slot dimensions should be set to a common multiple of the distance of the used wire resources and the size of possibly existing, dedicated FPGA resources. With these restrictions, modules can be freely placed in a two dimensional fashion while being able to carry out the communication via a regular mesh type network at runtime.

2.7 Experimental Results

For benchmarking the proposed ReCoBus and connection bar communication architecture, several tests have been performed that are organized in the following sections:

Section 2.7.1: Measuring the timing and performance characteristics of the ReCoBus architecture.

Section 2.7.2: Examining the logic cost for implementing the communication architecture.

Section 2.7.3: Evaluation of the impact of restricting the module routing on the achievable clock frequency and on the capability to fully route a module.

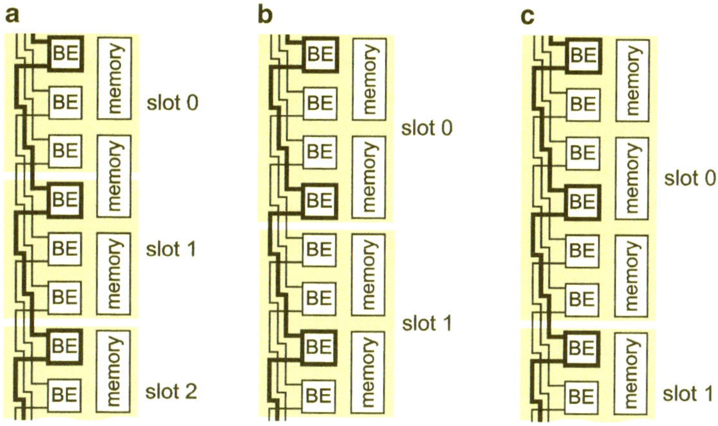

Fig. 2.50 Example of partitioning an FPGA into resource slots. In all examples, (*vertical*) online circuit switching is carried out via one fat drawn wire of a channel of span $S = 3$. (**a**) A height equal to the span, $H = S$, allows accessing the online circuit switching architecture at the same relative position inside each resource slot but may prevent the usage of dedicated resources (here the memory blocks). (**b**) Selecting the resource slot height to a multiple of the memory height provides full memory utilization but may restrict access to the circuit switching architecture. (**c**) By setting the resource slot height to a common multiple of the memory height and the span, it is possible to use the memories as well as to access the online circuit switching network at exactly the same relative position within the resource slots. On Xilinx Spartan-6 FPGAs, the memory and DSP blocks are four CLBs in height. This fits exactly the available quad line wires that route four CLB tiles further

Section 2.7.4: Investigation of transient effects on the communication architecture due to a reconfiguration process.
Section 2.7.5: Implementation example of a two-dimensional communication architecture featuring ReCoBus and connection bar connections, as well as a two-dimensional circuit switching network.

2.7.1 ReCoBus Timing Characteristics

The maximum throughput T of a ReCoBus is given by the data word width B of the bus, multiplied by the clock frequency f that can be achieved in the system ($T = B \cdot f$) and measured in the unit bits/second. Here it is assumed that a new bus transfer can start at any clock cycle and that the bus does not use multi-cycle paths in order to reach higher clock rates by the cost of multiple clock cycles per transfer. While B is typically set to the word width of a processor or memory, the value for the maximum clock frequency f is related to the reciprocal of the propagation delay of the critical path $t_{prop} = \frac{1}{f}$ to evaluate the combinatorial logic. For simplicity reasons, setup and hold times of the flip-flops will be omitted in the following. With respect to the proposed ReCoBus architecture, the critical

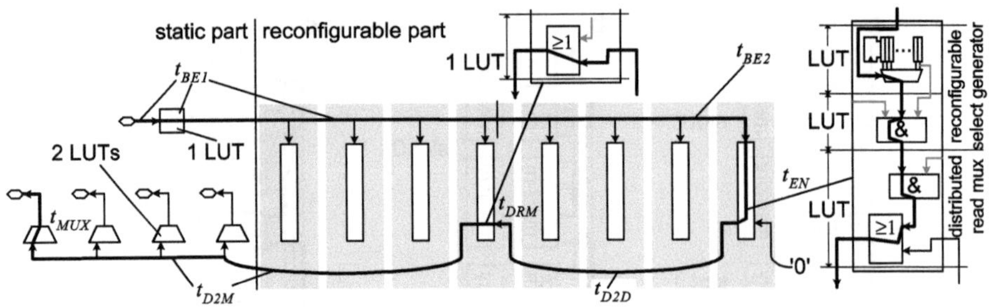

Fig. 2.51 Critical path of a transaction of a ReCoBus. Data read operations have the longest critical path consisting of a forward path including a reconfigurable select generator and a backward path containing a distributed read multiplexer. The path is measured between the master connection terminals driving the address signals and receiving the data through the final multiplexer. The model follows the example in Fig. 2.34 with four interleaved distributed logic-based bus chains ($N = 4$)

path will be taken on data read operations, in the case of a bus containing only slaves. Depending on the used features, the critical path consists of a forward path with the module select logic (see Fig. 2.28) and a backward path consisting of a distributed read multiplexer followed by a final multiplexer for aligning the correct byte order (see Fig. 2.34). A timing model for this path can be specified based on the bus architecture presented in Fig. 2.51 that is independent of the final FPGA implementation. The architecture is mainly parameterized by the resource slot width W and the number of interleaved read multiplexer chains N. Both values are given in terms of basic elements found on the reconfigurable target architecture (see also Sect. 2.3 for details on the assumed FPGA model). Then, the annotated delays in the architectural view have the following meanings:

t_{BE1} is the time required for the most critical address signal to pass a connecting resource (e.g., a look-up table) plus the propagation delay for the wiring of this address signal to the reconfigurable select generator, having the maximum signal propagation delay within the first N module connection points. These module connection points are located in the first N resource slots, located next to the static part of the system. In the case that the address signals are interleaved, the propagation delay of an additional alignment demultiplexer has to be considered. Note that all available reconfigurable select generators are connected to the address wires in order to take the delay for an increased fan-out into account.

t_{BE2} specifies the signal propagation delay for routing the most critical address signal to the next chained module connection point that is N resource slots further away.

t_{EN} gives the propagation delay for the reconfigurable select generator (see Fig. 2.27) plus the chunk of a distributed read multiplexer that feeds in the module data to the bus. The read enable signal (`read_en` in Fig. 2.27) is connected to $\lceil B/N \rceil$ chunks of a distributed read multiplexer, because it requires N resource slots to connect B individual data signals to the bus. The value of t_{EN} represents the most critical propagation delay over all $\lceil B/N \rceil$ paths.

t_{DRM} specifies the time to pass the look-up table of a distributed read multiplexer stage.

t_{D2D} denotes the propagation delay on the wiring between two chunks of a distributed read multiplexer.

t_{D2M} denotes the propagation delay on the wiring between the last chunk of a distributed read multiplexer chain and the multiplexers for adjusting the alignment of the N subwords. The value gives largest delay over all B data lines.

t_{MUX} specifies the time for the alignment multiplexers. These multiplexers have N inputs and ensure that the data signals are aligned in the same order, regardless of the position of a module at the bus.

These time values allow extrapolating the propagation delay to any size of the bus. The combined worst case propagation delay of the bus is calculated as follows:

$$t_{prop} = \underbrace{t_{BE1} + t_{EN} + t_{D2M} + t_{MUX}}_{fixed} + \underbrace{\lambda \cdot (t_{BE2} + t_{DRM} + t_{D2D})}_{variable} \tag{2.10}$$

The factor λ specifies the width in addition to the first N resource slots of the bus as a multiple of the number of interleaved distributed read multiplexer chains N. As shown in (2.10), the worst case propagation delay consists of a fixed and a variable part that can be used to extrapolate small reconfigurable buses to any size. The complete reconfigurable bus offers $R = (\lambda + 1) \cdot N$ resource slots, each one with an interface width of up to $\lceil B/N \rceil$ bits. With this timing model, the design space of a particular ReCoBus implementation can be accurately predicted without the need to fully implement the whole bus architecture on a particular FPGA.

As presented in Sect. 2.3, FPGAs offer routing resources, spanning over different numbers of basic elements that can be combined to build different regular bus structures. The resource slot width W may be chosen to a multiple of the grid of the basic building blocks of the respective FPGA architecture. Let H be the height, counted in basic blocks then the total amount of logic resources, a reconfigurable bus can provide to some modules, is $(\lambda + 1) \cdot N \cdot W \cdot H$ basic blocks.

For determining the impact of combining different wiring resources and resource slot widths, multiple variants of the bus architecture shown in Fig. 2.51 have been implemented on different FPGA devices. The experiments were carried out for different FPGAs from Xilinx Inc. that are all partially reconfigurable. As mentioned before in Sect. 2.3, a basic element of these FPGAs is called a *CLB*. And all examined FPGAs offer the same two different wire spans for building the local module interconnect. Double-lines connect a CLB with another one that is two CLBs far away. Analogously, Hex-lines connect a CLB with another one that is six CLBs far away. For all cases, the interleaving factor was set to $N = 4$ to match the interface size of a $B = 32$ bit data bus by using 8 look-up tables per

Table 2.3 Measured propagation delays for the proposed timing model. All values are in nano seconds. The table lists results for one Xilinx Virtex-II FPGA, two Virtex-IV devices of different speed grade, and a Spartan-3 FPGA. The values have been determined by a timing analysis of the different bus implementations

FPGA	Variant	t_{BE1}	t_{BE2}	t_{EN}	t_{DRM}	t_{MUX}	t_{D2D}	t_{D2M}
V-II-6	Double + double	1.095	0.486	1.988	0.347	0.92	0.639	1.148
	Hex − double	1.095	0.486	1.988	0.347	0.92	0.706	1.082
	Hex + double	1.581	0.972	1.988	0.347	0.92	0.706	1.179
V-IV-10	Double + double	1.178	0.445	1.46	0.195	0.779	0.581	1.025
	Hex − double	1.178	0.445	1.46	0.195	0.779	0.614	0.83
	Hex + double	1.673	0.89	1.46	0.195	0.779	0.624	1.345
V-IV-11	Double + double	1.828	0.374	1.244	0.166	0.664	0.492	0.867
	Hex − double	0.997	0.374	1.244	0.166	0.664	0.517	0.702
	Hex + double	1.415	0.748	1.244	0.166	0.664	0.525	1.137
S-3-4	Double + double	1.637	0.335	1.239	0.608	1.793	0.955	1.376
	Hex − double	1.637	0.335	1.239	0.608	1.793	0.927	1.376
	Hex + double	1.972	0.67	1.239	0.608	1.793	0.927	2.305

resource slot. For a resource slot width of $W = 1$, two cascaded wiring schemes have been manually implemented. In a first variant, two cascaded double wires (named *double + double*) have been used for routing a distance of $W \cdot N = 4$ CLBs. In another assembly, one hex wire in forward direction was cascaded with one additional double wire in backward direction (named *hex − double*). For a ReCoBus, featuring a resource slot width of $W = 2$ CLBs, one hex wire and one double wire, both cascaded in forward direction, have been selected (named *hex + double*). This leads to the required routing distance of $4 \cdot 2 = 8$ CLBs. The resulting timing values are listed in Table 2.3.

According to (2.10), the measured propagation delays have been extrapolated to larger bus sizes. The results are visualized in Fig. 2.52. It can be seen that the double + double approach leads to similar results as the hex-double one for all examined FPGAs. By defining a specific time budget for the bus propagation delay, and consequently defining the desired clock frequency of the bus, the possible parameters for the complete bus width $(\lambda + 1) \cdot N \cdot W$ and the resource slot width W of the ReCoBus can be determined. For example, if the specification demands a clock frequency of 50 MHz ($t_{prop} = 20$ ns) on a Spartan3 -4 FPGA, a ReCoBus can be implemented that is up to $(5 + 1) \cdot 4 \cdot 2 = 48$ CLB columns wide. This bus would be organized in 24 resource slots, each of them two CLB columns wide. Alternatively, $(7+1) \cdot 4 \cdot 1 = 32$ CLB columns, organized in 32 resource slots, would allow higher flexibility in smaller systems. Note that these values are best case values. Additional propagation delay may be added by some of the reconfigurable modules or the static part of the system. This will be the case if no registers are used at the module inputs or outputs.

So far, a discussion about heterogeneities found in the FPGA architecture was omitted. Such a heterogeneity can be a memory column or columns with dedicated IP blocks, like multipliers. For example, on Xilinx Virtex-II FPGAs a RAM column

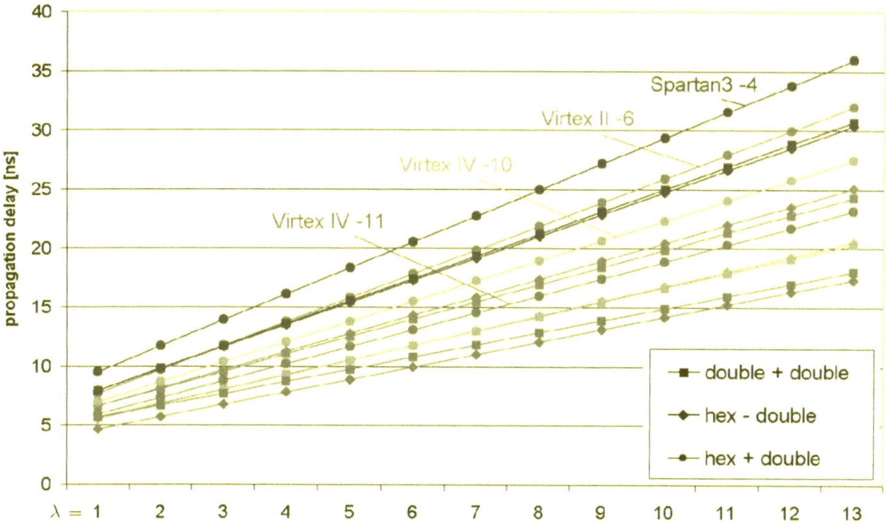

Fig. 2.52 Extrapolated propagation delays for different FPGAs and different cascading schemes over the number of attached interleaved resource slots λ. The resource slot width is $W = 2$ for the hex + double cascading scheme and $W = 1$ for the other schemes

is provided in the range of every 4–25 CLB columns. For all investigated devices, the additional propagation delay, found for bypassing a RAM column, was less than the delay required to bypass a CLB logic column. In addition, all RAM/multiplier columns provide the same routing resources as the logic columns. This allows to feed signals through heterogeneous columns without losing the regular interconnect structure of a ReCoBus. As a consequence, a RAM/multiplier column can be treated like any normal logic column with the exception that it cannot provide interface logic to a connected module.

Figure 2.52 points out that high system frequencies (with a 100 MHz or more) are only allowed for relatively narrow buses. Note that also static only designs require additional effort for the implementation of high speed systems. In order to overcome this limitation, pipelining the bus communication was examined. This still allows starting a communication transaction at any clock cycle, but data will arrive with a latency of one additional clock cycle at the output, while potentially transferring a new data word each cycle. Larger portions of data can be sent over the bus as one continuous transfer by using a burst protocol. Alternatively, split transaction protocols may be used that allow overlapping multiple bus requests in time.[10] It is preferable that the latency of the reconfigurable bus is independent of the placement position of a module. Therefore, pipelining is not suitable within a distributed read

[10]Both protocols are supported by the embedded hard IP processors (a PowerPC-405 Core) available on Xilinx Virtex2-Pro FPGAs. On these cores, a sequence of multiple bus transfers to different addresses can be started before the first transfer has been finished.

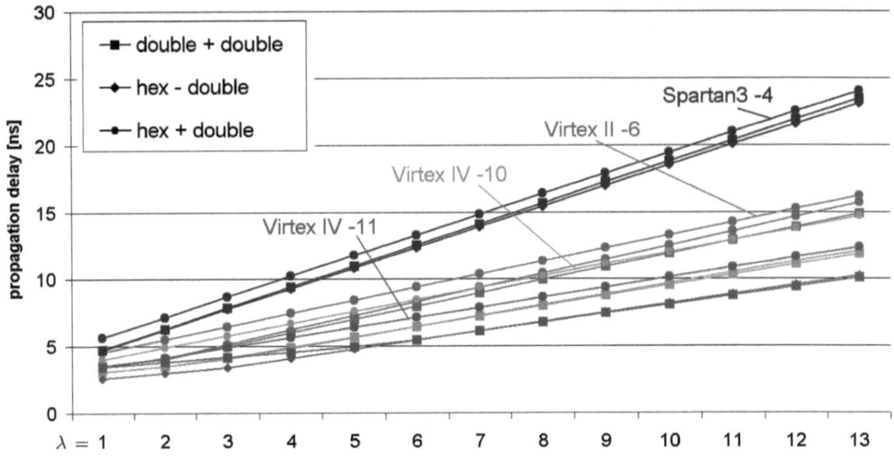

Fig. 2.53 Extrapolated worst case propagation delays for buses with one pipeline register in dependence of the number of attached interleaved resource slots λ

multiplexer. By analyzing the timing model in Fig. 2.51, it was determined to place a pipeline register between the forward path and the backward path, containing the distributed read multiplexer chains. In this case there exists the choice to place the pipeline register after or before the block, containing the reconfigurable select generator. As a consequence, t_{EN} will be considered either in the forward path or in the backward path. The worst case propagation delay of the reconfigurable bus with one pipeline register is then:

$$t_{BE} = t_{BE1} + \lambda \cdot t_{BE2}$$

$$t_{RM} = t_{D2M} + t_{MUX} + \lambda \cdot (t_{DRM} + t_{D2D})$$

$$t_{wc}^{pipe} = max\left\{max\{t_{BE}, t_{RM}\}, min\{t_{BE}, t_{RM}\} + t_{EN}\right\}. \tag{2.11}$$

The combined worst propagation delay when using pipelining t_{wc}^{pipe} is the maximum of the faster path plus the enable time t_{EN} and the slower path (in both cases the forward path and the backward path). The pipelined worst case propagation delay t_{wc}^{pipe} against the width in terms of λ is extrapolated as plotted in Fig. 2.53. The results reveal a substantial improvement as compared to the unpipelined version in Fig. 2.52. Now the Spartan-3 device allows implementing buses that are up to $(10+1)\cdot 4\cdot 2 = 88$ CLB columns wide at a speed of 50 MHz. These 88 columns would be 8 CLB columns wider than the amount that is offered by the biggest device within the Spartan-3 family and nearly double the amount, achieved for the unpipelined version.

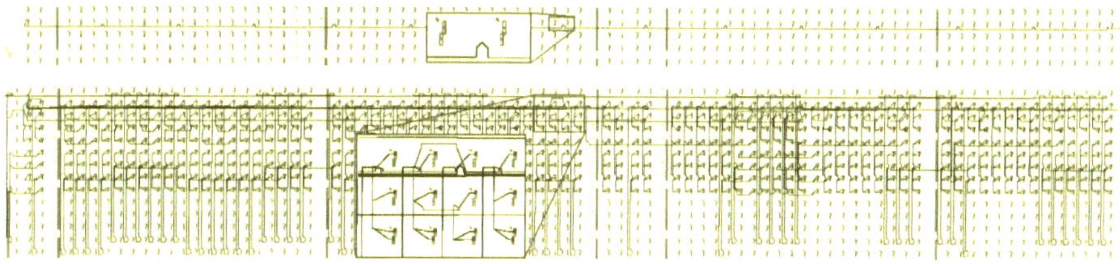

Fig. 2.54 Mapped test circuit for measuring the impact of the fan-out on the signal propagation delay of a Xilinx XC2V6000-6 FPGA. Starting from a regularly routed signal path (shown in the *top* for a hex line path) with one source at the left hand side of the FPGA and one sink at the opposite site, the design was successively extended with additional sinks. The figure shows the initial path and the path after extending it with 1,120 additional sinks. The time between the source and the first included sink was determined with the Xilinx timing analysis tool each time a new sink was included. The results are displayed in Fig. 2.55

2.7.1.1 Fan-out Impact on the Signal Latency

The signal propagation is not only related to the selected routing resources and the distance of a signal path, but also to the number of inputs connected to a particular wire. This number of destinations of a particular wire is called fan-out.[11] A high fan-out will lead to a considerable increase of the propagation delay on the critical path of a specific net. This increase must be known when the implicit connection scheme is used, where the module is connected to a ReCoBus or a connection bar without additional logic, and hence, without an additional driver. Connecting a module via an extra driver will decouple the module from the communication architecture, such that the module logic and routing has no influence on the signal timing of the communication architecture. The fan-out of internal bus signals can become very high, for example, in the case the partial modules contain a large register file that is directly connected to the ReCoBus.

When taking the impact of the fan-out into account, timing can be guaranteed, regardless of the current combination of reconfigurable modules, integrated into the system at runtime. Note that a full timing analysis is typically too difficult to be computed at runtime for an embedded system and that these effects have to be considered at design time. The impact of the fan-out on the signal propagation delay of a single signal wire was examined with the test circuit depicted in Fig. 2.54. The experiments have been carried out for a signal that is once routed regularly with the help of hex lines and once with double lines, as it would be for an implicitly connected shared write signal or the internal signal of a communication bar. The increase of the propagation delay on the initial path has been measured each time, after including and connecting a new sink to the test circuit. The results are plotted

[11] In general, *fan-out* states the amount of inputs a logic gate output can drive while still being able to reach the technology specific signal levels at the inputs. For FPGAs produced in CMOS technology, reaching signal levels is of less concern for static signals and the fan-out denotes the number of sinks a signal is connected to. Its value is related to the dynamic behavior of a signal.

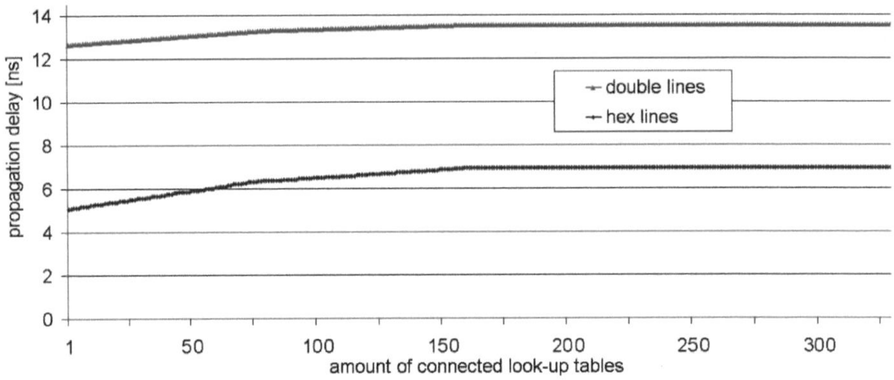

Fig. 2.55 Impact of the fan out (amount of connected look-up tables) on the propagation delay of a signal spanning the complete width of a Xilinx Virtex-II XC2V6000-6 FPGA

in Fig. 2.55 and point out that in the case of hex lines, the propagation delay can rise by up to 40%. For double lines, the increase was found to be 8%. For interpreting the results in Fig. 2.55, it must be mentioned that the additional sinks have been included consecutively one row after the other and that one row includes altogether 80 sinks. As can be seen, the propagation delay increases linear with the strongest slope for the first row and runs into a saturation after two to three rows in this experiment.

It was found that each wire segment has its own driver, thus electrically decoupling the wire output from the input. As the amount of possible destinations is limited for each wire segment, the impact on the propagation delay is limited, as well. While the first rows of sinks can directly connect to the initial signal path, this is not possible for all following sinks, because of the limited amount of direct connection terminals. Consequently, connections have to be provided via additional (decoupled) wire segments. As the sinks are regularly distributed among one row, the amount of connections to a particular segment of the initial prerouted signal path is about the same for each segment which explains the linear slopes in the result.

With this knowledge, an upper bound for the fan-out impact can be specified by determining the propagation increase after connecting all possible destinations of one particular wire segment to some sinks. For double lines, an increase of up to 12.7% in the propagation delay was found and an increase of up to 97% for hex lines on a Xilinx XC2V6000-6. In the latter case, the propagation delay increased from 0.596 to 1.175 ns when connecting all remaining taps of a hex line to some sinks. These values are upper bounds that will typically not be reached in practical designs. However, by reducing the slack on shared write signals with the maximal possible delay increase, partial modules can be implemented that will never violate the timing specification at runtime. Note that different wire segments of the same span differ with respect to their timing, because a higher driver strength requires more FPGA die area, while only the minority of the wire segments are used on timing critical signal paths [LAB+05]. This may be used for also implementing the communication architecture in a timing driven manner.

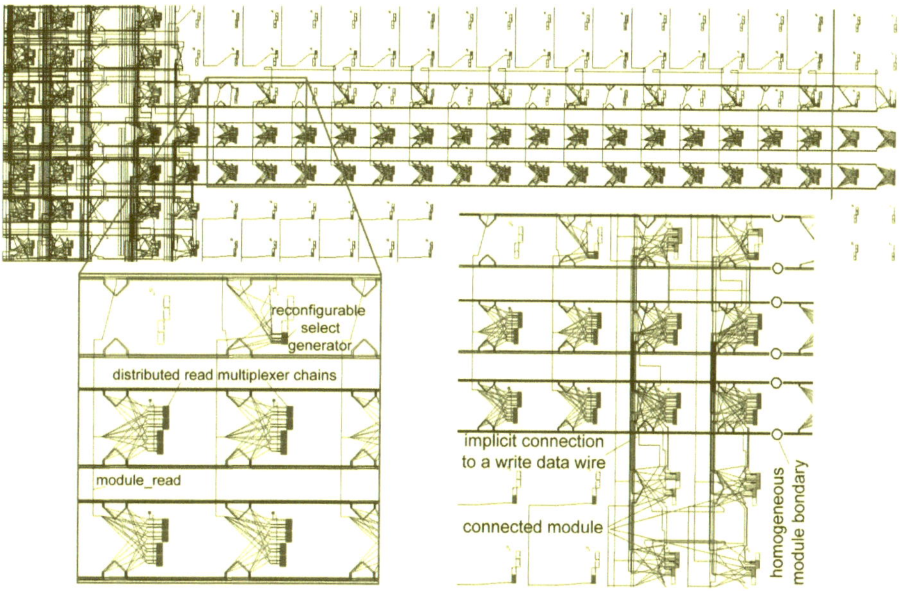

Fig. 2.56 Test system with a 32-bit ReCoBus connected to a stimuli pattern generator. In this example, each resource slot is $W = 2$ CLB columns wide and the bus is neither interleaved nor pipelined. The enlarged view in the right hand side of the figure shows a simple test module that is connected to the bus. The measured throughput of the ReCoBus is 800 MB/s on a Xilinx Virtex-2 XC2V6000-6 FPGA

2.7.1.2 High Throughput Test System

For demonstrating upper throughput limits on a ReCoBus, a test system was built, containing a stimuli pattern generator, an unpipelined 32-bit ReCoBus, and some reconfigurable test modules, as depicted in Fig. 2.56. The set of test modules contains only simple modules, such as Boolean functions. This allows keeping the achievable clock frequency of the system high. Whenever data is sent to a particular reconfigurable module by the stimuli generator, a copy is sent to an equivalent reference module. The reference modules are located in the static part of the system. This allows to verify proper bus operation by comparing all return values of read operations with the expected results from the internal test modules of the stimuli generator. The test pattern generator (a linear feedback shift register) and the error checker is pipelined such that the overall clock frequency is mainly related to propagation delay of the ReCoBus macro.

For this system, a requested clock frequency of 100 MHz was achieved for a ReCoBus with eight resource slots that are each $W = 2$ CLB columns wide on a Xilinx Virtex-2 XC2V6000-6 FPGA. The exact reported frequency was 104 MHz. By using only the address signals for controlling write operations to the test modules and respectively only the reconfigurable select generators for controlling read operations, read and write operations had been performed concurrently. This is possible as all modules contain only a single output register that is directly connected to the read data multiplexer chains. As a consequence, the bus is capable to transfer data at a throughput of 800 MB/s between the stimuli generator and the reconfigurable modules.

The stimuli generator contains two round robin schedulers, one for determining a module for the next read operation and another one for the next write operation. The schedulers ensure that modules are only accessed after proper module configuration. The partial runtime configuration and the schedulers are controlled by an external host PC. For this test setup, no errors have been found on the bus, while the system was continuously performing read and write operations at full speed, as well as exchanging test modules by partial runtime reconfiguration. Note that the configuration interface controller uses its own independent clock domain. This helps to detect transient effects due to the reconfiguration that may be masked otherwise, because of synchronizing the system with the reconfiguration process (also consider Sect. 2.7.4 that discusses issues on glitch-free reconfiguration).

2.7.2 Logic Cost of the ReCoBus Architecture

As revealed in the introduction of this chapter, communication architectures for reconfigurable systems must be implemented with a very low resource overhead in order to achieve an overall benefit due to partial runtime reconfiguration. While the communication bar approach for dedicated circuit switching connections omits any logic overhead in modules that are not connected to a particular link, some logic for providing access to a ReCoBus is required in all resource slots. The ReCoBus architecture is highly adaptable to the entire system requirements in terms of the bit widths for the data and address buses, the protocols (master or slave bus, pipelined or combinatorial), the granularity, or the placement style (one-dimensional slot based or a two-dimensional grid style). The aim of this section is to benchmark the different ReCoBus parameters for identifying the respective implementation cost and to take the best design decisions for a particular reconfigurable system. This covers also to evaluate the overhead functions for the most promising techniques as reported in Sect. 2.5. Note that the logic required to implement a ReCoBus is not completely overhead as it includes most of the bus multiplexing logic and some further logic. This logic would also be required when implementing a static-only system that does not use runtime reconfiguration.

For disclosing just a fraction of the diversity of the ReCoBus architectural parameters, three variants of a 32-bit slave bus are shown in Fig. 2.57. The ReCoBus implementations vary in the resource slot width W and the interleaving factor N. It can be seen that the implementation cost per resource slot shrinks substantially for highly interleaved buses (larger N values). Each resource slot requires one reconfigurable reset generator for providing a module select a read enable signal and a module reset. The reconfigurable area for the shown example buses include

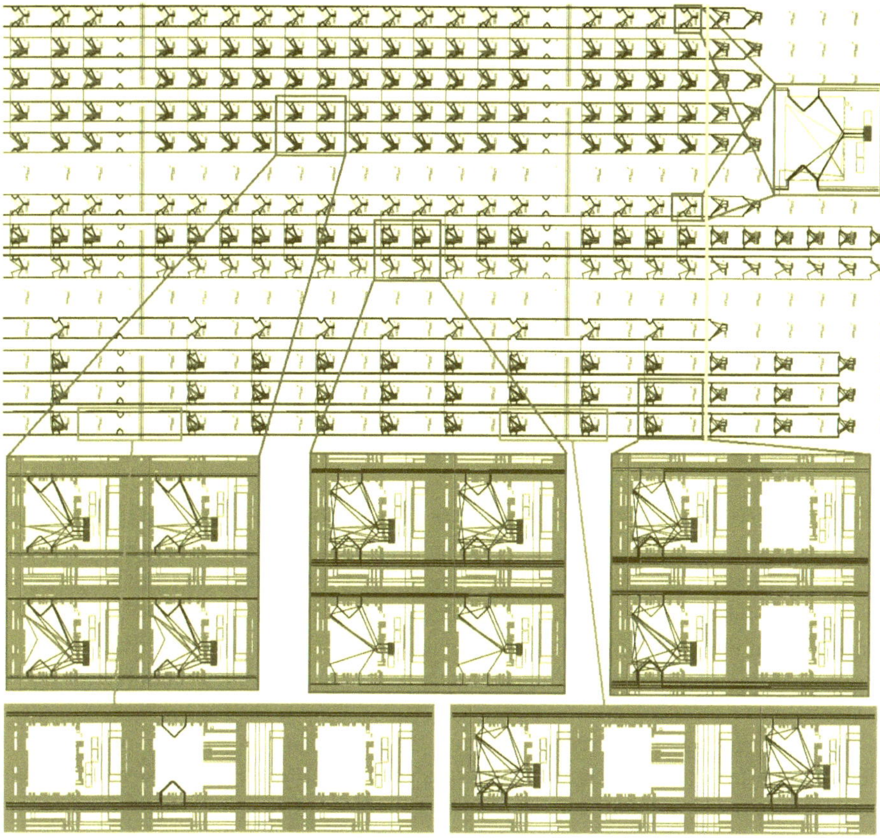

Fig. 2.57 Three ReCoBus implementation alternatives of a 32-bit slave bus. The bus on the top has a resource slot width of $W = 1$ and is $N = 2$ times interleaved. The bus in the middle is $N = 6$ times interleaved while the bottom one features a width of $W = 2$ and is $N = 3$ times interleaved. Write data lines and read data lines share the same slices (the *boxes* that provide two LUTs and two flip-flops). Each bus contains a chain of reconfigurable select generators on the top, for selectively enabling read access to one or up to N consecutive resource slots. The connections to the static part of the system are located at the right hand side of the figure. The *gray* wires in the enlarged boxes on the bottom are routing resources, available for the module implementation. For *the bottom bus*, it is highlighted that the signal routing at the two RAM columns is not identical. Each shown bus provides altogether 32 individual read as well as 32 write data lines, a module select, a module read enable, and a module reset signal, but no interrupt or any further address signals

two RAM columns that do not provide any connection logic to a ReCoBus. Note that the internal ReCoBus structure is identical for all logic columns in the case of a resource slot width of $W = 1$. For a resource slot width of $W = 2$, the layout at the RAM columns differs, thus restricting module relocation. In addition to the ReCoBus communication macros, some extra resources are required for the alignment multiplexing and some glue logic that is located in the static part of the

Table 2.4 Logic cost (in terms of number of used look-up tables) for different configurations of a horizontal 32-bit slave bus. The cases with ($W = 1, N = 1$), ($W = 1, N = 6$), and ($W = 2, N = 3$) are depicted in Fig. 2.57

Bus config.		Fixed cost								
		Static connection			Alignment multiplexer					
W	N	Data	Control	Sum	Read	Write	Sum	Glue logic	Sum fix.	Variable cost
1	2	64	12	76	32	32	64	15	155	18
1	4	96	12	108	128	96	224	19	351	10
1	6	96	12	108	128	144	272	23	403	8
2	1	64	6	70	0	0	0	5	75	34
2	3	72	6	78	64	66	130	19	227	14
3	2	64	12	76	32	32	64	15	155	18
4	2	64	6	70	32	32	64	15	149	18
6	1	64	6	70	0	0	0	5	75	34

system. This cost has been determined for a 32-bit ReCoBus by implementing this logic on a Xilinx Virtex-II FPGA. The results are listed in Table 2.4 as the *fixed cost* together with the *variable cost* within the resource slots. As can be seen, the resource overhead in the static part of the system increases remarkably with the interleaving value N. On the other side, interleaving also leads to a remarkable lower resource cost inside the resource slots. Consequently, interleaving is more suitable for larger systems.

When targeting more recent devices, such as Xilinx Virtex-6 or Spartan-6 FPGAs, the implementation cost in terms of look-up tables will be roughly half than what is listed in Table 2.4. This is because these devices provide 6-input LUTs that can be split into two individual 5-input LUTs. Then each individual LUT can provide roughly the same logic functions than implemented with the help of the 4-input LUTs used on the Virtex-II implementation.

In Sect. 2.5.6, it was proposed, how the one-dimensional interleaving scheme can also be applied for supporting a two-dimensional grid-based placement style of the reconfigurable modules. For this enhancement, the absolute logic cost within each resource slot will not rise and also the cost for the alignment multiplexers will be the same as compared to the one-dimensional case. Only the connection cost in the static part of the system has to be spent for each row of resource slots plus some additional glue logic (the OR-gates in front of the read alignment multiplexers shown in Fig. 2.44). This relationship is depicted in Fig. 2.58 that lists the total cost, required to implement 32-bit slave buses for different ReCoBus configurations. The values in the figure are based on Table 2.4 which lists the synthesis results determined for a Xilinx Virtex-II FPGA. In Table 2.4, the variable cost denotes the sum of the resource estimates denoting the cost in the resource slots. These estimates have been introduced in Sect. 2.5. For shared write and shared read signals (see (2.5) and Sect. 2.5.3.2), the logic cost within the slots is $\left\lceil \frac{S_{SW}}{N} \right\rceil$ and $\left\lceil \frac{S_{SR}}{N} \right\rceil$. Note that these two signal clases can share the same resources, because the ReCoBus

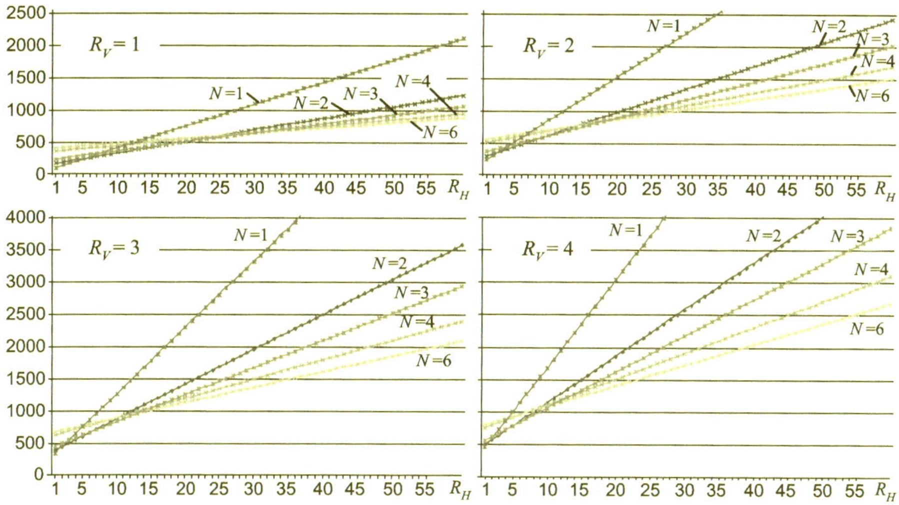

Fig. 2.58 Total implementation cost in terms of number of required look-up tables (the ordinate) in dependence of the amount of horizontal R_H (the abscissa) and vertical R_V resource slots for different interleaving factors of a 32-bit slave bus implementation

implementation uses the look-up tables for connecting shared read signals and the additionally available flip-flops for shared write signals at the same time. For the examined 32-bit slave ReCoBus implementations, this results in a variable cost of $\lceil \frac{32}{N} \rceil$ look-up tables for implementing these two signal classes. Table 2.4 denotes a variable cost that is always two look-up tables higher. This is the overhead required to implement the reconfigurable select generator as depicted in Fig. 2.27.

As the logic cost is not related to the resource slot width W, only the cost for the smallest resource slot width is plotted for a particular interleaving value N. The first example ($R_V = 1$) depicts the logic cost for a one-dimensional slot based placement style. Here, it can be seen that interleaving starts to pay-off for more than 6 resource slots if a ReCoBus is two times interleaved ($N = 2$) and for more than 13 slots in the case of $N = 6$. By dividing one large vertical slot into multiple smaller ones ($R_V = 2, \ldots, 4$), the fixed logic cost in the static part of the system increases (mainly because of the additional connection cost) and the gradient of the linear slopes steps by the amount of vertical slots R_V. The total amount of resource slots within the system is $R = R_V \cdot R_H$. For a Xilinx Virtex-II XC2V6000 FPGA, this results for a resource slot width of $W = 1$ to the following amount of look-up tables per resource slot:

R_V	1	2	3	4
#LUTs	786	386	256	192

Figure 2.1 on Page 46 shows the impact of the resource slot size and the communication cost on the overhead, caused by internal fragmentation. As can be seen in that figure, communication architectures, requiring 10 LUTs per resource slot, minimize the internal fragmentation if a single resource slots provides around 200–250 LUTs (for the assumed module distribution). Consequently, with the higher interleaved ReCoBuses ($N = 4$ and $N = 6$), this slot size can be implemented by tiling the full height of the XC2V6000 FPGA into three to four subcolumns without an explosion in the communication logic cost. Note that the ReCoBus that is $N = 6$ times interleaved requires 403 LUTs for $R_V = 1$ and 763 LUTs for $R_V = 4$ in the static part of the system. This difference of 360 LUTs is less than the average overhead stemming from internal fragmentation of just two modules (assuming a communication cost of 10 LUTs per resource slots for the example in Fig. 2.1). To summarize this observation: *minimizing internal fragmentation by using a fine module resource slot layout is more important than optimizing the logic overhead in the static part of the system.* In addition, a fine horizontal grid ($W = 1$) allows easier module relocation in the case of dedicated resource columns (e.g., a RAM column) as shown in Fig. 2.57. Finally, tiling the resource slots in vertical direction allows reducing the fragmentation of the dedicated columns, as multiple modules may use resources from the same RAM or multiplier column. As exemplarily shown in Fig. 2.64 on Page 138, it is not necessarily required that all used resource slots of a module are conjoined into one consecutive block.

So far, only a slave bus has been examined without dedicated read signals (e.g., interrupts) and addresses. As presented in Sect. 2.5.4, dedicated read signals can be implemented without additional logic resources inside a ReCoBus by manipulating tiny fractions of the configuration data. Alternatively, the time multiplexing approach can be used (see Fig. 2.36 on Page 95) that requires two LUTs within each resource slot. For providing address signals inside the resource slots, shared write signals can be interleaved as illustrated in Fig. 2.23 on Page 78. For instance, if the address space for the reconfigurable modules shall be 1 K words for a ReCoBus that is six times interleaved, it requires $\lceil 10/6 \rceil = 2$ additional LUTs per resource slot (see also (2.5) on Page 79).

A ReCoBus can also implement a bus master interface. In the easiest case, another instance of a macro, shown in Fig. 2.57, can be used to link the master output addresses towards the static part of the system. The resources of the read multiplexer chains can be shared with the address signals for the slave port. In the case of a 32-bit address bus, the cost in the static part of the system and inside the resource slots will double by this approach. Alternatively, addresses and data may be distributed in a time multiplexed manner on the same resources. This allows implementing the ReCoBus logic for a master interface inside the reconfigurable area at the same cost as compared to a pure slave interface. However, this approach comes along with additional latency and more complex logic within the static part of the system.

2.7.3 *The Impact of Modularization on the Physical Implementation*

Partial runtime reconfiguration requires to completely encapsulate the modules into a set of predefined resource slots. This means that modules are constrained to use logic and routing resources only within a predefined bounding box that can be adjusted in terms of resource slots. Consequently, this design methodology will influence the physical implementation of the modules. Only the communication via the module interface (through its entity) will pass the bounding box of a module. This might need of special communication primitives or macros (like a ReCoBus or a connection bar).

The impact on routability (the question, if all nets can be routed), power consumption, and speed, when constraining modules into fixed resource slots, was examined in [KPR04a]. The authors found that smaller modules pay a considerable penalty with respect to the achievable clock frequency for extreme aspect ratios. For example, a divider that is just one CLB column wide and that spans over the full height of the device has had a 94% increase in the propagation delay. For larger modules with more than 1,000 slices (equivalent to 2,000 4-bit LUTs), the timing overhead was in average below 10% for the results presented in [KPR04a].

Similar experiments have been performed in [KBT09a], and in addition, the impact of blocking specific wires for implementing the internal routing of a module was examined. This characterizes the cost of the ReCoBus and connection bar architecture in terms of routability and latency. Four different modules have been examined that all occupy more than 90% of the available logic resources within a bounding box that was set to a height h of 32 CLBs and width w of 5 CLBs. With eight look-up tables per CLB, this results in 1,280 LUTs in total that are available within the bounding box. Besides an FIR filter, a DES56 core, an FFT module, and a synthetic module was selected for the benchmark. The latter implements a huge congested ring shifter that cannot be fully routed with the selected logic utilization. The following code fragment lists the kernel (in VHDL) that is responsible for the congestion:

```
if clock'event AND clock='1' then
   for k in 0 to tmp_size-1 loop
      tmp(k) <= I(k MOD in_width) XOR
                tmp((k+distance_1) MOD tmp_size) XOR
                tmp((k+distance_2) MOD tmp_size) XOR
                tmp((k+distance_3) MOD tmp_size);
   end loop;
end if;
```

The shifter consists of a large internal register (`tmp`) where each single flip-flop is driven by a look-up table implementing a 4-input XOR-gate, hence, not allowing any functional optimizations. The flip-flop outputs are circular connected to the inputs of the XOR-gates. By increasing the `distance` between the connected

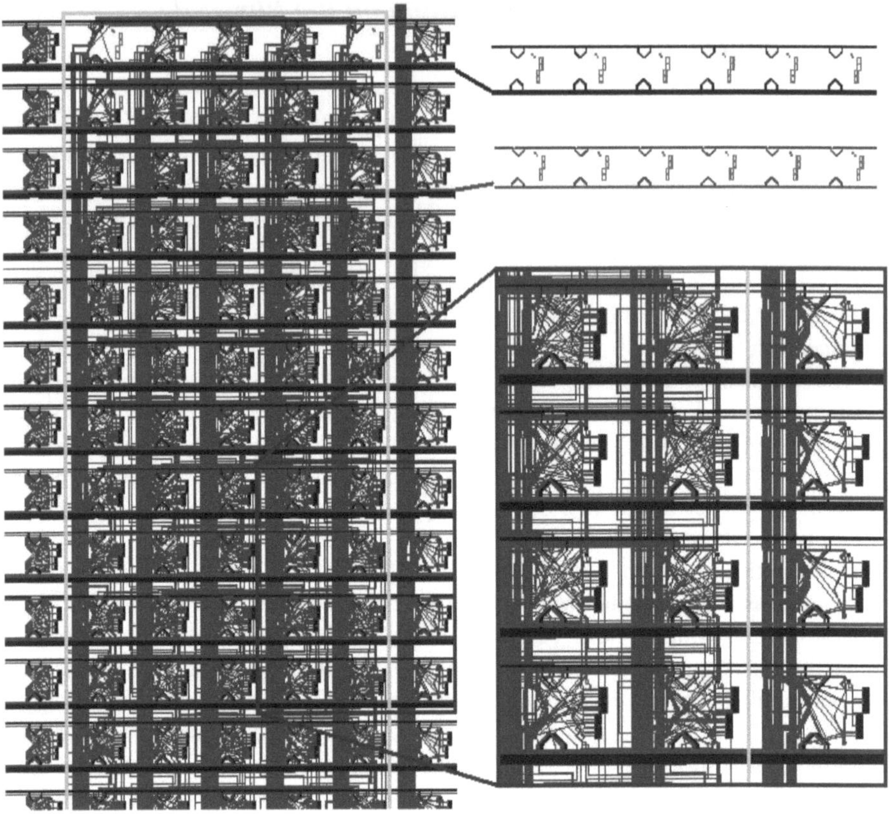

Fig. 2.59 FIR filter module constrained into a bounding box. In both horizontal directions, five double lines and five hex lines have been blocked for the module implementation. Special blocker macros, located around the bounding box, force the router not to use routing resources outside the module boundary (see also Sect. 3.2.4)

flip-flops highly congested circuits can be generated. These circuits can be placed on an FPGA but they are impossible to be fully routed.

The experiments have been carried out on a Xilinx Virtex-II device that has 10 double as well as 10 hex lines starting in each cell towards all horizontal and vertical directions. In each CLB from one to five double lines, hex lines, and double and hex lines together have been blocked. The blocking was only done for horizontal wires, but each time for both directions. This was chosen, because the proposed communication architectures occupy almost only horizontal wires. Figure 2.59 depicts an exemplary test design with the FIR filter where five double lines, as well as five hex lines are blocked. Consequently, these wires are not available for the module implementation, but are reserved for implementing the communication architecture responsible for integrating the partial modules into the system at runtime.

Table 2.5 lists the total number of wires that can be used to distribute data over the reconfigurable area. This implies that wires are interleaved and that a wire is extended if it ends in a CLB switch matrix within the module bounding box, as shown in Fig. 2.59. For example, in the case of blocking three hex and double lines in both horizontal directions, the hex lines will be interleaved in six independent chains, each containing three wires that are starting and respectively stopping in six horizontal consecutively aligned CLBs outside the module bounding box. Analogous, the double lines are interleaved in two independent chains. Therefore, one CLB row is blocking $(6 + 2) \cdot 3 = 24$ wires per horizontal direction in this example. These are 48 wires for both directions and 1,536 wires in total over the full module height of $h = 32$ CLBs.

In addition to the amount of blocked wires, the table lists the relative achieved latency after routing the module with respect to a routed module without area constraints. In the case of the synthetic module, the table also lists the number of signals that could not be routed at all. With the four marked exceptions, all place and route steps have been successfully completed for all other modules. Note that the router was forced not to cross the module border with the help of so-called blocker macros that are described in Sect. 3.2.4.

The first row of the table shows the influence when fitting a module into a fixed bounding box. The determined values confirm the results presented by Kalte et al. in [KPR04a]. All other values reveal the impact of additional wire blocking. The table contains two results for each module and wire blocking combination as two different methods for constraining the placement have been used. As expected, there was no strong performance decrease when blocking a larger amount of hex lines. The reason for this is that the module is only five CLBs wide. Therefore, only 40% of all hex lines can be accessed within the module bounding box via the mid-ports that allow to tap each hex line in the middle, after a routing distance of three CLBs. In other words, 60% of all horizontal hex lines could not be used for implementing inter-module communication because of these wires would leave the module bounding box. Because of the narrow, but tall aspect ratio that is best suitable for partial runtime reconfiguration on Xilinx FPGAs, and because of the carry chains for implementing arithmetic functions that have a predefined vertical direction, the horizontal routing is much less congested than the vertical one, as depicted in Fig. 2.60.

2.7.4 Glitch-free Reconfiguration

When fractions of a system are exchanged by partial runtime reconfiguration, it must be ensured that this can be accomplished without unpredictable side effects to other parts of the system. In the best case, exchanging partial modules can be performed completely transparent with respect to the rest of the system and without affecting the communication infrastructure. This enormously simplifies the interaction between the reconfiguration management and the execution of the rest of

Table 2.5 Impact of wire blocking on the latency and the number of unrouted nets (the values in brackets that are for the synthetic module) for different test modules. The latency is given relatively in % with respect to a physical implementation without any place and route restrictions. The first row demonstrates the impact of constraining the test modules into bounding boxes, while the subsequent rows are used to identify the additional impact of blocking horizontal routing resources. Two separate place and route runs have been accomplished per test module, one *only* with the help of special blocker macros, as described in Sect. 3.2.4 and another run that additionally applies a *bound*ing box constraint to the Xilinx place and route tool. The results are for the Xilinx ISE 8.2 design suite (using the timing driven option) and a Virtex-II XCV2-6000-6 FPGA

			FIR		DES		FFT		Synthetic	
	# blocked	Total wires	Only	Bound.	Only	Bound.	Only	Bound.	Only blocker	Bound. box
		0	102.8	113.1	101.7	103.1	105.0	102.5	130.4 (357)	123.6 (317)
Double	1	128	111.8	114.1	132.7	105.6	106.5	94.9	163.1 (136)	142.7 (392)
	2	256	114.8	113.2	129.8	106.2	115.3	98.0	135.1 (154)	156.5 (400)
	3	384	114.0	120.9	120.9	108.3	124.8	97.5	118.5 (236)	156.1 (450)
	4	512	119.7	121.1	144.0	111.1	114.1	106.4	137.1 (322)	128.1 (523)
	5	640	130.9	136.7	141.2[a]	114.3[a]	120.8	106.9	118.2 (328)	129.5 (580)
Hex	1	384	114.1	111.7	113.1	100.2	98.8	102.5	152.1 (55)	132.4 (333)
	2	768	118.0	113.9	127.5	102.1	101.4	95.5	143.4 (119)	131.1 (372)
	3	1,152	115.2	110.6	114.6	102.1	105.6	95.5	122.3 (82)	122.5 (372)
	4	1,536	113.3	113.2	122.6	102.1	102.9	95.9	156.4 (197)	138.2 (385)
	5	1,920	113.9	115.4	139.4	100.5	102.4	97.5	119.6 (134)	148.5 (365)
Hex and double	1	512	113.5	114.3	118.4	107.0	102.9	94.8	116.9 (138)	130.4 (393)
	2	1,024	109.0	121.7	122.2	112.3	105.8	103.1	128.9 (193)	133.2 (407)
	3	1,536	139.0	126.6	134.8	105.4	132.3	100.4	124.8 (290)	128.3 (505)
	4	2,048	131.0	116.6	135.3	117.6	112.1	103.4	117.7 (325)	131.3 (560)
	5	2,560	127.9	131.5	146.8[a]	145.7[a]	116.5	112.7	136.8 (458)	116.4 (645)

[a] Module was not completely routed (less than 5 nets)

Fig. 2.60 Magnified view of the partial FIR filter module that is constrained into a bounding box. The *gray colored wires* indicate unused routing resources and cyan is used for the module routing. While the vertical routing channels are almost fully utilized, there are still plenty of horizontal resources available

the system. For example, a transparent configuration process does not demand any synchronization between the reconfiguration and the bus schedule, for instance, to wait on a burst transfer completion.

This section is devoted to the testing of the behavior of the proposed ReCoBus and communication bar macros during reconfiguration. Firstly, it will be introduced how glitches can be measured on FPGAs and how this can be used to prove that the communication architecture allows to exchange modules at runtime without influencing other parts of the system. This will be shown on the ESM platform [MTAB07] that provides a Xilinx Virtex-II FPGA. After this, the proposed communication architecture will be extended, such that also Spartan-3 FPGAs can be used for implementing ReCoBuses glitch-free.

The disturbances due to a reconfiguration process can be classified into three categories:

1. *Glitch-free reconfiguration*: No influence to other modules or bus transactions if modules are exchanged.
2. *Glitch reconfiguration*: A transfer on a ReCoBus or communication bar may be disturbed seldom and for one predictable configuration clock cycle.
3. *Interrupted reconfiguration*: The bus may be disturbed or interrupted for multiple consecutive configuration clock cycles.

In almost all cases when the Xilinx documentation reports about glitches during runtime reconfiguration, this stands for the last category and disturbances may occur for many hundreds of configuration clock cycles (depending on the configuration data and the device). This categories have to be considered, when glitch-free reconfiguration has to be implemented on Xilinx Spartan-3 FPGAs, as revealed later in this section.

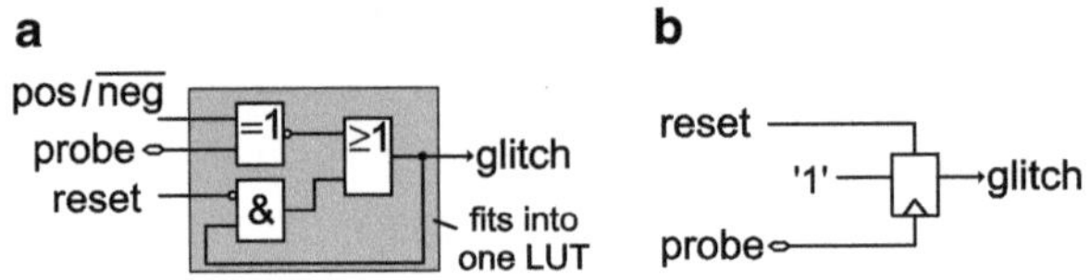

Fig. 2.61 Glitch detection circuits: (**a**) *glitch triggered latch*, (**b**) *glitch triggered flip-flop*. After releasing reset, the occurrence of a glitch event at the probe input will be indicated at the glitch output. With the $pos/\overline{neg}$ input signal in (**a**), the logic level of the idle state can be selected

2.7.4.1 Measuring Glitches on FPGAs

In order to prove that partial runtime reconfiguration can be performed glitchless, such events must be measurable. Measuring glitches on FPGAs can be performed by the use of an oscilloscope, if they occur at I/O pins. This has to be done with caution, because glitches are events lasting typically in the area of a nanosecond or even below, if they occur on-chip. Consequently, this requires an oscilloscope with a high input bandwidth in the gigahertz range. However, it is virtually impossible to connect an oscilloscope probe to an on-chip signal in order to detect glitches on this signal. Also, adding a connection from such a signal to an output pin by using some FPGA routing resources allows only a limited detection of glitches, because pin drivers typically have significant capacities and will consequently filter potential glitches. To overcome such limitations, the dedicated test circuits shown in Fig. 2.61 had been used. These glitch detectors can be added to existing systems in order to detect glitches on any wire.

This allows to prove that a partial reconfiguration process will not interfere with running data transfers on a ReCoBus or a communication bar. For demonstrating glitch-free runtime reconfiguration, when using the proposed communication macros, a test system was implemented on the ESM [MTAB07], as shown in Fig. 2.62. In the static part of this system, random test data has been applied on all macro inputs when exchanging some partial test modules. At the same time, the absence of glitches was checked with the circuits depicted in Fig. 2.61. Testing for glitches has been accomplished at the macro outputs, as well as on the internal signals of the communication architecture. As expected, no glitches have been detected during a reconfiguration process on a ReCoBus when using glitch triggered latches or glitch triggered flip-flops. This holds only for a ReCoBus architecture implemented on a Xilinx Virtex-II device. For Spartan-3 FPGAs the reconfiguration process follows more restrictions as revealed in the next section.

Connection bars are also not interfered due to a reconfiguration process as long as no module is included to or removed from a particular bar that also drives the bar. Exchanging a module that drives a connection bar will lead to a glitch reconfiguration and may also remove or include one or more registers in the data path. This is a short term event and occurs only on rearranging the data path. However, the point of time of this event is known, thus allowing to mask configuration effects to the rest of the system.

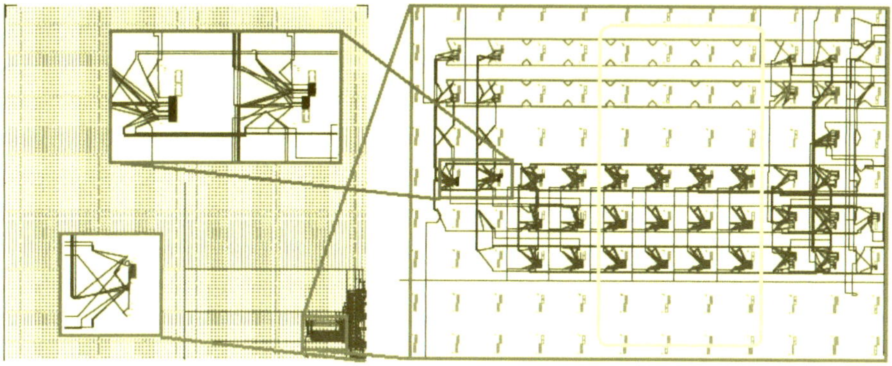

Fig. 2.62 Test setup on an ESM [MTAB07] for proving glitchless reconfiguration. The logic at the bottom left corner contains the interface between a host PC and the internal logic. In addition to the configuration interface (an ICAP port), the system contains a ReCoBus (colored brown in the center of a resource slot) and two connection bars in opposite direction (the green wires in the top of the resource slots). During reconfiguration of resource slots within the *yellow box*, static test data was applied to all macro inputs while observing the outputs with glitch detection circuits (highlighted in the *green box*)

2.7.4.2 Tolerating Glitches on Spartan-3 FPGAs

The ReCoBus architecture aims at building systems, where modules can be hot-plugged to a reconfigurable bus at runtime, by glitch-free partial reconfiguration. If a partial reconfiguration will influence a system depends mainly on the capabilities of the target FPGA architecture. For example, Xilinx Inc. guarantees glitchless reconfiguration modes on Virtex-II, Virtex-4, and Virtex-5 FPGAs for all logic, memory, and routing resources, while interruptions or disturbances are the normal case for Spartan-3 FPGAs. Note that it is only guaranteed that identical fractions of a configuration can be written multiple times to a Virtex FPGA without causing glitches. If a circuit is modified or even exchanged some side effects may occur. Note that the ReCoBus and communication bar macros will never change and that these macros are capable of masking effects from a reconfigurable module to the communication architecture. In the case, a reconfigurable system should be implemented on Spartan-3 FPGAs the ReCoBus or communication architecture needs to be enhanced for supporting glitch-free reconfiguration

Spartan-3 devices are *blockwise reconfigurable FPGAs* that require a clear operation to a complete logic column prior to the particular configuration data write operation. This clear operation also deletes the switch matrix setting within the logic column, hence potentially interrupting a ReCoBus or connection I/O bar. After the clear operation, a new configuration can be written in a frame by frame manner. A configuration frame is the smallest atomic piece of configuration data and a Spartan-3 FPGA has 19 frames that describe a complete logic column and the according switch matrices. The entire frame size depends on the FPGA and is in the range of a few hundreds to more than a thousand bytes. The delete operation is

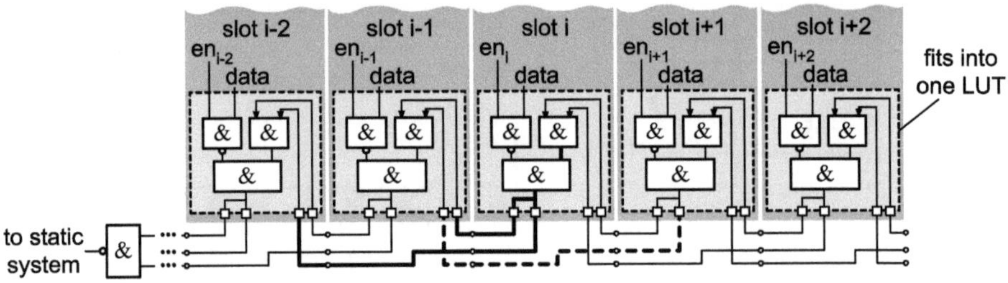

Fig. 2.63 Read multiplexer chain for glitch-less reconfiguration on Xilinx Spartan-3-FPGAs. During reconfiguration of slot i, the fat-drawn wire will be in a stable '1' state and the dashed wire will bypass slot i

performed by sending a command called *snowplow* to a Spartan-3 FPGA. With this command, the configuration data of one complete column will be reset to a logical '0' value. Consequently, communication macros will be interrupted if they use logic or routing resources of a reconfigured logic column.

In order to perform completely glitch-free reconfigurations on block erasable FPGAs, such as Xilinx Spartan-3-FPGAs, it must be possible (1) to exchange logic and routing of a column without influencing the surrounding system. In addition, (2) it must be possible to bypass the communication after any block erase. For achieving the first point, the behavior of wires during a column delete and configuration processes was examined. It was found that all horizontal double and hex lines carry a logical '1' value if the column is deleted. This is independent from the state of all possible sources to such a line. This behavior probably stems from a level restorer inside the switch multiplexers as revealed in [LAB+05].

For ensuring the second requirement, the bus has to introduce some redundancy to bypass deleted logic or routing. This is possible by arranging the routing in such a way that from each slot with a read multiplexer access point there will be paths to two different consecutive access points, as depicted in Fig. 2.63. Then, when only deleting one CLB column at a time, there will be always a functional bypass connection available. More details on partial reconfiguration using Spartan-3 FPGAs is presented in [KBT09a].

While Spartan-3 FPGAs introduce many difficulties in the implementation and management of runtime reconfigurable systems, Xilinx added full support for partial reconfiguration into the Spartan-6 series. These devices provide basically the same capabilities than Virtex-6 FPGAs. However, Xilinx has not enabled this support in their tools, despite that Spartan-6 provide the better cost and power efficiency. Anyway, there are academic tools available that support all recent Xilinx devices and that provide very powerful features that are missing in the Xilinx vendor tools. The most advanced tool is called GoAhead. This includes zero logic overhead integration of partial modules, module relocation, and support for island style and grid style reconfiguration.

2.7.5 Communication for Two-dimensional Module Placement

In the introduction of this chapter is was proven that the resource slots should be kept only a few hundred LUTs in size in order to optimize internal fragmentation. On a larger FPGA, such as the Virtex-II-6000 of an ESM [MTAB07] or most of the recent devices, this requires partitioning of the reconfigurable area into a two-dimensional grid for placing the reconfigurable modules. This also enhances the utilization of the resource columns providing dedicated resources including memory blocks and multipliers, because of the same dedicated resource column may be shared among multiple modules.

For demonstrating the two-dimensional communication capabilities of the proposed ReCoBus and connection bar communication architecture, the test system depicted in Fig. 2.69 on Page 143 was exemplarily implemented for the ESM platform [MTAB07]. This system provides a reconfigurable area with $R_h = 62$ (logic resource slots) $+ 4$ (additional memory resource slots) in horizontal direction. In vertical direction, the full height of the device was partitioned into $R_v = 4$ rows of resource slots that are each $H = 24$ CLBs high. This is an integer multiple of the block RAM height, and the vertical routing span of the hex lines. Therefore, the modules may access equivalent resources (a particular hex line, a block RAM or some logic resources) at exactly the same relative position inside a resource slot, regardless of which row the module is currently placed in. Altogether the system provides $R = R_h \cdot R_v = 264$ resource slots, whereof all 248 logic resource slots can directly connect a reconfigurable module to the communication architecture. With a resource slot width of $W = 1$ and eight LUTs per CLB, a slot provides $\sigma = 24 \cdot 1 \cdot 8 = 192$ LUTs in total per logic resource slot.

This fine slot grid allows implementing the reconfigurable modules by using different resources as well as by fitting modules into different shaped bounding boxes, as depicted in Fig. 2.64. In the example, an FIR-filter module is once implemented without dedicated RAM/multiplier resources and three times with these resources. The logic-only module uses $w \cdot h = 15 \cdot 2 = 30$ slots, whereof only four slots are used for connecting the module. With a connection cost of $c = 14$ LUTs, the module wastes $(30-4) \cdot 14 = 364$ LUTs because of the unused ReCoBus connections. This is 6.3% of the logic available within the module bounding box. The internal fragmentation, in terms of the number of unused LUTs within the module bounding box, is 45 LUTs or 0.8%. Therefore, the total overhead of logic to be paid for fitting this particular module into a bounding box and for providing communication is about 7.1% more as compared to a static only implementation. For the last two implementation alternatives shown in Fig. 2.64, the waste of logic resources is $(6 - 4) \cdot 14 = 28$ LUTs for unused ReCoBus connections and another 75 LUTs because of the internal fragmentation, which results in altogether 8.9% overhead with respect to the total amount of LUTs available in the six logic resource slots of the module. It must be mentioned that only focusing on the logic utilization does not express that in some cases the logic density has to be reduced for carrying out the routing. The reason for this is that the switch matrix resources

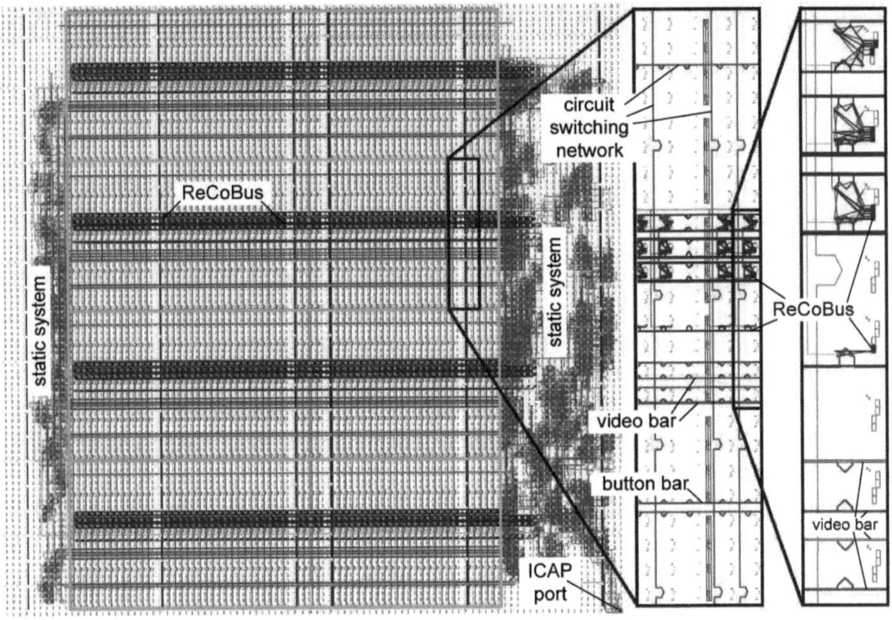

Fig. 2.64 Four implementation alternatives of the same FIR filter module. All variants connect the module within four resource slots to a 32-bit two-dimensional ReCoBus. The first example uses only logic resources and is $w = 15$ resource slots wide and $h = 2$ slots high. The second uses dedicated RAM/multiplier resource slots and requires only $(3_{\text{logic}} + 1_{\text{RAM}}) \times 2$ slots. The two implementations at the right hand side are based on the same netlist as the second variant but have been implemented with a different form factor of $(6_{\text{logic}} + 2_{\text{RAM}}) \times 1$. The last alternative provides a specialty, as it relaxes the requirement that the resource slots have to be continuously arranged. The module skips four logic resource slots and connects a single outlaying RAM/multiplier resource slot without using any routing resources within the switch matrices of the skipped slots. The gap can accommodate a module that may also access the full 32-bit of the ReCoBus without interfering with the surrounding FIR filter module

consume considerably more area on the FPGA die as the logic cells itself. Therefore, shrinking the switch matrices by leaving some logic resources unused results in an overall benefit [Deh99].

Reconfigurable modules placed in the tile grid of the system, depicted in Fig. 2.65, may communicate arbitrary via (1) a ReCoBus, (2) two connection bars, and (3) a two dimensional circuit switching network, as listed in the following. In this example system, various instances of different modules have been configured at runtime and without interference to other partial modules or the static system. Because of the full column-wise reconfiguration scheme of the used Virtex-II FPGA, modules have been implemented that do not use distributed memory. This restriction was removed by Xilinx for all recent FPGAs. At maximum, 60 individual partial modules were integrated into the system at the same time. All these modules contained a register file that could be accessed by the static system via the ReCoBus.

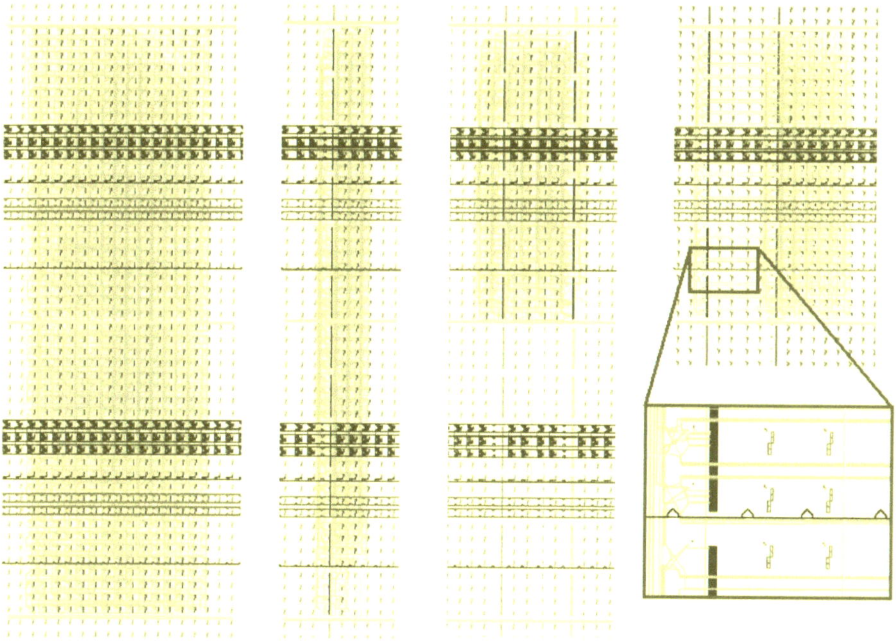

Fig. 2.65 Example system featuring a two-dimensional grid style placement on an ESM [MTAB07]. The reconfigurable area is tiled into $R = [R_h] \cdot R_v = [(62\,(\text{logic}) + 4\,(\text{RAM})] \cdot 4 = 264$ individual resource slots. A resource slot is one CLB (one RAM / multiplier column) wide and 24 CLBs (6 block RAMs / multipliers) in height. This results in $\sigma = 192$ look-up tables. In each resource slot, containing logic, modules can connect to a ReCoBus and to connection bars for video data (24 + 3 bit bidirectional) and push buttons. In addition, modules can connect to a circuit switching network for directly linking input pins (the ESM debug pins from the top) to the freely placeable modules. With a connection cost of $c = 14$ LUTs in each slot, the bus provides access to a slave bus interface with up to 8-bit read data, 8-bit write data, two address bits, a module select, a read enable signal, a module reset, and a shared interrupt wire per slot. The static system at the left hand side of the FPGA generates a VGA background data stream that can be linked to any of the four horizontal video bars. The output of a particular bar is passed to a pipeline register within the connection macro and directly send backwards to the left from where it can be send another time via any other video bar towards the right part of the static system. A register controlled multiplexer inside the right static part selects the particular output video bar. This system is controlled by a host PC for performing bus transactions and for sending configurations to the internal configuration access port (ICAP) on an XC2V6000 FPGA

Furthermore, all modules manipulated the video stream (e.g., to overlay an object or to perform some video filtering or processing). The video stream was routed via the connection bars. Moreover, some modules had a direct connection to external push buttons. This was, for example, used to control an arcade-like video game (Pong). Finally, some modules could receive data from the external debug pins of the ESM. These pins were routed using the circuit switching network of the system.

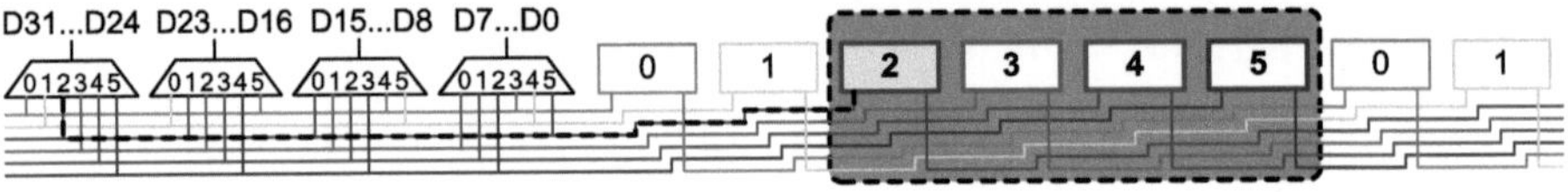

Fig. 2.66 Data bus interleaving of the example shown in Fig. 2.69. Only up to 4 out of 6 independent interleaved chains are used for implementing ReCoBus data lines. The numbers inside the resource slots on the right hand side of the figure denote the multiplexer select value that is determined by the most left logic resource slot as shown for one module

2.7.5.1 ReCoBus Communication

Figure 2.65 depicts the mapped ReCoBus logic. In the right hand side of the illustration, it is shown that the communication cost per slot for implementing a bidirectional 8 bit slave ReCoBus interface is $c = 14$ LUTs (inside a slot, the seven depicted slices that are providing two LUTs each). The four ReCoBus macros are horizontally $N_h = 6$ times and vertically $N_v = 4$ times interleaved. By using six interleaved chains for the ReCoBus macros, the bus can be implemented with the help of hex lines that are substantially faster for larger buses as compared to the alternatively available double lines. Highly interleaved buses can be used for implementing ReCoBuses with a lower amount of possible consecutive bus connections than specified by its interleaving factor as shown in Fig. 2.66. This scheme is used for this example setup and it logically matches the system illustrated in Fig. 2.44 on Page 104 with a two-dimensional ReCoBus that is four times interleaved in both, horizontal and vertical direction for providing 32-bit data access. The address signals use the full horizontal interleaving scheme. This is used for extending the 2 bit address range of a single resource slot to up to $2 \cdot 6 = 12$ bits or 4 K in this example.

2.7.5.2 Communication Bars

In addition to a ReCoBus macro, each row of resource slots provides two horizontal connection bars (see also Fig. 2.65). One bar connects the same four push buttons in parallel to all four rows of resource slots, hence, these signals can be tapped in all 246 slots. The second bar distributes a VGA video signal with 24-bit color depth to all resource slots. All video connection bars allow modules accessing the data that is streamed from the left to the right of the system. In addition to this, each connection bar provides a backward path for returning the stream again to the left hand side of the system. As revealed in Fig. 2.67, a multiplexer located at the beginning of each video connection bar macro (located at the left hand side of the figure), can selectively chose between an initial signal from a background module or a loop-back signal from another macro. This allows streaming the video data over all rows of resource slots in all (2^8) possible permutations. Note that the video background signal or the data from a backward path can be sent to multiple video connection

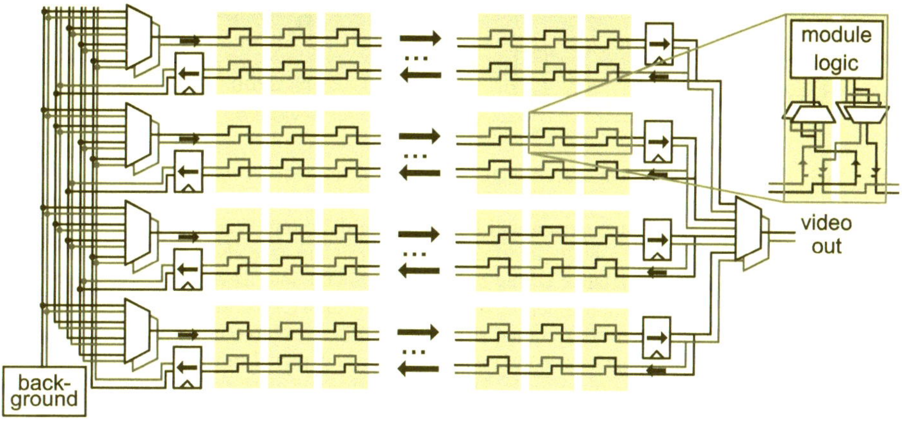

Fig. 2.67 Architecture of the video bar routing via connection bars. For each row of resource slots, a multiplexer at the left hand side of the system can freely select between an initial video stream from a background module or from a loop-back signal of another bar. The video signal is $N_{video} = 2$ times interleaved, as depicted by different colored chains. Consequently, a module must be at least two slots wide and including alignment (de)multiplexers for accessing the stream adjusted with respect to the current start position of a module in a read-modify write fashion

bars in parallel. A multiplexer in the right hand side of the system selects one video connection bar to be linked to the output pins.

The complete set of video signal is divided in two subsets that are two times interleaved. This requires an alignment multiplexer/demultiplexer pair for adjusting the input stream according to the module position (odd or even resource slot index) and also for adjusting the output stream according to the initial format. This is the only logic required within the reconfigurable area for the video bar and this logic is only necessary if the module connects to a video bar. The alignment can be controlled in multiple ways:

1. Using identity functions instead of multiplexers in the according LUTs that are exchanged by modifying the configuration bitstream.
2. For modules, only requiring a video input interface, it is possible to implement the alignment directly within a switch matrix, because on Virtex FPGAs the signals of both chains can be tapped within the same switch matrix. This requires bitstream manipulation at runtime.
3. Including extra signals into the connection bar may be used to automatically control the alignment (de)multiplexers. For example, by setting one additional wire within the first interleaved chain to a constant '`0`' value and a corresponding wire within the second chain to a constant '`1`' value. Within a reconfigurable module, sensing one wire indicates if a module is placed at an even or an odd horizontal resource slot index.
4. Instead of using extra wires within the connection video bar, existing signals may be used to determine if a module is placed at an even or odd slot position index.

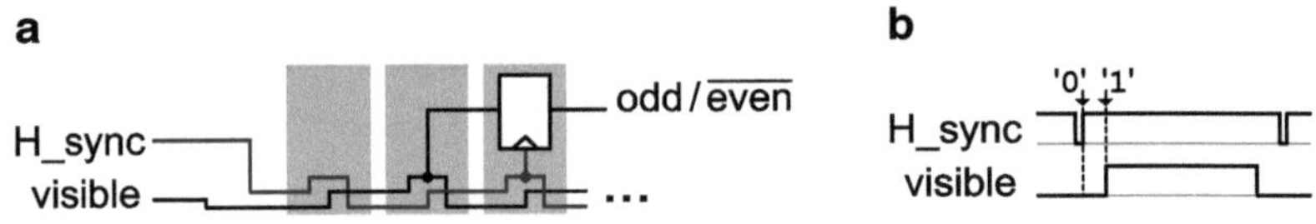

Fig. 2.68 Automatic alignment control signal generation for video connection bars. The VGA video signal contains a horizontal synchronization signal (H_sync) and another signal indicating if the current pixel of the video stream will be visible on the monitor or not. These signals will be distributed in separate horizontally interleaved connection bars (see also Fig. 2.67) in such way that the signals use the same relative wiring resources within the interleaved bars. These wires are connected within a module to a flip-flop as shown in (**a**). If the H_sync signal is connected to the clock input, this indicates an even resource slot index and if the visible input is connected to the clock input, this will be recognized as an odd slot index as depicted in (**b**) that shows the fixed phase relationship between the two signals. The odd / $\overline{\text{even}}$ signal can be directly connected to the alignment (de)multiplexers

This approach is chosen for the implemented example system. By connecting the horizontal synchronization signal (H_sync) and the visible signal that is provided together with the VGA color information in corresponding wires, located in separate interleaved connection bar chains, a simple flip-flop is sufficient to automatically detect if a module is placed at an even or odd slot position, as illustrated in Fig. 2.68.

2.7.5.3 Two-dimensional Circuit Switching Network

As a third alternative in the demo system depicted in Fig. 2.65, modules can directly connect to some I/O pins via a two-dimensional circuit switching network. Figure 2.69 reveals details of the network implementation that is carried out directly within the switch matrices of the FPGA fabric. All network inputs (here the debug pins available at the top of the ESM) are routed vertically with hexlines down to the bottom of the device. At two vertical positions within each resource slot, these hexlines can be directly connected to horizontal channels that are implemented with double lines towards both directions, eastwards and westwards. Alternatively, the horizontal double lines may route through the incoming signal. As the network must be regularly arranged (for allowing free module relocation) such that all resource slots contain an identical internal logic and routing layout, the horizontal double lines have to be interleaved with $N = 2$. Note that a connection to a vertical pin wire automatically segments the horizontal routing track such that multiple pins may be routed to a set of modules within the same horizontal track.

A module connection is provided via a flip-flop connection macro. This flip-flop is either wired to the endpoint of a horizontal double wire routing track, to a middle tapping point of a double line, or to a vertical pin wire.[12] Consequently,

[12]The pin wire connection requires an additional routing segment within the module connection switch matrix on a Xilinx Virtex-FPGA. This is considered in the connection macro design.

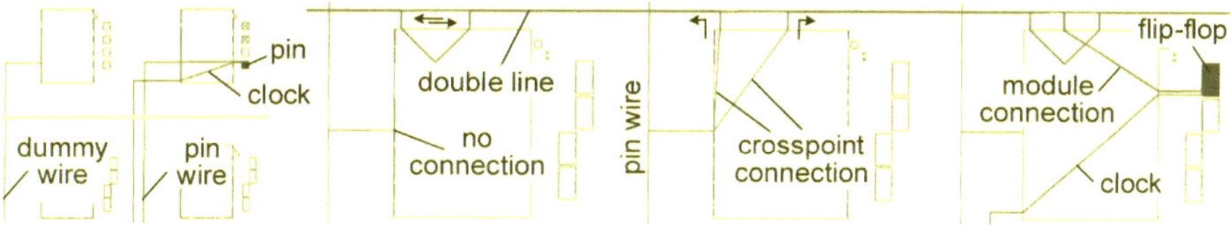

Fig. 2.69 Details of the circuit switching network. Some input pins are passed through an input register and directly connected to a vertical pin wire that spans the full height of the reconfigurable area. In columns that do not provide a pin, the equivalent wire is instantiated as a dummy antenna. Two bidirectional routing channels per horizontal row of resource slots may route through the horizontal wires, connect it to a vertical pin wire, or connect it to a flip-flop for interfacing a module

modules can access a particular horizontal track independent from the horizontal placement position. Both, the pin wire connection to the horizontal routing track and the module connection is set via the configuration interface and controlled by the external PC.

The whole network can be seen as a large crossbar that is scattered regularly over the resource slots. For minimizing internal fragmentation, the full horizontal logic columns should be actually tiled in three commensurately rows, which results in 256 LUTs per resource slot, when considering the results in Fig. 2.2 on Page 47 for a communication cost of $c = 14$ LUTs. However, with this tiling, a slot would be 32 CLBs high, which is not divisible without remainder by six, which is the routing span of a hex line. In other words, this tiling is improper for the proposed two-dimensional circuit switching network implementation. An alternative implementation, using double lines for the vertical pin wires, would straightly allow a tiling in three rows but would substantially rise the propagation delay of the circuit switching network.

2.7.5.4 Evaluation of the System

The example system runs with a clock frequency of 100 MHz on the ESM platform [MTAB07]. This high speed is only possible by passing all inputs and outputs of the communication architecture through pipeline registers within the ReCoBus, the connection bar macros or the input and output terminals of the circuit switching network. For the latter one, the signals at the input pins to the circuit switching network must be synchronized with the system clock for preventing metastability. The vertical routing to the bottom horizontal routing track consumes 6.1 ns out of the 10 ns timing budget when the system is clocked at 100 MHz. This allows a horizontal routing distance of up to ±26 resource slots. For the top row of resource slots, the horizontal routing track is fast enough to distribute a signal from any input pin above the reconfigurable area to any of the 66 horizontal resource slots. Note that longer routing distances or data rates can be achieved by removing the flip-flops from the routing path and by synchronizing the input data stream within the reconfigurable modules.

The following list denotes the logic cost of the implemented communication architecture depicted in Fig. 2.65, in detail:

- Logic within the resource slots: $R \cdot c = 248 \cdot 14 = 3,472\,\text{LUTs}$
- Connection of the ReCoBus macros within the static part costs 528 LUTs in total and is composed of as follows:

  1. For connecting 8 read data and 8 write data signals in each resource slot, 16 flip-flops are instantiated (which is counted as 16 LUTs), multiplied by N_h and $R_v = 384\,\text{LUTs}$.
  2. For the address signal connection $4 \cdot N_h \cdot R_v = 96\,\text{LUTs}$ are used as the two required connections per ReCoBus macro and horizontally interleaved chains are not shared within the same slice (a slice contains two LUTs and two flip-flops on a Xilinx Virtex-II FPGA). As a half-occupied slices cannot be used to further implement additional logic by the Xilinx tools, this logic overhead was considered twice.
  3. The control signal connection requires $12 \cdot R_v = 48$ LUTs

- Alignment (de)multiplexers: 272 LUTs that are broken down as follows:

  1. Data to static system (see Fig. 2.66): $8 \cdot 4$ 6-input multiplexers = 128 LUTs (see Table 2.2 for the logic costs of a certain multiplexer),
  2. Data from static system: $8 \cdot 6$ 4-input multiplexer = 96 LUTs, and
  3. Address alignment: $6 \cdot 2$ 6-input multiplexer = 48 LUTs.

- Static connection of the communication bar macros: $R_v = 4 \cdot [64\ \text{(video)} + 12\ \text{(push buttons)}] = 304$ LUTs,
- Multiplexers for the video data routing, as depicted in Fig. 2.67: 280 LUTs
  The video routing architecture consists of:

  1. Video connection bar input multiplexers: $R_v = 4 \cdot 28$ (# used wires of the video bar) = 112 4-input multiplexers requiring 224 LUTs and
  2. The video output multiplexers with 28 4-input multiplexers, which are implemented with 56 LUTs.

- ReCoBus and communication bar glue logic: 154 LUTs.

The total amount of LUTs required in the static part of the system is 1,538, thus resulting in 5,010 LUTs for the complete communication architecture. This is 7.4% of all available LUTs of the ESM main FPGA. As compared to the results presented in Sect. 2.7.2, the system presented here requires 40% more logic within the resource slots as it provides an additional address port and an interrupt input. At first glance, the logic cost within the resource slots seem to be unlikely large, but it amortizes easily because of the reduced internal fragmentation. For example, a module with a size of 1,000 LUTs can be implemented in 6 resource slots that consume $6 \cdot l = 1{,}152$ LUTs on the ESM main FPGA. [MTAB07] reports an alternative implementation of a one-dimensional reconfigurable system with a resource slot width of 4 CLB logic columns on the same ESM. The resource slots of that system provide 3,072 LUTs and the 1,000 LUT example module will leave

2,072 − 152 = 1,920 more LUTs unused due to internal fragmentation as compared to the example discussed in this section. In this comparison, the waste of logic due to internal fragmentation is reduced by a factor of 13.6 with the help of the ReCoBus communication architecture. Moreover, the vertical tiling further allows a higher utilization of the dedicated memory and multiplier resources.

2.7.5.5 Competitive Analysis of the ReCoBus Architecture

The best related system, providing an interface to reconfigurable modules via a backplane bus [HKKP07b], is based on a Xilinx Virtex-II-4000-4 FPGA with 16 resource slots that each are 4 CLB columns wide and one-dimensionally aligned. The authors report a resource consumption of 4,277 Slices (equal to 8,554 LUTs) for the full communication architecture providing a 32-bit master backplane bus. Due to the large resource slot size of 2,560 LUTs, that system will be similarly vulnerable to internal fragmentation as the ESM system in [MTAB07]. The communication architecture in [HKKP07b] includes only the backplane bus and is not interleaved. Consequently, the throughput on that bus is rapidly decreasing down to a peak of only 280 MB/s [HKKP07a] for a bus with 16 slots, despite that the bus is equally pipelined as the ReCoBus that achieves 400 MB/s peak. The better performance of the ReCoBus has two main reasons. First, due to interleaving, the routing is carried out via fast hex lines and the signal path of one interleaved chain is only 10 LUTs deep, as compared to 16 LUTs in [HKKP07b]. The second reason is that the ReCoBus is implemented much more densely, hence, reducing the signal propagation delay within the bus.

The here proposed ReCoBus system only provides a slave interface. However, by joining two vertical resource slots together, one ReCoBus macro can be used for connecting data and the other one for connecting the address signals. This would result in about the same logic cost and half the amount of resource slots, but would allow integrating modules with a master interface into the system. Such a system would then provide a 1.4 times higher throughput, and 8 times more resource slots within the same reconfigurable area, while requiring about half the amount of look-up tables for the implementation as opposed to [HKKP07b].

2.7.5.6 Competitive Evaluation of the Connection Bar Architecture

In [BHB$^+$07], a network for linking video data in a reconfigurable system is proposed that is related to the two-dimensional connection bar example depicted in Fig. 2.67. The approach includes multiple one-dimensional resource areas that each include multiple modules in a daisy chained fashion, as shown in Fig. 2.11 on Page 57. This routing is similar to the here presented connection bar approach.

However, within a one-dimensional resource area, all signals crossing a module bounding box in [BHB$^+$07] demand connection macros. This results in an overhead of two LUTs per signal wire that pass a module and the overhead is also required for the case when a module is not accessing a particular signal. As [BHB$^+$07] does not consider variable module sizes, a fine grained, two-dimensional resource slot layout (as the one presented in Fig. 2.67) would result in an overhead of $2 \cdot (24 + 3) \cdot 248 = 13{,}392$ look-up tables for the video signal example (24-bit RGB plus control signals). These are 20% of all available LUTs within the main FPGA of an ESM. Opposed to this, a communication bar that is not accessed by a particular module is completely transparent to the module interface and does neither include any logic overhead nor demand module modification of the interface for looping unused signals through a module. And even in the case that a module requires read or write access to a signal, one LUT is sufficient per signal wire to connect the upstream together with the downstream at the same time. This results in half the logic cost of [BHB$^+$07].

2.8 Chapter Contributions and Discussion

The ReCoBus and connection bar methodology, presented in this chapter, provides mechanisms for dynamically integrating bus-based communicating modules into a system at runtime. The contributions of this chapter are as follows:

- Overview of state of the art communication techniques, allowing to integrate reconfigurable modules into FPGA-based systems at runtime.
- Investigation of the internal fragmentation due to partitioning the reconfigurable area into resource slots of fixed size [KBT09c].
- Design and analysis of the novel ReCoBus communication architecture that is capable of directly mapping partial master and slave modules into the address space of an embedded system [KHT08]. This architecture is capable of reducing the resource slot size – and consequently able to increase the amount of slots – by at least an order of magnitude as compared to the best existing approaches. This progress in placement flexibility is achieved at very low logic cost while still allowing high throughput on the bus.
- Extension of the ReCoBus architecture for allowing to place modules in a two-dimensional grid style with even smaller resource slots [KBT09c, KSHT08a]. So far, there is no other bus-based communication architecture available that is capable to directly connect modules in such manner – despite the necessity to minimize internal fragmentation. Besides enhancing logic utilization, the grid style further allows to use dedicated resource columns, including block RAM and multiplier columns by multiple different modules.
- Design and analysis of the novel connection bar technique for directly linking reconfigurable modules with other parts of the system. This includes I/O pins, modules in the static part, or even other partially reconfigurable modules [KBT09c, KSHT08b]. Connection bars are highly adaptable and can be arbitrarily segmented into multiple sub bars at runtime with the capability of providing point-to-point connections as well as one-to-many multicast streams.

The connection bar approach was further extended for grid style systems and allows high throughput links, even if modules are placed in two dimensions on the FPGA.

- Design of a two dimensional circuit switching network by reserving a set of wires for carrying out changeable module connections directly within the routing fabric of an FPGA at runtime. This technique builds a logical routing layer on top of the FPGA and it is the only available two-dimensional routing architecture allowing a runtime system to freely route signals throughout the reconfigurable modules. Consequently, the reconfigurable modules are not obstacles for the proposed circuit switching network.
- Experimental evaluation of the proposed communication architecture for determining its latency and its logic cost [KHT08]. Furthermore, the impact of bounding modules into fixed regions and restricting the routing on the physical implementation was examined. For the ESM platform, it was proven that hundreds of wires can be reserved for implementing a communication architecture for including partially reconfigurable modules into a system at runtime, before noticeably reducing the achievable clock frequency or the routablility of the partial modules [KBT09a].
- Implementation of the ReCoBus architecture on low cost, blockwise reconfigurable FPGAs, such as the members of the Xilinx Spartan-3 family [KBT09a]. This paves the way for exploiting runtime reconfiguration in very cost sensitive applications. With the extensions of the ReCoBus architecture for supporting glitch-free reconfiguration on Spartan-3 FPGAs, it was possible to provide multiple consecutive resource slots for the first time on these devices with the capability of swapping modules without interrupting communication with modules not involved in the reconfiguration process.

So far, partial runtime reconfiguration as an integral part of an FPGA-based system has not remarkably found its way into commercial applications. Moreover, this has not been widely exploited despite its potential benefit, including cost and static power reduction (that is related to the FPGA size). This has two major reasons. Firstly, the overhead due to internal and external fragmentation and for connecting reconfigurable modules with the static part of the system is mostly that high when using other approaches that typically no overall benefit may me achieved (see also Fig. 2.3).

Secondly, present tools and design methodologies have not reached a sufficient stable state and designing partially reconfigurable systems is an error-prone process. The first issue is tackled with the ReCoBus and connection bar approaches that limits the worst case resource overhead per reconfigurable module to a few hundred look-up tables. This is in most practical cases more than an order of magnitude better than existing approaches presented for Xilinx FPGAs. The fine-grained placement grid is further helping for reducing external fragmentation. The tool and design issues will we considered in the next chapter.

Chapter 3
Building Partially Reconfigurable Systems: Methods and Tools

This chapter covers all components required to specify, verify, and implement runtime reconfigurable systems based on the ReCoBus and connection bar communication architecture. This includes offline design aspects, as well as functionality required to load partial modules onto an FPGA at runtime. The goal of this chapter is to enhance a typical system design flow in such a way that fractions of the hardware (implemented on an FPGA) may be exchanged by partial reconfiguration. This comprises technical aspects that are required to provide the communication to reconfigurable regions on an FPGA, as well as controlling and managing the reconfiguration of the modules at runtime. This Book puts focus on highly optimizing these components and integrating them into a framework, such that sophisticated reconfigurable systems can be built easier and faster, with a higher level of flexibility, and with the lowest possible overhead in terms of logic resources and reconfiguration time, as compared to many pioneering and related approaches in this field.

Throughout this chapter, the following key components for building partially runtime reconfigurable systems will be examined in detail:

Designflow and Tools: Designing runtime reconfigurable systems introduces new aspects into traditional FPGA design that demand a suitable *designflow* to deal with the increased complexity. The flow should provide a high level of abstraction with the help of *tools* for automating the FPGA device specific low level implementation aspects. This includes the specification and synthesis of the communication architecture that is responsible for integrating the reconfigurable modules into the system, as well as guiding through all physical design aspects. This includes floorplanning, constraining the place and route process, and assembling the bitstream.

Reconfiguration Control and Management: The *module and resource management* is responsible for deciding *where* and *when* a reconfigurable module is loaded to the FPGA device. Therefore, this component includes a placer and a scheduler that is typically provided as an operating system service. The *reconfiguration control* comprises all technical aspects, required for loading partial configuration bitstreams

D. Koch, *Partial Reconfiguration on FPGAs: Architectures, Tools and Applications*, Lecture Notes in Electrical Engineering 153, DOI 10.1007/978-1-4614-1225-0_3,

to a specific area on an FPGA. This implies online modifications to the configuration that is stored on the device or sent to the FPGA (e.g., for modifying fractions of the routing at runtime).

High-speed Reconfiguration: The reconfiguration speed has a major impact on the feasibility of applying runtime reconfiguration within a system. For example, it is only possible to allow redundant cold spare hardware modules in a fault tolerant system, if the entire reconfiguration time is lower than the allowed time to repair. Or reconfiguring compute kernels in a multimedia signal processing system will also demand to accomplish strict real-time constraints. For addressing configuration speed issues, configuration prefetching, bitstream decompression, and defragmentation techniques are presented.

With these components, it is possible to enhance FPGA based systems with the capability of partial runtime reconfiguration. The reconfiguration capabilities are closely related to the target FPGA features. This means that if a system must provide partial reconfiguration, also the FPGA must allow exchanging only portions of the currently loaded configuration. And if the system should continue execution during reconfiguration, which is mandatory in the case of partial *self reconfiguration*, this must be supported by the target FPGA architecture as well. However, as revealed in [KT04], a system may compose a new complete bitstream by linking together multiple module configurations that are then loaded as a *full configuration* to the device. As this method is different to the one used for active partial reconfiguration, this technique is called *pseudo partial reconfiguration*.

The remainder of this chapter reveals new contributions on the offline and online aspects of implementing partial reconfigurable systems on FPGAs, capable to exchange fractions of the configuration, while keeping the rest of the system active.

3.1 A Framework for Component-Based Reconfigurable System Design

The goal of this section is to expand typical FPGA design flows for implementing complex runtime reconfigurable systems. For enhancing design productivity while reducing error-prone manual design steps, this expansion is highly automated and supported with novel tools as illustrated in Fig. 3.1. These tools have been seamlessly integrated into the Xilinx ISE design suite, resulting in a powerful framework for easily designing runtime reconfigurable systems for Xilinx FPGAs. This framework is targeted on the design, test, synthesis, and bitstream assembly of complex reconfigurable systems on a high level of abstraction. In particular, the following additional steps, as compared to completely static only systems, have to be accomplished in this framework (see also Fig. 3.1):

1. *Static/dynamic partitioning* for splitting the system into a static part and a set of dynamically reconfigurable modules.

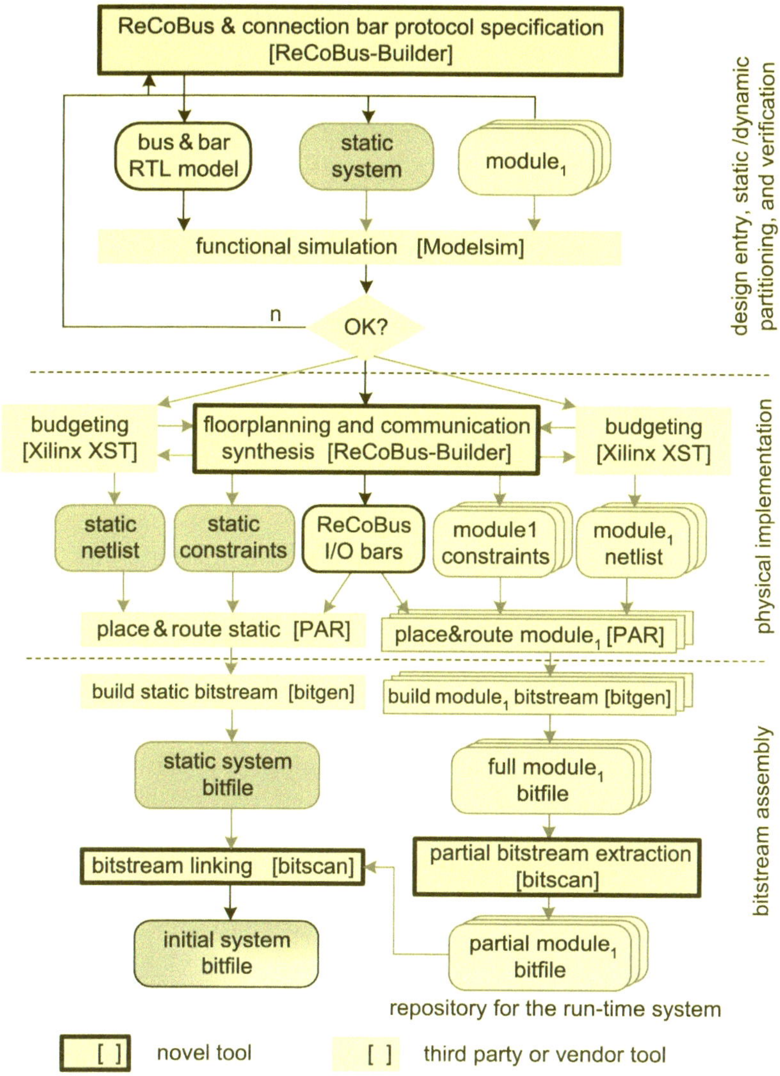

Fig. 3.1 Design flow for building reconfigurable systems with the framework ReCoBus-Builder. The system designer has to partition the system into a static part and the dynamically reconfigurable modules. According to the port requirements, the communication architecture will be specified in the following. After passing simulation, the physical implementation will be carried out respecting area and location constraints, determined in the floorplanning step. Finally, the static and partial modules bitstreams are assembled

2. *Specifying the communication architecture* that has to provide all communication demands of the reconfigurable modules. The partial modules will be completely encapsulated and any connections to other parts of the system have to be accomplished via this architecture.

3. Extending the simulation for *simulating runtime reconfiguration.*
4. After passing the functional simulation, the *physical implementation* can be carried out. This requires:
   a. *Budgeting* the resource requirements of the static system as well as for the individual partial modules.
   b. This information is then used for *floorplanning* the FPGA. Here, the device is partitioned into a fixed static region and an area, tiled into resource slots, for accommodating partial modules.
   c. In addition, this step requires to *freeze the placement of the communication architecture* that is implemented with the help of dedicated communication macros, located at fixed positions.
   d. With this information, *module bounding boxes* and corresponding *module origins* can be defined in terms of resource slots.
   e. Next, place and route will be performed separately for the static system and all partial modules according to the placement constraints specified during floorplanning.
5. Finally, the *configuration bitstreams* for the static system as well as for all reconfigurable modules are generated. For the latter ones, the partial module bitfiles have to be extracted from the full bitstreams.
6. If required, an initial bitstream may be composed out of the static bitstream and the partial modules.

The static/dynamic partitioning has to be done manually by the designer, as it is in general impossible to automatically derive information about mutually exclusive function blocks from a system. However, as these function blocks are typically encapsulated within separate branches of the design hierarchy, this partitioning step is very easy to perform. The result of the static/dynamic partitioning is a static design, typically containing all modules that are required at any time, for example, a CPU, the memory controller, the configuration interface logic, and dedicated I/O interface modules. Beyond the static design, the partitioning will define the total set of partially reconfigurable modules $\mathcal{M}$. All reconfigurable modules and the components of the static system possess a port and the full set of ports is $\mathcal{P}$.

Each module $m_j \in \mathcal{M}$ may possess different port requirements and to each module m_j a set $P_j \in \mathcal{P}$ is associated. The communication architecture has to fulfill all these port requirements and must provide at least the union of all requirements: $P_{\cup} = \bigcup P_j, \forall j = \{1, \ldots, |\mathcal{M}|\}$, as exemplarily depicted in Fig. 3.2. This port set has to be provided by ReCoBus and connection bar macros. With these two types of macros, virtually all practical communication demands can be fulfilled. The macros are instantiated at the systems level as an interface module, like for example, a bus. The designer has to identify the port requirements and to specify the protocol on the physical layer for the macros. For connection bars, this involves a specification of the number and the direction of the signal wires and if the wire connections include pipeline flip-flops at the macro inputs, the outputs, or at the module side. For a ReCoBus, this is similar, but here, not only the direction has to be specified

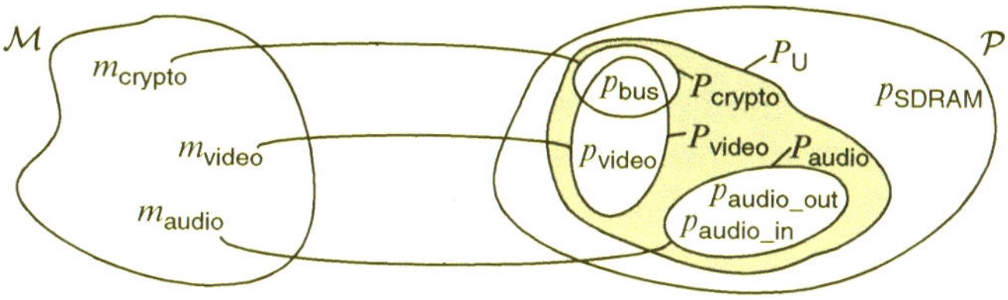

Fig. 3.2 The communication architecture has to provide the port requirements $P_{\cup}$ for all partially reconfigurable modules $\mathcal{M}$. In this example, three modules (m_{crypto}, m_{video}, and m_{audio}) require a bus and a video connection as well as an audio I/O interface, which has to be implemented with ReCoBus and connection bar macros

but also if a particular wire is dedicated or shared among multiple modules (see also Sect. 2.5). Furthermore, the maximum number of simultaneously loaded master and slave modules can be specified, as well as the signal leveling of the reset and control signals. Note that only the physical aspects (mainly in terms of wires and registers) are specified, hence, being virtually compatible with any protocol implemented on top of the communication architecture, for instance, a Wishbone or CoreConnect bus protocol. A physical specification of a ReCoBus or some connection bars can be reused by multiple designs. Such specifications are independent to the target FPGA, and the same specification may be used to build systems for different FPGA families. Currently, the framework supports all Xilinx Virtex-II/II Pro FPGAs as well as all Spartan-3 devices and its derivates including Spartan-3E/-3A/-3AN. The ReCoBus specification process itself is supported by a GUI inside ReCoBus-Builder tool and the process can be carried out within a few minutes.

The here considered systems are strictly modularized and allow dynamically exchanging hardware in a way that exceeds the capabilities of standard ASICs or static only FPGA designs. Consequently, this demands additional tools and techniques for implementing such systems. The following sections are devoted to tools and methodologies required for physically implementing and verifying (Sect. 3.2.4) runtime reconfigurable systems, as well as techniques required to integrate modules at runtime (Sect. 3.2.6).

3.2 ReCoBus-Builder: A Builder for Reconfigurable Systems

The central tool of the proposed framework for building complex partially reconfigurable systems is the *ReCoBus-Builder*. As depicted in Fig. 3.1, the tool is assisting and automating all further steps required for physically implementing reconfigurable systems. ReCoBus-Builder provides a comfortable GUI (see Fig. 3.3) that helps in specifying and synthesizing a communication architecture fulfilling the requirements for all reconfigurable modules. The tool is seamlessly integrated into the Xilinx ISE design flow and helps floorplanning the FPGA. The coupling

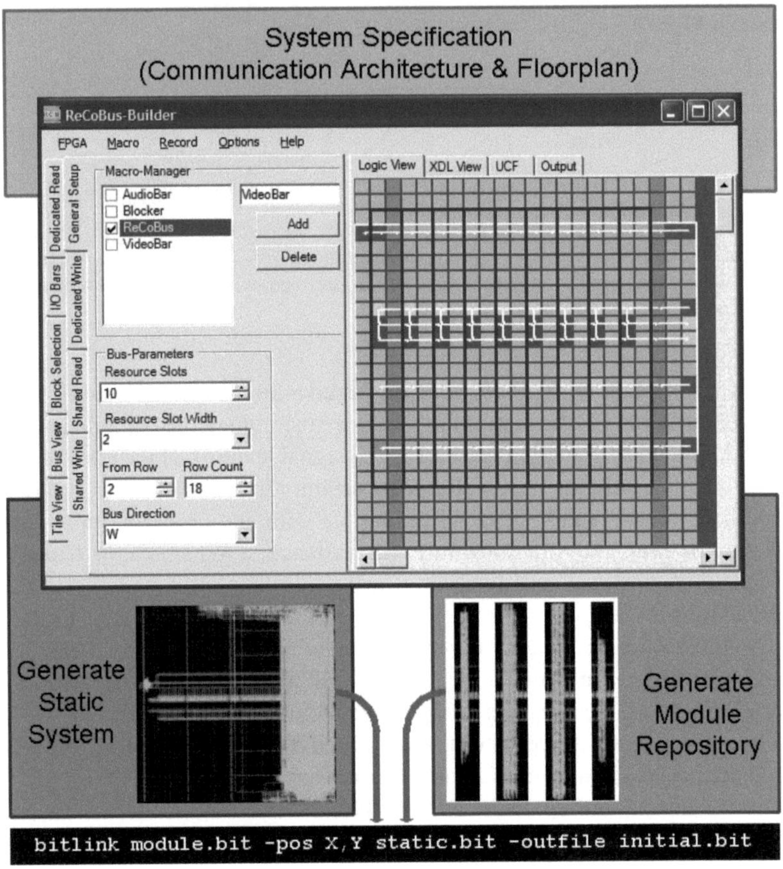

Fig. 3.3 ReCoBus-Builder GUI. The tool assists in specifying the complete ReCoBus and connection bar communication architecture. In addition, the tool provides a geometrical view on the target FPGA for floorplanning the design. This includes the definition and tiling of the reconfigurable region, placement of the communication infrastructure, and adjusting module bounding boxes. The synthesis of the communication architecture is a push-button process and ReCoBus-Builder generates all instantiation templates and constraints required for easily implementing the system with the Xilinx ISE design suite. This results in a static system and a module repository. Finally, the tool assists in the integration of partial modules by bitstream linking (see also Sect. 3.2.6)

to the Xilinx vendor tools is achieved by synthesizing ReCoBus and connection bar communication macros that can be directly instantiated inside a VHDL specification of the system. The tool supports this process with an included template code generator.

The floorplanning leads to a set of location constraints as well as bounding box and prohibit region definitions. ReCoBus-Builder is capable of compiling these specifications into constraints that guide all place and route processes. Furthermore, the tool can generate so-called *blocker macros* that allocate all logic and an

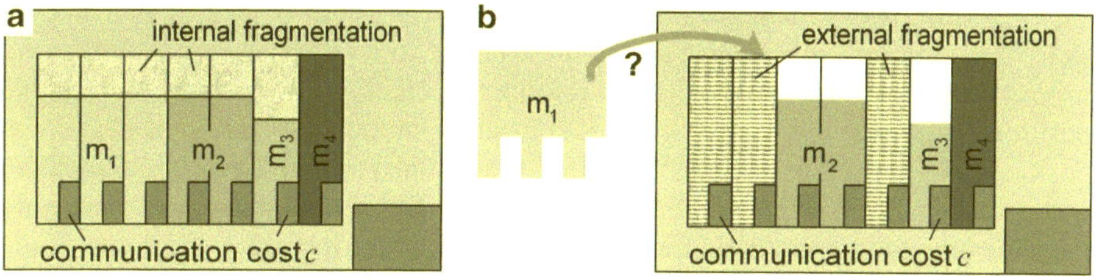

Fig. 3.4 (**a**) Optimal module packing of an assumed peak module set. (**b**) If the free space is not available as a *continuous* block, we might not be able to place a new module, despite that there are sufficient free resource slots available

adjustable set of routing resources within user definable regions of the FPGA. These macros circumvent a lack in the current versions of Xilinx's router that is not strictly following bounding box definitions.

3.2.1 Floorplanning

While static-only FPGA-based systems may be implemented without additional area constraints in a completely automatic fashion with the help of a placer tool, floorplanning is essential when designing runtime reconfigurable systems. Here, all partial modules have to be constrained into bounding boxes and during the implementation of the static part of the system, the assigned reconfigurable area must be prohibited for implementing any static logic or routing. This is illustrated in Fig. 3.3.

Floorplanning typically starts with a preliminary synthesis run for budgeting the resource requirements of the static system and of all reconfigurable modules individually. Based on this initial resource budgeting, the target FPGA can be selected. On this target, one or more reconfigurable areas have to be defined that will be tiled into a grid of resource slots. In this step, the system is split into a static part and a dynamically reconfigurable part containing the FPGA resources that are shared by multiple modules over time. These slots will accommodate the partial modules and the grid must provide sufficient resources to fulfill all logic (look-up tables) and memory (block RAMs) requirements. The resource slot grid must be balanced between the communication cost and the overhead for internal fragmentation. For a given set of modules, this can be computed using (2.1).

In Sect. 2.2, we introduced the term peak module set $\widehat{\mathcal{M}}$. This is the set of modules requiring the most resources at runtime among all possible module layouts. This definition is fine for many floorplanning cases, but in general, we have to consider that the reconfigurable resource slots get fragmented. This might require to reserve more resources for hosting reconfigurable modules than what is requested by $\widehat{\mathcal{M}}$. The example in Fig. 3.4, for instance, would need an additional resource slot to host the new incoming module, if the currently running nodules cannot be interrupted for a compaction of the module layout.

After tiling the reconfigurable area into a linear array or a grid of resource slots, the communication architecture must be set to fixed positions such that the infrastructure uses the same logic and routing resources at always the same relative position within the slots. This enables relocating modules to different placement positions at runtime or when assembling the initial system. The communication architecture will be implemented by fully placed and routed macros that are only instantiated and that maintain their internal structure during place and route of the static system or the partial modules. This allows that the individual reconfigurable modules can be also implemented in individual synthesis processes. These macros will then force the place and route steps to use exactly the same resources for the communication across the module interface. From this perspective, a ReCoBus acts like a Xilinx bus macro. But however, a ReCoBus connects multiple tiles and not only a single reconfigurable island. Moreover, a ReCoBus hardmacro implements the multiplexer and module select logic. Consequently, a ReCoBus macro is not pure resource overhead like a Xilinx bus macro.

Note that the design process of different modules can be easily delegated to multiple design teams that can work in parallel. And each team has access to its own individual resources without worrying about what other teams.

With the tool PlanAhead [Inc08c], Xilinx offers a vendor tool that also guides designers during floorplanning for partially reconfigurable systems. However, PlanAhead (in combination with the present partial design flow [LBM$^+$06]) can only assign one module exclusively to one reconfigurable area within the FPGA at a point of time. Thus, it is not simply possible to replace one large module by multiple small ones within the same reconfigurable area at runtime. The main reason for this restriction is a lack of sophisticated communication techniques that provide a flexible communication infrastructure for dynamically partial reconfigurable systems. PlanAhead only provides point-to-point communication between a static part and a dynamically reconfigurable part of the system by so-called *bus macros*.[1] Most newer related projects on building runtime reconfigurable systems are based on PlanAhead. Opposed to PlanAhead, ReCoBus-Builder uses the flexible ReCoBus and connection bar communication architecture that allows integrating multiple reconfigurable modules into one shared area at runtime.

3.2.2 *ReCoBus Generation*

After specifying the system specific and communication related parameters, ReCo-Bus-Builder generates a ReCoBus *hard macro*, containing the complete logic and routing of the ReCoBus that provides the communication between the static part of

[1]The name *bus macro* is misleading, as these macros provide only point-to-point links. In this book the phrase *ReCoBus* is used for a monolithic macro containing a real bus with connections to multiple (reconfigurable) modules at the same time. In addition, the flexible technique providing multiple point-to-point connections will be named with the phrase *connection bar*.

the system and the reconfigurable modules. Once generated, hard macros will not require any logic synthesis and they will be placed on the FPGA prior to the final place and route step. In this step, the connections between the monolithic ReCoBus hard macro and the rest of the design will be accomplished. The complete hard macro is generated using the *Xilinx Design Language* (XDL) that gives designers, among other low level details, full access to all of the FPGA logic resources (e.g., the look-up tables and routing wires). In addition, FPGA device descriptions, containing all device specific information, can be generated in the XDL language format. This information permits building device specific ReCoBuses as dedicated monolithic hard macros. For instance, this allows automatically deriving the positions of the block RAM columns in order to generate regular structured hard macros even if the reconfigurable area contains such dedicated resources. Note that partially reconfigurable modules may use such RAM columns. Another example for device specific information that can be derived from the XDL device description, is the set of available routing wires provided by a particular FPGA. It was found that available routing wires vary even within the same FPGA family. For instance, the largest Xilinx Spartan-3-FPGA (XC3S5000) provides wires called 'hex lines' within all CLBs that are not available in the CLBs of the smallest device (XC3S50) within the same device family. ReCoBus-Builder considers all this information in order to build an optimized monolithic ReCoBus hard macro.

A ReCoBus macro is considered to be optimized if it uses the minimum amount of slices within the resource slots that are connected with minimum routing cost. Thus ReCoBus-Builder packs as much logic inside the slices as possible to provide the bus logic and the connection to the reconfigurable modules. For instance, the two look-up tables inside one slice will be used for implementing the logic for two read multiplexer chains (see also Sect. 2.5.3). Independently to this, the two additional available flip-flops will provide a connection for two write data signals from the static part of the system to the reconfigurable modules connected in the same slot. With this packing, double the amount of signals per slice, as compared to the slice macros proposed in the Xilinx partial design flow [LBM+06], can be provided.

A ReCoBus is built by first placing the slices according to the desired resource slot width and specified density within the area assigned for the resource slots. The slices are placed in a regular fashion along the resource slots such that their relative positions are the same for all slots. Then, a breadth-first search is carried out for finding the routing paths between two consecutive resource slots and the internal routing within a resource slot. The search will continue in order to find all possible paths that require the minimum number of wires (hops) between the allocated slices. Out of all found paths, the one leading to the lowest propagation delay will be selected. In case that multiple paths have the same delay, the one containing wires providing the lowest connection grade will be chosen. This leaves the wires with higher connection grade for implementing the partially reconfigurable modules. ReCoBus-Builder is capable to solve routing conflicts that may occur if the ReCoBus is densely built. The paths found between the first two resource slots and inside the first resource slot are then copied to all other resource slots in order to provide a homogenous routing over all resource slots and the borders to the static

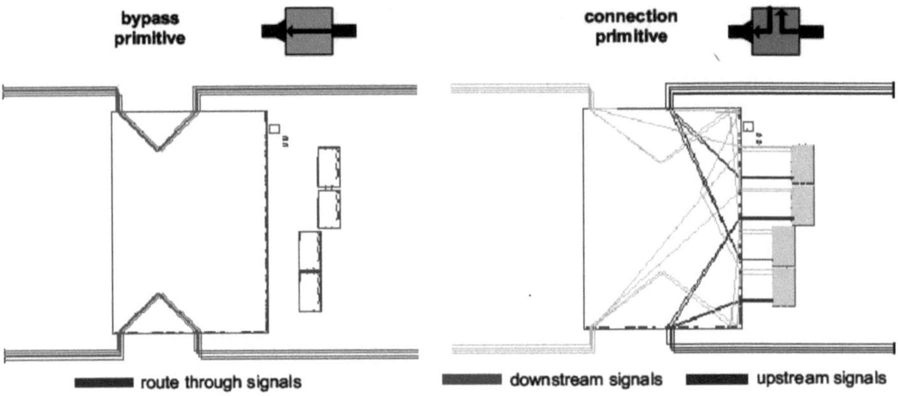

Fig. 3.5 Connection bar generation. Module connections points are only generated where required

part of the system. After this, the slice logic is set according to the determined routing. This includes a shuffling of the look-up table inputs according to the found routing tracks. Finally, a macro interface for connecting the static part of the system is generated. The complete ReCoBus hard macro generation runs automatically as a push button process.

For implementing a fully PLB bus compatible ReCoBus macro, it is sufficient to define a start point in the floorplanning GUI, the number of slots to generate, and the direction (towards west or towards east). By default this generates a hardmacro that can integrate master and slave modules via a 32-bit PLB bus. However, by modifying the default settings, virtually any bus protocol can be implemented (e.g., WISHBONE or AVALON).

Most commonly, the ReCoBus will map slaves directly into the system address space and in the case of masters, the ReCoBus will allow these modules to directly access the address space. In the case of multiple masters, an arbiter within the static part of the system is required that directs the access to the bus. ReCoBus-Builder does not generate an arbiter together with the ReCoBus, but it can provide all dedicated signals between multiple masters and an arbiter. Consequently, any policy for granting masters access to the bus can be implemented.

3.2.3 Connection Bar Generation

Besides the ReCoBus hard macro, ReCoBus-Builder assists in specifying the routing and placement of the connection bars. Multiple bars can be included into the system at different horizontally aligned positions that are located parallel to the ReCoBus as depicted in Fig. 2.46. It can be selected if and at which position a connection bar has to provide a module connection point. Connection points are only required for modules that demand access to a particular bar. If no bar connection is required, no logic resources will be wasted, as shown in Fig. 3.5. Each bar is exported as an individual hard macro, as described in the last paragraph for the ReCoBus. In each resource slot, the ReCoBus-Builder tool ensures that the routing is carried out by using the same equivalent wire across the slot boundary, regardless of whether a connection bar contains a connection or not. Consequently, modules can be placed freely within the reconfigurable resource area.

For older Xilinx devices (Virtex-II and Virtex-4) connection bars with up to eight wires per CLB can be implemented. In these devices, the number of LUTs per CLB is limiting the connection bar. By interleaving multiple wire segments (most useful is to interleave double lines with an interleaving factor of $N = 2$), the total number of wires per CLB row can be increased. However, in practice, not more than four wires should be used per CLB. First of all, this is the maximum number of wires that are suitable on Virtex-6 or Spartan-6 FPGAs for connection bar implementations (here limited by the available wires). Then occupying many wires may cause congestion or even unroutable situations.[2] Moreover, on all Xilinx FPGAs, there are eight LUTs per CLB available whereof at maximum four LUTs belong to the same carry chain. In other words, in most cases, we will not be able to process more than four input wires (= signal bits) per CLB. By default, the ReCoBus Builder proposes four wires that would allow to change the signal direction (e.g., from east to west) of a connection bar with the help of bitstream manipulation at runtime. This means, we could implement a connection bar towards east – and in the same CLBs another one towards west. We can then plug a module into the system and can define for this module whenever to access the first or the second connection bar.

3.2.4 Routing Constraints (Blocking)

While the Xilinx vendor placer properly follows area constraints, this is not strictly performed for the routing. For the standard partial design flow proposed by Xilinx [LBM$^+$06], this has no side effects, because partial modules are always placed to exactly the same reconfigurable island as illustrated in Fig. 3.6a. However, this causes problems if modules have to be relocated to different positions on the FPGA. This could be solved by merging routing violations from multiple reconfigurable islands into one island that is then used for implementing the reconfigurable modules as depicted in Fig. 3.6b, c. This is not applicable for systems with plenty of resource slots and the merging may fail as shown in Fig. 3.6d.

As a workaround, ReCoBus-Builder can generate so-called blocker macros that occupy all logic and routing resources within a user defineable region of the FPGA.

---

[2] While most wires look equivalent to each other, some have special purpose. For example, on Virtex-II or Spartan-3 FPGAs, the wires with the index 0 should be left unused for connection bars. This is recommended as these wires can connect also the clock and flip-flop enable inputs of the logic primitives. Other wires cannot provide the same connections. Such information is added to the ReCoBus-Builder for guiding the resource allocation and in most cases the default settings will lead to the best results.

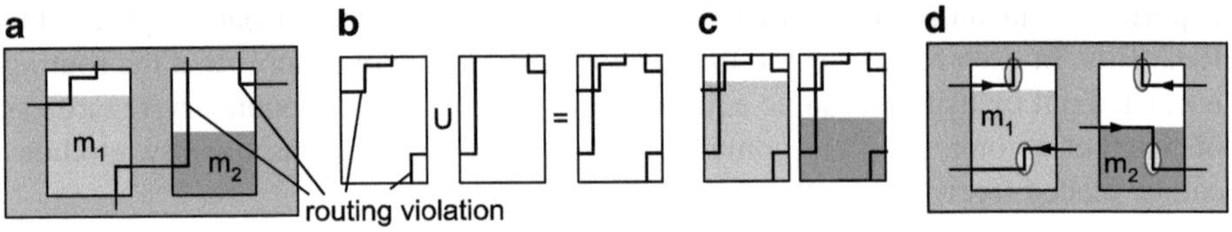

Fig. 3.6 Enhancing relocatability for modules including static routing. (**a**) Module m_1 cannot be swapped with module m_2 as this will cut the static routes through the reconfiguration island. (**b**) By merging the static routes of both islands that are then (**c**) included into all reconfigurable modules, relocation becomes possible. (**d**) However, merging may fail on wire conflicts, as a wire track can only implement one static routing path through the reconfigurable area

Blocking can be performed for all CLB tiles as well as for block RAM or I/O tiles, as depicted in Fig. 3.7. This process incorporates that the logic and routing resources of the communication architecture have to be skipped by the blocker. The blocker includes antennas into the switch matrices such that all outgoing wires will be occupied with such antennas. Consequently, the Xilinx vendor router will be forced to use alternative wire segments in unblocked switch matrices. The blocking antennas demand a logic tile (e.g., a look-up table) driving such an antenna to be compatible with the router. As the ReCoBus uses logic tiles that are then not available for blocking, fully blocking all outputs may be impossible by only using resources of a particular switch matrix. ReCoBus-Builder solves this issue with a blocking algorithm that can flood switch matrix outputs via neighboring CLBs. The designer can select within ReCoBus-Builder if any combination of (1) local interconnect lines, (2) horizontal long lines, or (3) vertical long lines should be blocked or left for implementing the static or module routing. For example, this allows using horizontal long lines for bridging some static signals across the reconfigurable area. Independently, vertical long lines may carry out the routing inside modules that are multiple resource slots high.

As illustrated in Fig. 3.3, the router must be prevented from using routing resources within a reconfigurable region during the implementation of the static part of the system. This is implemented by instantiating look-up tables that drive blocker antennas. In addition, these look-up tables will be connected to one or more global clock trees. This activates all necessary clock drivers within the reconfigurable region. The configuration of the clock drivers is stored in a dedicated part within the initial bitstream. The configuration of the clock network is not touched at runtime and the blockers will consequently ensure that the partial modules will be clocked, regardless of the placement position within the partial region. Again, this is because all primitives (LUTs, block rams, multipliers) within a reconfigurable region will be connected to global clock signals.

The blockers are included after synthesis and placement but before the final routing step. This is achieved by modifying the netlist after the placement with the help of the XDL language [BKT11]. The process is fully automated and carried out in seconds. Note that blocking is not limited to rectangular regions and

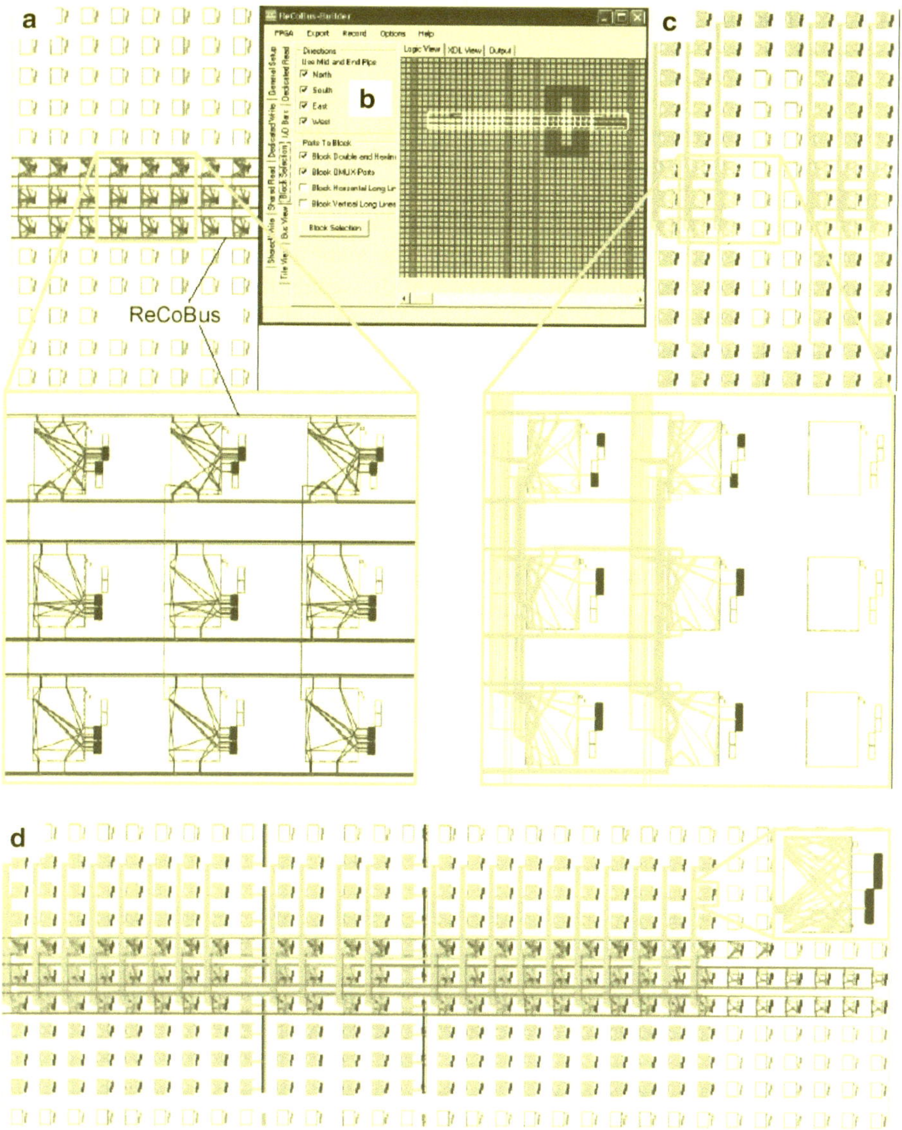

Fig. 3.7 Blocker macro generation for constraining the routing. (**a**) On top of the communication architecture, (**b**) blocker macros can be defined and (**c**) automatically generated that will wall the module routing into bounding boxes. Note that the blocker uses resources that are disjoint to the communication architecture as depicted in the enlarged boxes. (**d**) Blocking is also used for defining prohibited routing regions which is required when implementing the static part of the system (see also Fig. 3.3)

bounding boxes may be defined possessing any shape. Furthermore, modules are not necessarily required to be bound to a single continuous area, as shown in Fig. 2.64 on Page 138.

3.2.5 Timing Specification

Apart from constraining the physical wires, the communication architecture is further defining the timing behavior of the ReCoBus and the connection bars. This allows generating the different partial modules completely independent from the static part of the system. Accordingly, the static part can be implemented independent from all reconfigurable modules. This allows design iterations without interfering with other parts of the system. The static and the module timing can be decoupled by placing flip-flops between all signals that cross partial module boundaries or connect to the static system. Alternatively, the Xilinx vendor place and route tool allows defining slack offsets for a routing net. With this constraint, the total available slack (typically the reciprocal of the clock frequency) can be split into a fraction for the static system and another fraction for the partial modules. During the place and route phase of the partial modules, they will be implemented at the most critical position that is furthest away from the static part of the system. This ensures a valid timing when placing the module to any position within the reconfigurable area.

The ReCoBus hard macro can include flip-flops for pipelining the communication (which is typically used). However, the Xilinx router will not ensure timing to these flip-flops if no further constrains are defined. This means that defining a target clock frequency will (by default) not force the router to meet this timing for the signals connected the ReCoBus. This issue is removed by adding timing constraints to the user constraint file (UCF):

```
# specify the target clock frequency
NET ''clock'' TNM_NET = ''clock'';
TIMESPEC TS_clock = PERIOD clock 10000 ps; # 100 MHz

# add all flip-flops of the ReCoBus to a time group
TIMEGRP ''ReCoBusTimingGRP'' = FFS ''*inst_ReCoBus*'';

# define the ReCoBus routing latency
TIMESPEC ''TS_to_bus'' = FROM ''clock'' TO
                                ''ReCoBusTimingGRP'' 9900 ps;
TIMESPEC ''TS_from_bus'' = FROM ''ReCoBusTimingGRP'' TO
                                          ''clock'' 9900 ps;
TIMESPEC ''TS_internal'' = FROM ''ReCoBusTimingGRP'' TO
                              ''ReCoBusTimingGRP'' 9900 ps;
```

3.2.6 Bitstream Assembly and Hardware Linking

ReCoBus-Builder assists in forcing modules into bounding boxes, as well as defining and implementing swappable interfaces including their timing. Thus, partial modules can be cropped out of the complete bitstream for *linking* them together to an initial system or for loading them to an FPGA at runtime with the help of partial runtime reconfiguration. For this purpose, a *bitstream linker* was developed that can compose together partial modules in a two-dimensional fashion. This linker provides all functionality required to produce a module repository and an initial system at design time. The capabilities of the bitstream linker are as follows:

- *Extraction of modules from a full bitstream* with the capability to extract any logic tile, BRAM/multiplier tile, the BRAM content, and the I/O tiles. The latter one is in particular required for implementing partial systems on an ESM platform [MTAB07] where reconfigurable modules include external pins that are directly connected via an additional external crossbar to some external peripherals.
- *Merging of extracted module bitstreams* into a *full bitstream* where the target placement position can be selected in a two-dimensional fashion. The linker will check if the module resource requirements match the provided resources within the target region. Besides generating full bitstreams for an initial system, it is alternatively possible to generate *partial bitstreams* that contain only the configuration data for a particular module.
- *Modifying compare values inside the reconfigurable select generators* that decode dedicated signals for the reconfigurable modules inside the ReCoBus hard macro. This includes select and grant signals. With this feature, address ranges can be assigned for each module instance separately.
- *Analyzing or remapping of clock domains* within a definable region. The linker can scan a region for all used clock domains (global and local clock nets). Additionally, all clock input multiplexers of the FPGA fabric can be adjusted to any one of the global clock nets. This is helpful when porting a partial module from one design to another or for adapting the speed of a module, and hence, the power consumption.
- *Adjusting connection bars* for selecting within each resource slot to either bypass or connect a data stream of a particular connection bar. In detail, the output multiplexer, selecting between the preceding resource slot or the module output, can be adjusted.

Compared to related approaches [BJRK$^+$03, HLK02, KJdlTR05, BBH$^+$06, MMP$^+$03], the bitstream linker is not only capable of extracting and merging configuration bitstreams but also able of performing small bitstream manipulations for exchanging look-up table functions as well as modifying the routing. The extraction and merging allows two-dimensional module relocation (as opposed to [HLK02, MMP$^+$03]) and supports all logic, dedicated memory, and I/O cells. JBits [GLS99], providing a library of Java classes for manipulating configuration bitstreams, includes all functions required for bitstream linking. However, JBits supports only Xilinx Virtex-II devices and no Virtex-II Pro or Spartan-3 FPGAs. The developed bitstream linker supports all these devices and the binary code of the runtime bitstream linker requires less than 10 KB when implemented on a standard 32-bit softcore CPU. When alternatively using JBits, a java runtime environment will be required.

Hardware linking may be carried out alternatively on the netlist level. This may be helpful for debugging the system, but is inapplicable for runtime reconfigurable systems, as the configuration bitstream generation from a netlist is unlikely to be performed at runtime.

3.2.7 Simulating Runtime Reconfiguration

An essential requirement to correctly build any type of complex digital system is the capability to simulate it before implementing the actual design. This is required for debugging purposes, performance analysis and all kinds of diagnosis. For example, for verifying arbitration or handshaking mechanisms during swapping a reconfigurable module, it is necessary to simulate the reconfiguration process on a *cycle-accurate* base with the rest of the system.

In order to find errors, resulting from the integration of a partial module into a system, it is necessary to simulate the system at the register transfer level on a true cycle-accurate model that behaves exactly like the real model running in the FPGA. For instance, this allows verifying if a module follows a bus protocol during and after a configuration process.

3.2.7.1 ReCoBus Simulation

Introducing runtime reconfiguration adds additional functionality and complexity to a system and we need techniques to verify the correct behavior of the system. Rather than testing this behavior in the target system, it is more appreciate to simulate the system at design time. This allows a better observation of the system and more advanced test cases. The simulation has to include the capability of partial reconfiguration. The ReCoBus communication macros allow plugging-in or disconnecting modules by partial reconfiguration such that the rest of the system can continue its operation. And even more important, modules may be swapped without influencing running transfers on the ReCoBus at any time. Consequently, the reconfiguration process itself is completely transparent to the reconfigurable system. This property is exploited for carrying out cycle-accurate simulation for ReCoBus-based reconfigurable systems. As the reconfiguration is a completely deterministic process[3] that is transparent to the runtime system when using the ReCoBus communication architecture, it is possible to activate/deactivate modules, without considering the entire configuration data itself. This helps enormously to

[3]For Xilinx Virtex-II FPGAs, this holds true for a configuration speed of up to 50 MHz [CMZS07]. Beyond this speed, the internal FPGA configuration state machine may request some unpredictable wait states.

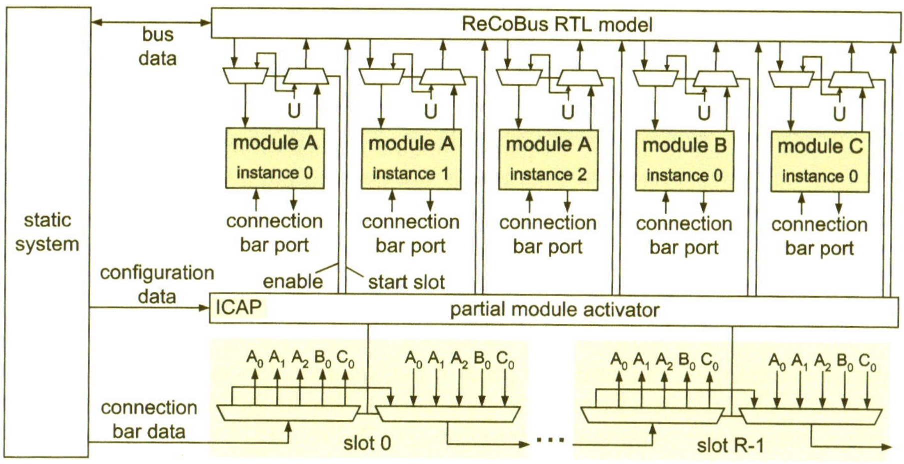

Fig. 3.8 Simulation of a reconfigurable system connecting partial modules via a ReCoBus and a connection bar. A *partial module activator* interprets the configuration data that is sent to the configuration port of the FPGA (ICAP). This data stimulates a state machine that keeps track of the current slot configurations and the module instances. This automata controls the module activation, the alignment multiplexing and the settings of the connection bar multiplexers

simplify the simulation of the reconfiguration process and it is sufficient to control the activation of the partial module instances without considering all internals of the FPGA fabric, as shown in Fig. 3.8.

Here, a module, called *partial module activator*, senses the data sent to the configuration interface (typically the internal configuration access port ICAP). This data stimulates a state machine that stores for each slot if it is idle or accommodating a particular module type. In the latter case, the system will further remember the instance that is executing the partial module for each slot. This information is evaluated for setting the activation of the instances. Within the simulation model, a ReCoBus can connect any number of module instances in parallel and the activator will reflect the resource limitations of the runtime system. All ReCoBus sockets are basically identical. However, if the ReCoBus is interleaved, the interleaved signals will pass an alignment multiplexer/demultiplexer pair for each ReCoBus socket. The alignment is controlled by the activator that stores the start slot position. The designer has to specify the maximum amount of simultaneously integrable instances for each module. During simulation, the first available instance of the requested module will be activated. If a module is deactivated, its inputs and outputs are forced to an undefined state. This allows determining if partial modules consume random data or if the static system tries to read data from a terminated module instance during reconfiguration.

3.2.7.2 Connection Bar Simulation

Connection bars are simulated by a demultiplexer/multiplexer pair for each logic resource slot. This allows connecting the connection bar input data either to the next consecutive slot or to any of the partial module simulation instances. Note that in the case the connection bar is interleaved, not the next but a slot that is located N positions further will be connected.

3.2.7.3 Controlling the Module Activation

Within the activator, the particular state of the resource slots and consequently the activation of the particular modules is set according to the amount of bitstream data written to the configuration interface. For Xilinx FPGAs, the configurations are written in a frame by frame manner where a frame corresponds to the smallest atomic piece of configuration data that can be accessed in the FPGA fabric. For configuring a complete logic column (including the routing), an FPGA family-specific amount of frames has to be written to the device that is 21 frames in the case of Virtex-II FPGAs. The size of a frame depends on the height of the device in terms of basic elements (CLBs). A Virtex-II XC2V6000 FPGA (the main FPGA of an ESM) holds a frame size of 984 bytes ($2 \cdot 12$ bytes for the top and bottom I/O tiles, plus $96 \cdot 10$ bytes for the CLBs). A configuration bitstream may include some header information that will be ignored by the FPGA configuration state machine. This is used to simplify the module identification in the activator by placing a module identifier in the bitstream header. The following lines denote a header example of a partial ESM bitstream:

```
0000: AD D0 00 02    // add instance of module 02
0004: AA 99 55 66    // start FPGA configuration
0008: 30 00 80 01    // reset CRC register
000C: 00 00 00 07    // reset CRC register
0010: 30 00 20 01    // write column address register
0014: 00 0C 00 00    // start column address is 6
0018: 30 00 40 F6    // write configuration command
001C: 00 00 00 00    // first frame data
```

When configuring the FPGA, the data of a full frame is at first shifted into an internal shadow register. From this register, a copy is stored in an additional hold register after finishing one frame. This data is then transferred to the logic cells on the arrival of a new frame in the hold register. Consequently, an extra dummy frame has to be written for flushing the last frame from the hold register to the logic cells. This is the procedure following the documentation [Inc07]. A slightly faster alternative was proposed by Claus et al. in [CMS06]. In that work, flushing the last frame is triggered by applying a so-called *multiple frame write command* instead of sending an extra dummy frame to the FPGA. While this technique is very advantageous on writing single frames in a random fashion, the benefit is marginal when reconfiguring larger consecutive parts of a Xilinx FPGA.

For a Xilinx Virtex-II XC2V6000 device, the standard procedure results in an idle offset of 984 frame data write cycles for filling the configuration pipeline. After this, a new column configuration is accomplished after $21 \cdot 984 = 20{,}664$ write cycles. In other words, out of the complete bitstream information, the configuration activator will only interpret the beginning module identifier and the start column address. Thereby, the start column address is set by the placer and is determined by the system (e.g., by some control software running on an embedded CPU) at runtime system or respectively during simulation. All further reconfiguration effects are now controlled by counters that are triggered by writing data to the configuration port.

The simulation model requires some module specific parameters, including the module size (in terms of CLB, memory, or DSP columns), the ReCoBus interface width, and relative location information about the connection bars. This information has to be annotated to the simulation model for each module individually. With this approach, a cycle accurate simulation of the reconfiguration process is accomplished and a reconfigurable module, activated in the simulation, behaves exactly like the real module after a partial reconfiguration process. The activation time of a module is set to the point of time when the real system will finish the configuration. The simulation model will generate a module reset itself after activation, like the real module running on an FPGA (as described in Sect. 2.5.2.3). This reset is released by the system level reconfiguration step that sets the base address of the last reconfigured module (see also Sect. 2.5.2.3).

As the configuration data content is completely ignored, simulating runtime reconfiguration can be carried out with the help of placeholder modules, long before physically implementing a module. In this case, the module size has to be estimated and a bitstream containing dummy data will be sent to the module activator.

3.3 Core-Based System Design

As the communication infrastructure requires only low resource overhead by providing high flexibility and low latency, the ReCoBus-Builder tool flow is also suitable to implement static-only systems. As the interface of the partial modules is completely encapsulated, a re-synthesis step of a particular module will not influence the timing of other parts of the system, and consequently, does not require a timing re-verification of the untouched parts. In addition, this tool flow allows to parallelize the synthesis as well as the place and route process. Furthermore, the partitioned design flow may help to massively reduce the memory consumption of the design tools, because focus has to be put only on fractions of the design. If a module is once encapsulated and fitted into a bounding box, it can be integrated into a system by instantiating the module as a fully routed hard-IP core.

Gupta et al. postulated in [GZ97] to compose on-chip systems with the help of hard-IP cores. In that work, focus is put on ASIC design and the motivation for the core-based system design is identical to designing static systems based on the ReCoBus-Builder tool flow. In both cases (ASIC and FPGA), the system is composed of fully routed modules that provide a standardized interface for carrying out the top level communication. Modules are verified against this standardized interface, hence simplifying the integration of such IP cores into a system. This allows to skip costly verification processes and enhances design productivity. While in the ASIC case, dedicated metal layers can be used for wiring the top level modules, routing resources have to be preserved in the FPGA case. These resources form a logical layer on top of the FPGA routing fabric that allows to link modules without touching the internal routing of any module. As the reserved routing resources are not available for the internal module interconnect, the communication architecture has to be carefully designed to meet the communication requirements of both, the top level interconnect as well as of the internal module routing. So far, the ReCoBus and connection bar communication architecture is the only available technology that is capable of efficiently composing on-chip systems with hard-IP cores for FPGAs.

Besides academia, the EDA industry is also working on core-based system design and user groups such as the Virtual Socket Interface Alliance (VSIA) [vsi07] or the Spirit Consortium [spi07] develop standards for providing compatibility among tools and IP cores. This includes vendor neutral IP core generation as well as interface standardization, which in particular, includes bus specifications. These user groups target ASIC designs and the FPGA vendors Xilinx and Altera address the need for component based system design with a concept called partitioning. In a partition, the internal look-up table placement is fixed and the module internal routing is kept as far as possible during design iterations. In order to maintain the speed of a particular partition, critical paths are identified and labeled as critical. These critical paths will be not touched when performing any top level routing between the different partitions. However, as the internal module routing and the top level routing share the same FPGA routing fabric resources, there is a risk that the partition concept fails to maintain the partition timing or to accomplish the routing. Consequently, the partition design flow may require time consuming design iterations. Opposed to this, the strict separation of FPGA routing fabric resources for implementing the top-level routing and the internal module routing guarantees to maintain timing as well as to accomplish all top level routing demands.

3.4 Runtime Reconfiguration

The last section has covered aspects on building reconfigurable systems, including specification, simulation, floorplanning, and partial bitstream generation that have to be considered at design time. This section puts focus on runtime aspects. When fractions of an FPGA-based system have to be exchanged, the configuration stored inside the FPGA has to be exchanged as well. This is carried out by sending a configuration bitstream to the FPGA that updates only the changed parts inside the FPGA fabric. This process is related to the used FPGA architecture that defines how configuration data can be accessed and also how the FPGA primitives are encoded

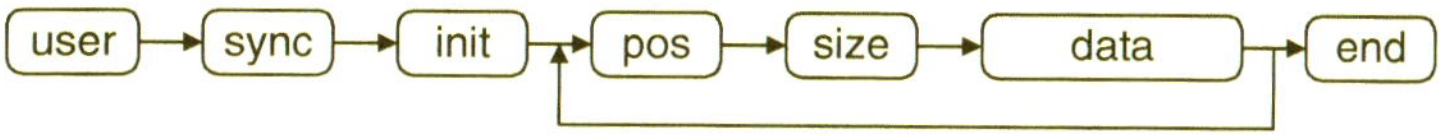

Fig. 3.9 Configuration bitstream format for Xilinx FPGAs. The configuration state machine scans the input data for a syncronisation (sync) sequence. Any preceding optional user data before this sequence will be ignored. After this, some initialisation commands follow. Next, the frame start position (pos), the size of the configuration data, and the configuration data itself is send to the FPGA. The last three steps may be repeated arbitrary times. At the end further commands are required for activating the configuration or for finishing the configuration process. The frame start position is automatically incremented if multiple consecutive frames are written to the FPGA

inside the configuration bitstream. For instance, while the configuration of Atmel AT40KAL FPGAs can be performed via an SRAM memory like interface [Inc03] in a very fine grained manner in portions of 8 bit, Xilinx Virtex-II FPGAs can only be configured in frames. Such a frame stores a fraction of the configuration data for a full logic or memory column of the FPGA including the attached top and bottom I/O cells. Consequently, the frame size depends on the device height and is, for example, 7,872 bit in the case of a Virtex-II-6000 FPGA. This amount of data is the smallest atomic piece of configuration data that can be individually accessed on these devices.

For all recent devices, Xilinx bounded the configuration frames to the height of a clock region. The clock region height is identical within an FPGA family and the different devices differ in the number of clock region rows. For a Spartan-6 FPGA, the heigth of a configuration frame is spanning 16 CLBs, 4 BRAMs, or 4 DSP48 primitives. The corresponding configuration frames consist of 128 bytes (plus a 2 byte checksum). More device specific information can be found in the corresponding user guides.

3.4.1 *Horizontal Bitstream Relocation*

In reconfigurable systems, providing a very flexible module placement, it is inapplicable to store alternatives for all possible positions inside a module repository of the runtime system. Consequently, a partial module bitstream should be able to run at multiple possible placement positions. For relocating modules in horizontal direction on a Virtex-II FPGA, the start position of a logic or RAM column has to be adjusted inside the configuration bitstream that is depicted in Fig. 3.9. As it is possible to write the module configuration data as one single block,[4] it is sufficient to set the start position once at the beginning of the configuration process. This can be easily carried out in software. Alternatively, it is possible to manipulate the module

[4]This holds true for consecutive modules that contain no block RAM or dedicated multiplier resources. For non-consecutive modules (see Fig. 2.64 for an example) and for dedicated resources, multiple blocks, each containing another start position, have to be written to the FPGA.

start position within the bitstream in hardware. This approach was firstly presented by Kalte et al. for Xilinx Virtex-E FPGAs in [KLPR05] and later for Virtex-II / II Pro FPGAs in [KP06]. Manipulating bitstreams in hardware will not show an advantage as compared to a manipulation in software, if modules densely fit into resource slots, which is possible when using a fine grained resource slot tile grid. The reason for this is that for a densely packed module, typically all configuration frames within the area of a swapped module have to be written to the FPGA. Consequently, it is sufficient to set the bitstream position once at the beginning of the configuration process. The partial bitstream generation, presented in Sect. 3.2.6 will ensure that all columns of a module are included inside the resulting partial module bitfile.

For one-dimensional slot based reconfigurable systems, based on Xilinx Virtex-II FPGAs, such as the ESM, a horizontal bitstream relocation scheme is sufficient and a partial module bitfile can be directly written to the target position on the FPGA. This requires that modules span the full height of the device including the I/O pins located above or below a particular module. This typically prevents using these I/O pins and all communication has to be carried out with I/O pins located at the left or right border of the device, or I/O pins that are not located within the reconfigurable region of the system. Consequently, designing slot based reconfigurable systems does also include considerations for the PCB (printed circuit board). An platform dealing with these issues is the ESM that actually provides an external crossbar switch for allowing partial modules to connect to external peripherals. This simplifies the configuration process at runtime but is costly bought with an increased latency plus the cost for an additional crossbar FPGA. The ReCoBus and connection bar communication architecture can fulfil all communication demands on a single FPGA.

3.4.2 Vertical Bitstream Relocation

For saving an external crossbar and enhancing I/O utilization, partial module bitstreams have to be mergeable with the static part of the system or other reconfigurable modules. This means that a partial module may not span the full height of the device and that parts of the static system or even other reconfigurable modules share the same CLB column. For Xilinx Virtex-II FPGAs, this is obligatory if modules are placed in a two-dimensional grid style. As mentioned before for these devices, the atomic reconfigurable fraction of configuration data that can be directly accessed is a full frame. However, Xilinx guarantees that no glitches or other side effects occur when overwriting any configuration with exactly the same content. This holds true as long as no look-up tables are overwritten that contain a state. Look-up tables may contain a state if they are used as (1) shift registers or (2) distributed memory. The Xilinx synthesis tool can be constrained not to use these two options. Note that it is not necessarily prohibited to use distributed memory within reconfigurable modules. If, for example, the configuration of a reconfigurable module will never be overwritten as long as the module is running, because it spans

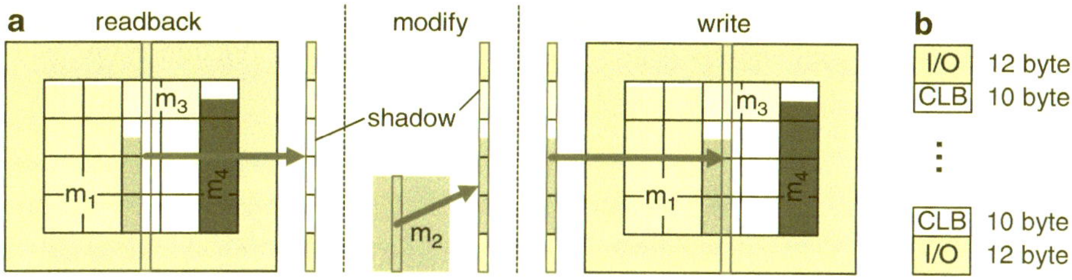

Fig. 3.10 (**a**) Readback-modify-write bitstream manipulation. For changing fractions of a particular full column on a Xilinx FPGA, the columns can first be stored in a shadow memory using the configuration readback capability. This shadow can then be accessed freely for performing bitstream manipulations. Finally, the modified configuration data is written to its original location. (**b**) Vertical inner-frame addressing for Xilinx Virtex-II FPGAs. The configuration data for I/O tiles and CLBs is strictly separated and the data of all tiles is concatenated from top to bottom and over all frames within the configuration bitstream

the full height of the reconfigurable area, the state inside the look-up table is not required to be touched by any reconfiguration process. This is the case for one-dimensional slot based systems and for such systems, a state stored inside some look-up tables must not be further considered.

Merging configuration bitstreams for Xilinx FPGAs was first proposed in a patent of the Xilinx Inc. [CKL+]. Sedcole et al. used the same approach for relocating modules on Xilinx Virtex-4 FPGAs in a two-dimensional manner [SBB+06]. In both approaches, the bitstream merging was carried out in a configuration *readback-modify-write* operation, as illustrated in Fig. 3.10a. In the presented prototypes in this book, the bitstream manipulations are not performed directly within the FPGA but in an extra shadow memory, hence, hiding the configuration readback process. Figure 3.10b depicts the inner-frame addressing scheme of Xilinx Virtex-II FPGAs. As can be seen in the figure, the configuration bitstream is similarly regularly structured as the FPGA architecture. A single block RAM is sufficient to store a full configuration frame even for the largest Virtex-II device. However, a single frame buffer will dramatically reduce the configuration speed, because of Virtex-II FPGAs require one extra frame to be written to the device for each configuration read or write operation. This is independent of the amount of configuration data that is transferred during read or write operation. Let c_r and c_w be the data transfer rate for configuration read and write operations and o_r and o_w be the overhead required for reading and writing a portion of configuration data via the configuration interface, then the time for readback-modify-write bitstream manipulation method is:

$$t_{rmw} \approx f_m \cdot \left(\frac{f_h}{c_r} + \frac{o_r}{f_b \cdot c_r} + \frac{f_h}{c_w} + \frac{o_w}{f_b \cdot c_w} + \frac{m_h}{p_m} \right) \tag{3.1}$$

Where f_m denotes the number of configuration frames, involved in the modification process, and f_h and m_h specify the total amount or respectively the amount of configuration data for a module inside one frame. f_h and m_h are related to the height of the device and the module. The buffer size f_b is given in terms of frames and the actual bitstream manipulation is processed with the rate p_m. Equation 3.1 extends a description by Sedcole et al. [SBB+06] by a parametrizable buffer size and the initial overhead required for each access to the configuration port. The equation gives an estimation for the upper bound, as it is not assumed that the processes readback, modify, and write are performed in parallel. For instance, by using a double buffer for the readback data, the bitstream modifications can be performed in parallel with the readback process. The maximum configuration read and write transfer rate is in most cases the same and the bitstream manipulation in the buffer can be typically carried out considerably faster ($c_r = c_w \leq p_m$). The overheads o_r and o_w are slightly more than the amount of data stored inside one frame (f_h) in the case of Xilinx Virtex-II FPGAs. The configuration time is therefore at least four times longer than the time required to write al f_m frames to the configuration port, if the buffer can store one frame. For larger buffers, the configuration time converges towards double the amount of time to write all f_m frames. However, by using a buffer that can store all frames of the reconfigurable area, the readback process can be completely omitted. In this case, the data rate of the configuration interface can be fully utilized. A two-dimensional reconfiguration process is then carried out by first updating the configuration image in the buffer followed by writing all modified frames to the FPGA in typically one block. This will enormously speed-up the configuration process for the full buffer-based bitstream manipulation to roughly:

$$t_b \approx f_m \cdot \left(\frac{f_h}{c_w} + \frac{m_h}{p_m} \right) \tag{3.2}$$

Note that the configuration image manipulations can be performed per DMA and concurrently with the write process to the configuration port. Claus et al. presented in [CMZS07] how DMA can be used for transferring data to the internal configuration access port (ICAP) of a Xilinx Virtex-II Pro FPGA at very high speed. The reported configuration speed is about 95 KB/ms which allows to exchange a 32 bit softcore processor in less than 2 ms. However, this comes along with a considerable memory burst during the configuration process that may interfere with the rest of the system. Section 3.5.3 presents compression techniques for reducing such bursts.

This reconfiguration speed enhancement of the full buffer-based bitstream manipulation is bought at the cost of an additional amount of RAM that is, in the worst case, identical to the size of a full FPGA configuration bitstream. However, the monetary cost for this additional memory will typically be negligible as compared to the cost of the FPGA itself. Alternatively, the shadow memory can be saved by concatenating the configuration data on the fly by directly merging bitfiles from the initial configuration storage and the partial module repository.

Despite that recent Xilinx FPGAs are configured in sub columns, it can still be an option to apply the here presented bitstream merging technique to these devices. This is for example usefull, if the floorplanning does not fit to the predefined sub column layout.

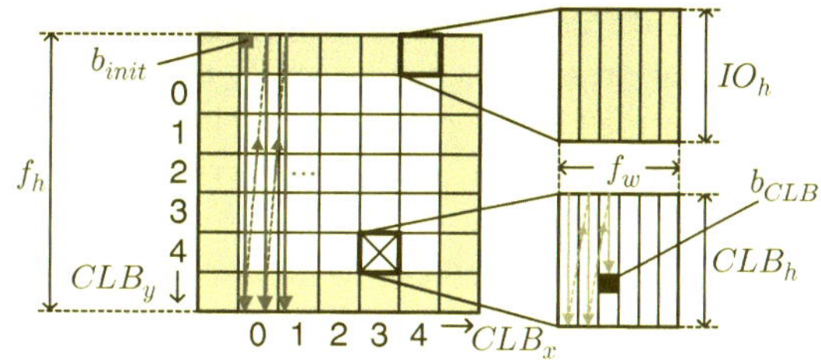

Fig. 3.11 Addressing an individual bit inside a CLB

3.4.3 Small Bitstream Manipulations

Besides modifying larger parts of the configuration when loading new modules to the FPGA, some additional small manipulations may be required at runtime. This allows, for example, to change the address of a reconfigurable select generator (see also Sect. 2.5.2.3) or to connect the output of a partial module to a connection bar after configuration and initialization of a partial module. An individual configuration bit of a CLB can be addressed inside the configuration bitstream by its CLB position (CLB_x, CLB_y) and the bit offset (b_{CLB}) within the CLB:

$$b_{ADR}^{CLB}(CLB_x, CLB_y, b_{CLB}) = b_{init} + f_h \cdot f_w \cdot CLB_x + f_h \left\lfloor \frac{b_{CLB}}{CLB_h} \right\rfloor + IO_h + CLB_h \cdot CLB_y + b_{CLB} \ mod \ CLB_h \quad (3.3)$$

With the device specific constants b_{init} for denoting the initial offset for the first CLB column, f_h and f_w for stating the frame length and the number of frames per CLB column, and IO_h and CLB_h for defining the configuration frame length related to a single I/O cell or a single CLB. All parameters are illustrated in Fig. 3.11. The BRAM configuration data is separated from the CLB configuration data in Xilinx Virtex-II FPGAs and located at the end of a configuration bitstream [Inc07]. The bit addressing inside the BRAM columns is analogous to the CLB bit addressing scheme but with a different parameter set. Equation 3.3 can also used for recent FPGAs. However, except the IO_h term we have to add a row offset term.

While the configuration bitstream data for a particular LUT is stored well-ordered in one single frame,[5] all switch matrix configuration bits are spread irregularly over a CLB configuration bit array of size $f_w \cdot CLB_h$. However, the internal bitstream structure of one CLB was found identical throughout the complete FPGA fabric. Within the software driver, all bits, belonging to a particular elementary reconfigurable resource (e.g., a clock input multiplexer), are stored in a list. This list first comprises the number of configuration bits n_c followed by a set of CLB configuration bit offsets: $\{n_c, b_{CLB,0}, \ldots, b_{CLB,n_c-1}\}$. With this data structure,

[5]This holds for 4-input LUT Xilinx FPGAs. Newer FPGAs (e.g., Virtex-6 or Spartan-6) require to write multiple frames to (re)configure a LUT function. However, LUT values are strictly separated from configuration bits describing the routing.

the specific configuration encodings for the different possible settings of the elementary resources can be easily assigned in software. At first, all modifications are performed on the configuration image stored in the RAM memory of the system and then updated by selectively writing all modified frames to the FPGA fabric. Note that selectively updting the routing using the configuration image approach will cause no sideefects to running modules. This is in particular usefulr for implementing circuit switching networks directly inside the FPGA routing fabric.

3.5 High Speed Reconfiguration

In some systems, reconfiguration may be frequently triggered and the configuration time can, in some cases, dominate within the system. For example, in the DISC-system [WH95], 71% of the overall time is spent on updating instructions in a reconfigurable processor.

Consequently, techniques for speeding-up the reconfiguration process are highly relevant for many runtime reconfigurable systems. These techniques include *configuration prefetching*, *caching*, *configuration port overclocking*, *bitstream decompression*, and *defragmentation*. It is possible to apply these techniques individually or in combination with others. For example, bitstream decompression can be used for accelerating the defragmentation of the reconfigurable area, which, in turn, may be accomplished in the background before a new task reconfiguration is requested.

3.5.1 Configuration Prefetching

Whenever possible, configurations should be *prefetched* such that the configuration and the execution of different modules overlap in time. The goal of this technique is to fully hide the reconfiguration time by swapping modules in the background. However, while this may be possible for periodically activated tasks, this is difficult if activations are unpredictable. In this case, speculative prefetching can be applied. Hauck proposed in [Hau98] to trigger the configuration prefetch with the help of dedicated preload instructions that are automatically included in a software binary. In the proposed system, some kernels of a software program have been replaced with reconfigurable accelerator modules and the configuration is triggered with respect to the current program state (defined by the program counter) and prior to the kernel invocation. Other related work includes reconfiguration tasks belonging to a particular hardware task that will be scheduled to a reconfiguration interface. In [DF07], Dittmann and Frank presented deadline monotonic scheduling of such reconfiguration tasks with fixed priorities. The work assumes to preempt a reconfiguration task to push the configuration of higher prior tasks while allowing to seamlessly resume the preempted reconfiguration at a later point in time. The same model is used by Sim et al. in [SWT09] for a latency-optimal list scheduling

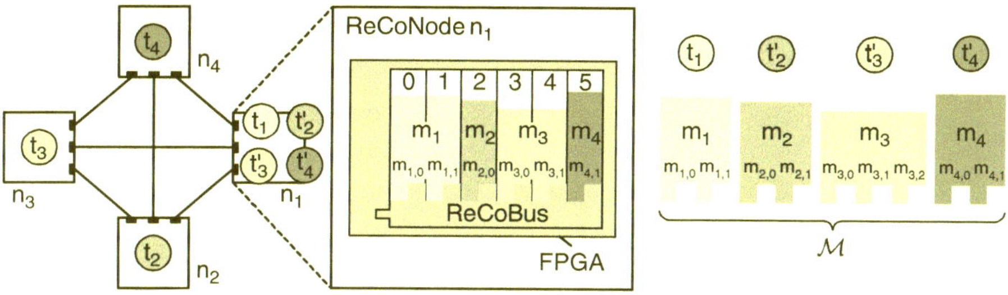

Fig. 3.12 Configuration prefetching. Assuming a set of replicas (t'_2, t'_3, and t'_4) that each possess a repair slack sufficient to configure one resource slot before activating a replica in case of a fault, it is possible to bind all replicas as a cold spare to the same node n_1. In this case, any single node failure of n_2, n_3, or n_4 can be compensated

that runs in polynomial time. The algorithm runs in two phases, with the first one identifying the reconfiguration task dependencies (e.g., a reconfiguration process can only start if the allocated FPGA resources are free or the configuration process must complete before activating the corresponding hardware task). In a second phase the schedule is computed using preemptive reconfiguration.

Rather than prefetching compleete modules, prefetching can alternatively be applied on fractions of a module. The idea is to limit the configuration latency by prefetching parts of multiple modules at the same time. If then one of these modules has to be activated, only the remaining configuration of this module will be written to the FPGA. Consequently, this can be acompished in less time than writing a complete partial module to the device. This can, for example, be used to increase fault tolerance or to improve repair times in a fault-tolerant distributed control system, such like in a ReCoNet (see Chap. 4 for more details).

A corresponding example is sketched in Fig. 3.12. In that system, the FPGA on node n_1 hosts the active task t_1. In the case of a node failure of any other node (n_2, n_3, or n_4), the FPGA on node n_1 should start a copy of the entire destroyed task (a replica task t'). If we now assume that the time to start a replica does only allow to configure a single slot, we could use prefetching as shown in the figure. In this scenario, we save three configuration slots while being able to finish the configuration process of all possible replica modules in less time than starting a reconfiguration from scratch.

3.5.2 *Configuration Port Overclocking*

While the improved CMOS processes of FPGAs resulted in many faster features over time, the specified configuration speed had been kept rather low. As has been revealed in the introduction on Page 22, the configuration speed of a modern Virtex-5 or Virtex-6 FPGA is limited to only 100 MHz of operational clock speed.

This holds even for the internal ICAP port that provides a direct connection to the user logic implemented on the FPGA fabric. As the on-chip ICAP port requires no pin drivers, it should be considerable faster when assuming the 65 or 40 nm CMOS process of Virtex-5 or Virtex-6 devices. Moreover, the FPGA internal configuration logic consists basically of a shift register that should provide significant faster operation than 100 MHz.

In order to provide ultra fast reconfiguration and for exploring the configuration speed limits of Virtex-5 FPGAs, a hardmacro for enhancing the ICAP port was proposed in [HKT11]. As shown in Fig. 3.13, this macro multiplexes the internal 32-bit ICAP port to a 64-bit wide ICAP port. This results in an enhanced ICAP port that provides – except the 64 bit wide operation – the same protocol than the original ICAP primitive. The connection of the multiplexer output to the ICAP primitive has a critical path delay of only 1.2 ns. Due to the multiplexing to the double word size, the user logic can work at moderate speed for sending configuration data to the enhanced ICAP port. With this macro, configuration at 550 MHz at the original ICAP port (which is 5.5 times the specified 100 MHz) was possible. Together with a 32-bit wide configuration interface, this results in a 2.2 GB/s sustained configuration throughput. This speed is sufficient fast to reconfigure a fully featured Microblaze softcore CPU in less than 100 μs. Providing that high data rates is not trivial and many systems might not be able to configure even close to this value. This might be if they suffer of a low throughput to read configuration data from external memory. Such a bottleneck can be widened by transfering compressed bitstream, as investigated in the next section.

3.5.3 *Hardware Accelerated Bitstream Decompression*

With progress in FPGA evoluation, two trends, related to reconfiguration speed, can be observed: (1) higher possible configuration speeds at the FPGA configuration interfaces and (2) an increase in functionality (FPGA size), and consequently, an increase of the configuration data to be stored and transferred within the system (see also the chart on Page 22). Not only the functionality of the device is rapidly increasing but also the complexity of typical modules that are integrated in a system. For instance, instead of integrating a simple 8 bit CPU, present systems may provide a complex 32 bit core.

The first trend results in an increasing gap between the possible FPGA reconfiguration speed and the throughput of non-volatile configuration memories. While many present FPGAs can be configured with data rates of up to 400 MB/s (e.g., Xilinx Virtex 4 / 5 or Altera Stratix IV), a typical NAND-flash memory device provides only a few tens of megabytes per second. Similarly, the gap between the network throughput in distributed embedded systems and the configuration speed has rapidly increased, hence, inhibiting remote configuration, where one node may configure another one via the network at full FPGA configuration speed.

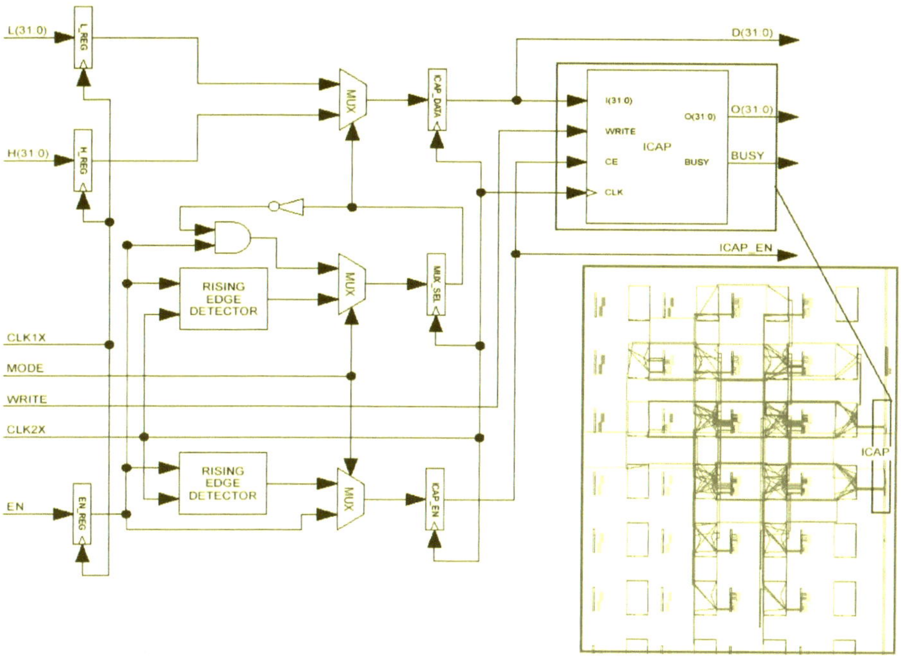

Fig. 3.13 ICAP-64 hard macro for high speed reconfiguration

The trend towards an increased functionality leads to larger configuration memory requirements as well as to an increased amount of configuration data that has to be transferred for loading some modules to an FPGA.

These effects can be diminished by *storing and transferring compressed configuration data*. This data can then be decompressed just in front of the relatively fast configuration interface. Because of the required throughput, this decompression has to be carried out in hardware (Fig. 3.14). If we assume, for example, that the configuration data can be compressed to 50% of its original size, only half the data is required to be stored and transferred. Hence, the configuration speed may be potentially doubled. Bitstreams may contain fractions of data that are selectable to any value (don't care values) that can be exploited for achieving higher compression ratios [LH99]. However, as all bitstream formats of today's FPGAs are confidential, exploiting don't care values is not used and only lossless compression algorithms will be examined. Throughout this section, different hardware accelerated decompression techniques will be evaluated according to the following objectives:

- *Compression ratio*: for allowing considerable memory savings and throughput enhancements,
- *Throughput*: the decompressor has to emit at high continuous data rate. This includes the achievable clock frequency of the hardware decompressor as well as the compression algorithm.
- *Resource overhead*: The decompression hardware should be as small as possible.

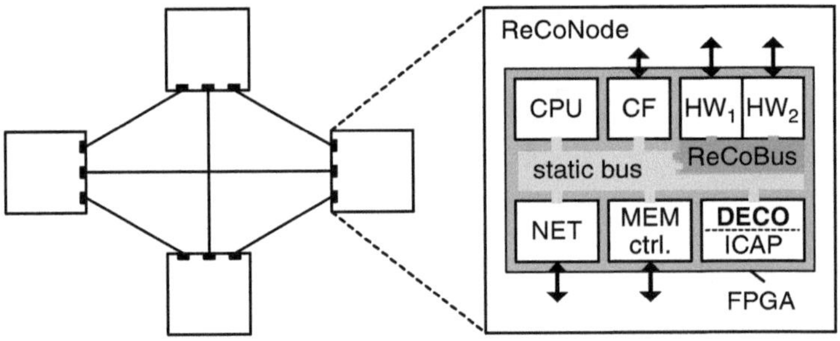

Fig. 3.14 Optimizing reconfiguration speed and configuration storage with hardware accelerated decompression in a ReCoNet. Each ReCoNode of the ReCoNet is equipped with an FPGA for implementing a CPU, a CompactFlash controller (CF), some reconfigurable modules, a network interface, a memory controller, and an interface for the FPGA internal access configuration port (ICAP). A decompressor (DECO) in front of the ICAP port can optionally decompress configuration data stored on a CompactFlash card, the local memory, or data located on another node. For tolerating node failures, configuration data has to be stored redundantly on different ReCoNodes

3.5.3.1 Impact of the Compression Ratio on the Configuration Speed

Bistream compression is the process of encoding FPGA configurations using fewer bits than the original unencoded bitstream. Data savings are expressed with the *compression ratio* $\eta = \frac{size(compressed\_bitstream)}{size(original\_bitstream)}$. The main goal of bitstream compression is to widen the throughput bottleneck of slow configuration storage devices or slow network links. For the following, it is assumed that a bitstream decompressor is capable to emit one output token of word size $|w_{deco}|$[6] per clock cycle. With $|w_{FPGA}|$ being the size of the configuration interface, the decompression accelerator has to operate with at least $\frac{|w_{FPGA}|}{|w_{deco}|}$ times the clock frequency of the configuration interface in order to allow full configuration speed.

For accelerating the configuration process, it is not only important to put focus on the compression ratio, but also to consider the limited bandwidth at the decompressor input and output. Considering an FPGA with the peak configuration data rate d_{FPGA} from a configuration memory providing the peak read data rate d_{MEM}, then the configuration data rate is:

$$w_c = \begin{cases} d_{FPGA} & if \;\; d_{MEM} \cdot \frac{1}{\eta} \geq d_{FPGA} \qquad \text{full speed} \\ d_{MEM} \cdot \frac{1}{\eta} & else \qquad \text{limited by memory bandwidth} \end{cases}$$

Note that for any lossless compression algorithm, it is possible to construct cases where the compression ratio is larger than one [Heu03], thus, degrading the

---

[6]The vertical bars denote the word size of the surrounded word that is the number of bits used to code a word. This notation is taken from Salomon [Sal04] and will be consistently used throughout this section.

configuration time. For example, assuming a configuration bitstream where 10% of the bitstream contains worst case compression data that may be compressed by a ratio of $\eta_1 = 4$, while the 90% remaining bitstream data may be compressed at $\eta_2 = 0.1$. Then, the average compression ratio is $10\% \cdot \eta_1 + 90\% \cdot \eta_2 = 49\%$. When further assuming that the configuration memory provides half the throughput of the FPGA configuration interface. Then, at first view, it seems possible to achieve the optimal configuration time t_{opt}, that is only related to the size of the configuration bitstream divided by maximum configuration interface speed d_{FPGA}.

However, this holds true only if a decompression accelerator reads at a continuous data rate from the configuration memory. With respect to the assumed example, the 10% worst case configuration data produce a configuration memory burst that idles the configuration interface, while for the remaining 90% of the bitstream, the maximum configuration speed is limited by the configuration interface data rate. In the latter case, it is not possible to take full benefit of the high compression ratio and throughput is wasted at the configuration memory interface. As shown in Sect. 3.5.5.3, these bursts cannot be efficiently hidden with additional buffers between the configuration memory and the decompression accelerator. When taking the bandwidth of the configuration memory into account, the configuration time for the assumed example is $\frac{10\% \cdot \eta_1}{0.5 \cdot d_{FPGA}} + \frac{90\%}{d_{FPGA}} = 1.7 \cdot t_{opt}$. This means that the configuration time is reduced by only 15% as compared to an uncompressed configuration process that would take two times t_{opt} for the assumed memory data rate. Despite the good average compression ratio of below 50%, this result is far away from the optimal configuration time t_{opt}.

If the bitstream compression algorithm will be designed, such that the worst case compression ratio is limited at the cost of a reduced best case compression ratio, the configuration time can be further enhanced. For instance, by limiting the compression ratio of the 10% of the bitstream containing worst case configuration data to, for example, $\eta'_1 = 2$, while having a compression ratio of $\eta'_2 = 0.4$ for the remaining 90%, the average compression ratio will result in $10\% \cdot \eta'_1 + 90\% \cdot \eta'_2 = 56\%$. The configuration time for this adjusted algorithm is then only $\frac{10\% \cdot \eta'_1}{0.5 \cdot d_{FPGA}} + \frac{90\%}{d_{FPGA}} = 1.3 \cdot t_{opt}$.

These numerical examples point out that looking only at the overall compression ratio will not necessarily enhance the configuration speed. This observation reflects Amdahl's law. As compared to the execution time of an algorithm that cannot be reduced below its sequential fraction, the configuration process cannot be accelerated beyond the configuration interface limit (t_{opt}) regardless of the achieved compression ratio.

3.5.4 Algorithms for Bitstream Compression

In this section, existing data compression algorithms will be analyzed with respect to their suitability for hardware accelerated bitstream decompression. Various modifications to standard algorithms will be discussed in order to fulfill the objectives *compression ratio*, *throughput*, and *resource overhead* at the same time and with high quality. Here, high quality means implementations that achieve compression ratios that can compete with state of the art software programs, such as gzip, implementations that allow throughputs of several hundreds of megabytes, and implementations with small resource requirements. In the following, we will examine configuration ratio issues together with implementation aspects for variations of well known compression algorithms, such as run length encoding, Lempel Ziv, and Huffman encoding.

3.5.4.1 Run Length Encoding

Run length encoding [Sal04] is a simple technique to compress a sequence of identical tokens belonging to a data stream. Thus, run length encoding is a good candidate to compress configuration bitstreams that are sparsely populated with binary `1` values and that offer lots of consecutive identical tokens. The recurrences will be encoded by tuples (τ, l) with τ being the next token emitted by the decompressor and l being the run length denoting the number of times this token is repeated in the original data. For example, the sequence `A B B C C C` is encoded as (`A`,1) (`B`,2) (`C`,3). The data width of a token is $|\tau|$ bit and the amount of bits used to encode the run length is $|l|$ bit. Run lengths larger than $2^{|l|}$ will be encoded by multiple tuples that allow to recompose the original input data. When τ and l are encoded with a fixed amount of bits, the compression ratio η_{RLE} is in the following range ($\widehat{\eta}_{RLE}$ denotes the worst case compression ratio):

$$\frac{|\tau| + |l|}{|\tau| \cdot 2^{|l|}} \leq \eta_{RLE} \leq \frac{|\tau| + |l|}{|\tau|} = \widehat{\eta}_{RLE}. \tag{3.4}$$

The cost to encode a run length tuple is always $|\tau| + |l|$ bits. These bits encode in the best case $|\tau| \cdot 2^{|l|}$ bits and in the worst case only $|\tau|$ bits. Note that for the latter case, the compressed data will be larger than the original input data.

The token bit width $|\tau|$ (and consequently, the amount of different tokens) and the amount of bits to encode the run length $|l|$ have to be determined with care, because when random data is compressed using run length encoding, the compressed data will be larger than the original input data. This is illustrated in Fig. 3.15 on the basis of a probabilistic state machine. The probability that the next token in a random data stream is identical to the present one is $\widetilde{p}(1) = \frac{1}{2^{|\tau|}}$. The probability for longer recurrences i decreases exponentially: $\widetilde{p}(i) = \frac{1}{2^{|\tau| \cdot i}}$. The compression ratio for an infinite random data stream is therefore:

$$\widetilde{\eta}_{RLE} = \sum_{i=1}^{2^{|l|}} \frac{|\tau| + |l|}{|\tau| \cdot i} \cdot \frac{1}{2^{|\tau| \cdot i}}. \tag{3.5}$$

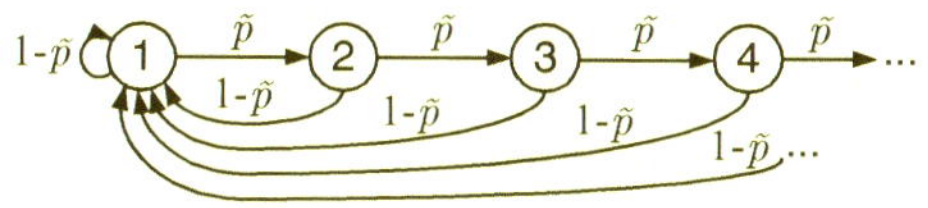

Fig. 3.15 Probabilistic state machine for the compression of random data with run length encoding. The states represent the run lengths and the probability $\tilde{p} = \frac{1}{2^{|\tau|}}$ denotes the chance that the next token is identical

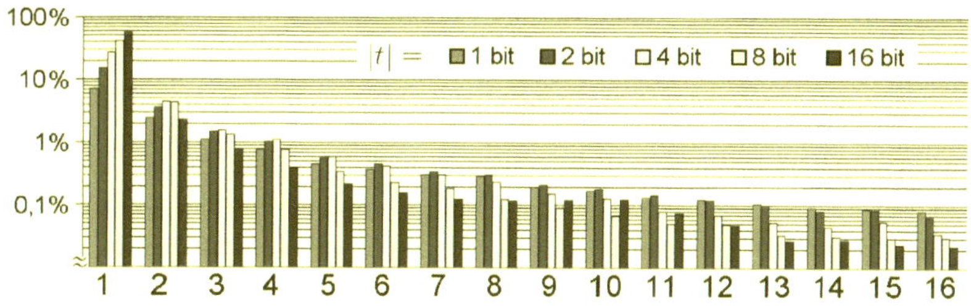

Fig. 3.16 Histogram of the first 16 most occurring run lengths found inside all benchmark bitstreams (see Sect. 3.5.5). The abscissa enumerates classes of run lengths by their occurrence that is not necessary the run length itself. However, class 1 also denotes the occurrence for a run length of 1. The columns quote the portion of a particular run length occurrence class for different token bit widths (1, 2, 4, 8, 16) with respect to the total amount of codewords required to encode the benchmark using RLE

Even for relatively small token bit widths, the probability that just two consecutive words are identical is about zero (e.g., $< 1\%$ for bytes). Therefore, $\widetilde{\eta}_{RLE}$ converges against the worst case compression ratio $\widehat{\eta}_{RLE} = \frac{|\tau|+|l|}{|\tau|}$ for increasing $|\tau|$.

The values of $\widetilde{\eta}_{RLE}$ and $\widehat{\eta}_{RLE}$ indicate the behavior when the bitstream contains fractions of high entropy data. This can happen especially for initialization data of internal RAM blocks. As the relative amount of RAM initialization data is not negligible,[7] the compression algorithms should have an $\widetilde{\eta}$ and an $\widehat{\eta}$ as small as possible, with $1 < \widetilde{\eta} < \widehat{\eta}$.

For parameterizing the run length algorithm, a histogram of the probability distribution of the run lengths for different token bit widths $|\tau|$ of the complete benchmark corpus (see also Sect. 3.5.5.2) has been analyzed.

The histogram in Fig. 3.16 reveals that higher rates of consecutive sequences exist only for small token bit widths. For instance, for a token bit width of $|\tau| = 16$, the probability is 55% that the run length is 1, which means that a sequence of identical tokens contains only one token. Furthermore, the histogram points out the potential for data compression through run length encoding. By adding, for example, the probabilities of the first ten run lengths ($\sum_{i=1}^{10} p(i)$), it can be observed that for all token bit widths except for the width $|\tau| = 16$, the sum of the probabilities is below 50%. Thus, for these token bit widths, the majority of run lengths is larger

[7] According to the data sheets for the chosen benchmark FPGAs [Alt09, Xil07] in Sect. 3.5.5.3, the relative amount of RAM preload data is in average 14% of the overall bitstream size.

than 10, thus, pointing out feasibility for run length compression. However, when using the described encoding, the run length 1 will lead to a relatively large portion of data σ in the compressed data stream that is larger than the original one: $\sigma = \frac{|\tau|+|l|}{|\tau|} \cdot p(1)$. For instance, if the token bit width is set to $|\tau| = 4$ bit, 27% of two consecutive words differ (see class 1 in Fig. 3.16).

In order to overcome these issues, two RLE variations have been examined for hardware accelerated bitstream decompression. The first one works on a bit width of a single bit, while the second one uses a flag to distinguish between the case of a recurrent sequence or the case that a sequence contains only one word.

3.5.4.2 Toggle RLE

If the token bit width is set to $|\tau| = 1$ bit, the run length value specifies the distance between 2 bit flips. By further defining a fixed start bit value, the complete bitstream data is compressible by encoding only run lengths l while omitting the next token τ. By reserving some codes for specifying run lengths without toggling at the end, it is possible to compress recurrent sequences of any length. For encoding the run length, different prefix free encodings have been tested for balancing between the compression ratio and the hardware required to decode the prefix free code words. In addition, the encodings ensure a bounded compression ratio for random data as well as for worst case examples. For an example, the following encoding is assumed:

Run length	1	2	3	4	4[a]	8[a]	16[a]	64[a]
Code	`00`	`01`	`100`	`101`	`1100`	`1101`	`1110`	`1111`

[a] No toggling after this run length

If the start bit value is chosen to `0` then the sequence `1 00 111 000000` is encoded as `00 01 100 1100 01`. In this example, a run length of 1 will double the amount of data required to be transferred from the configuration memory. However, a run length of 1 exists in only 7% of the used benchmark bitstreams as depicted in Fig. 3.16 for class 1 and $|t| = 1$.

As illustrated in Fig. 3.17, the toggle RLE hardware decompressor requires only an input buffer (containing the decoder for the prefix free code), two counters, a control state machine, and a decoder for the run length values. The toggle RLE technique is suitable for FPGA as well as for CPLD implementations. The major drawback of this decompression technique is its sequential operation.

3.5.4.3 Flag RLE

For enhancing the throughput, a variation of the traditional run length encoding scheme was examined that uses a flag to distinguish between the case that consecutive tokens differ and the case of a recurrent sequence of identical tokens

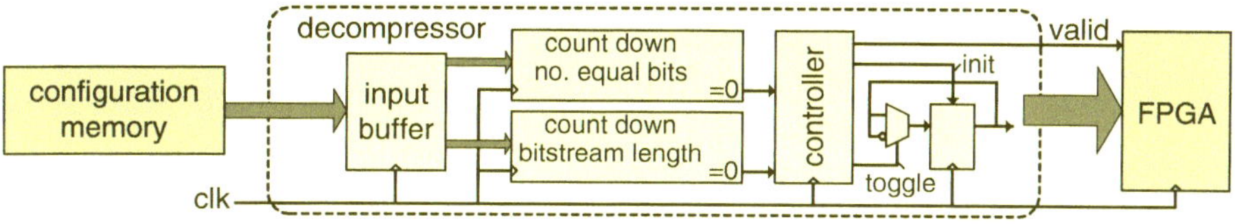

Fig. 3.17 Toggle run length decompressor. At the beginning of a decompression process, the output flip-flop is initialized. Each time when the sequence of equal bits ends, the flip-flop is toggled and the equal bit counter will be loaded with the next run length until the complete bitstream is decompressed

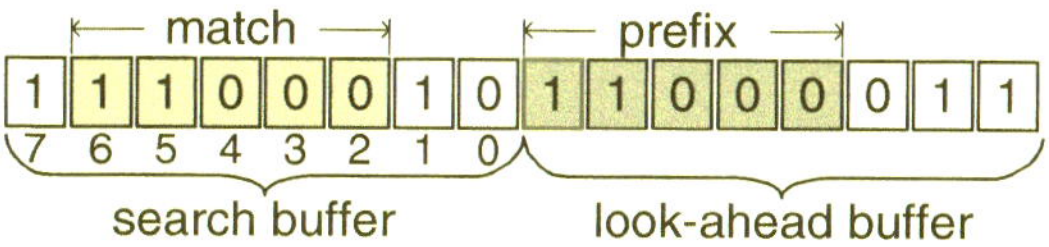

Fig. 3.18 LZSS compression. A search buffer holds already compressed data and a look-ahead buffer contains the next portion of uncompressed data. The example shows a 5 bit prefix of the look-ahead buffer that is also found in the search buffer at position six. This match will be coded as $('1', 6, 5)$ and denotes a reference to the search buffer

(run length equal or larger than one). The corresponding encodings are $('0', \tau)$ and $('1', \tau, l)$. Thus, the worst case compression ratio is $\widehat{\eta}_{RLE} = \frac{|\tau|+1}{|\tau|}$. Note that the flag RLE approach is a special case of a prefix free encoding of the run length l. When running at the same clock frequency, the flag RLE approach possess a $|\tau|$ times higher throughput as compared to a toggle RLE decompressor. The flag RLE algorithm is suitable for FPGA as well as CPLD implementations

3.5.4.4 LZSS Encoding

The LZ algorithms (named after their inventors A. Lempel and J. Ziv) are string substitution schemes. A well known LZ variant is LZ77 [ZL77]. String substitution schemes replace uncompressed data by references to already compressed data. During compression, as shown in Fig. 3.18, a search buffer keeps track of the last already compressed tokens and a look-ahead buffer allows accessing the next tokens of the uncompressed data. When compressing data, the search buffer is scanned for matches of all prefixes within the look-ahead buffer. The longest prefix of the look-ahead buffer, found in the search buffer, is then replaced by its starting *offset* and its *length*. If no prefix is found, the first token (*nextToken*) of the look-ahead buffer will be emitted.

The LZSS approach [SS82], as an improvement of LZ77, that uses a flag to distinguish between these two cases (that a match was found in the search buffer or not). This has similarities with the flag RLE methodology presented in the last section. Let o, l, and τ denote the *offset*, the prefix *length*, and the *nextToken* and $|o|$, $|l|$, and $|\tau|$ the number of bits to code these values. Then, during LZSS compression,

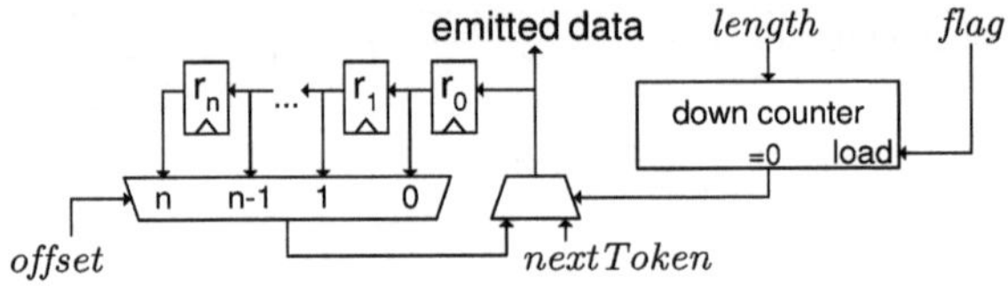

When assuming the following LZSS encoding with $|\tau| = 8$, $|o| = 5$, $|l| = 3$ for encoding some ASCII characters:

$$(0,\underbrace{01000001}_{nextToken\,\mathtt{A}})\ (0,\underbrace{01000010}_{nextToken\,\mathtt{B}})\ (1,\underbrace{00001}_{offset},\ \underbrace{010}_{length^{*}})\ (0,\underbrace{01000011}_{nextToken\,\mathtt{C}})$$

The decompressed sequence of this encoding is: A B A B C

*In the final implementations, the shortest length is 2 (encoded as 000) because a length of 0 is futile and a length 1 is encoded with a new token.

Fig. 3.19 LZ77 HW decompressor proposed in [HW99], and an accord example for LZSS decompression

either the codeword $('1', o, l)$ or $('0', \tau)$ is emitted (see also Fig. 3.19). The best compression ratio is achieved if the longest possible prefix match of size $l_{\max} = 2^{|l|}$ is found. In this case, $l_{\max} \cdot |\tau|$ bits are encoded with $1 + |o| + |l|$ bits. If no match was found, a token of $|\tau|$ bits is encoded by $1 + |\tau|$ bits. This leads to the following lower and upper bound for the compression rate η of LZSS:

$$\frac{1 + |o| + |l|}{l \cdot |\tau|} \leq \eta_{LZSS} \leq \frac{1 + |\tau|}{|\tau|} = \widehat{\eta}_{LZSS} \tag{3.6}$$

The upper bound is equal to the upper bound of the flag RLE algorithm. Thus, LZSS does not lead to a drastic downgrading of the compression ratio if fractions of the configuration bitstream contain data with a high entropy.

3.5.4.5 LZSS Hardware Decompression

During decompression, offset and length refer to already emitted data. Thus, the decompressor must keep track of the last n emitted tokens, with n being the search buffer size of the compression program. For the LZSS decompression, an existing LZ77 hardware decompressor [HW99] was enhanced with a state machine that decodes LZSS encodings as illustrated in Fig. 3.19. During decompression, the emitted data is fed into a shift register that represents the mentioned search buffer. If there is no match in the shift register, the next token *nextToken* will be emitted. In case of a match, the offset field o specifies the position of the shift register containing the first token of an l token long match.

As CPLDs only provide a relatively small amount of memory bits, LZ decompressors are inapplicable for CPLD implementations, because of the required shift register. For implementing the shift register on FPGAs, the following four options exist:

1. Look-up table flip-flops (the flip-flops available behind each look-up table),
2. Frame Data Input Register (FDRI) provided in Xilinx FPGAs [Inc07] (this register stores the configuration data of a complete frame and is only accessible via shift operations),
3. Dedicated RAM-Blocks, and
4. Dedicated shift register primitives (SRL16) in Xilinx FPGAs [Inc05a].

The first solution consumes an unfavorable amount of hardware resources. The second one is advantageous when implementing hardware decompression on the FPGA itself as a fixed part of the chip. Accessing data stored inside the FDRI is very slow when using the configuration interface. However, [PMW04] and [LH01] present compression techniques that are based on the FDRI, but both papers omit a discussion of a possible hardware decompressor implementation.

The third variant of implementing the LZ shift register uses dedicated on-chip RAM resources and allows implementing very large shift registers. However, as huge shift registers require more bits $|o|$ for specifying the offset o, no benefits have been found by using larger shift registers. By analyzing how circuits are typically mapped onto an FPGA, many regularities can be found within the resulting configuration primitives and consequently within the corresponding configuration data. For example, many arithmetic functions use carry chains, hence, resulting in identical configurations among a set of consecutive basic elements. In the case of Xilinx FPGAs, the basic elements of a carry chain are vertically aligned; and the configuration for the basic elements is stored directly adjacent within the bitstream. Consequently, short shift registers perform better on LZSS for decompressing configuration bitstreams for Xilinx FPGAs than larger ones.

Instead of using dedicated RAM blocks, the last option for implementing the LZ shift register with the help of dedicated shift register primitives is most advantageous for bitstream decompression. It was discovered that even small search buffers, as little as 32 tokens, result in good compression ratios. Such shift registers are appropriate to be implemented with the help of *SRL16* primitives on Xilinx FPGAs. This primitive allows to use a 4-bit look-up table alternatively as a 16 bit shift register with random access to the individual register contents. An l deep shift register of token bit width $|\tau|$ can be created by cascading $|\tau| \cdot \lceil \frac{l}{16} \rceil$ SRL16 primitives.

The best parameter set found for LZSS encoding was 1 bit for the flag, $|l| = 3$ bit, $|o| = 5$ bit, and $|\tau| = 8$ bit. This results in a 9 bit long codeword which does not match a typical 8-bit interface. To simplify the connection to 8-bit wide flash memory interfaces, an extra register is used for the flags. This register stores the flags for a block of eight consecutive codewords, thus, leading to 8-bit wide remaining codewords.

Recent FPGAs offer high speed configuration interfaces that are up to 32-bit wide and that operate with up to 100 MHz [Alt09, Xil07]. In order to match these interfaces, the token bit width can be extended to 32 bits. This would influence the compression as follows: With an increased token bit width, long prefix lengths become more rarely. Therefore, the subset of encoded prefix lengths has to be adapted. In addition, the worst case compression rate improves with higher word widths. However, it is also possible to enhance the throughput by increasing the clock frequency of the decompressor in combination with a serial-to-parallel converter at the output. The LZSS decompressor consists of a few simple basis elements and the whole design can be easily pipelined. Consequently, we can easily build decompression modules that run at least at 200 MHz (e.g., when implemented on a Xilinx Virtex-II device). This speed in combination with a datapath width of 16 bit allows decompressing at a rate of 400 MB/s. Thus, the full configuration interface bandwidth of today's FPGAs can be utilized under optimal conditions.

3.5.4.6 Huffman Encoding

As an entropy encoding scheme, Huffman encoding [Huf52] assigns short code-words to frequently occurring tokens and longer code words to infrequently occurring tokens. Huffman encoding is based on a Huffman tree, which in turn is created upon a probability distribution of tokens. Creating an individual Huffman tree for each bitstream file is called *dynamic Huffman encoding*. Such a Huffman tree is part of the compressed data and has to be reconstructed before decompression. Such decompressors turn out to be costly in terms of hardware usage [WL05].

Applying the same Huffman tree for all bitstreams is called *static Huffman encoding*. The static Huffman tree is based on the probability distribution of a benchmark corpus and will never change. The higher the variation in the probability distribution, the better the compression ratio achievable with Huffman encoding.

Let $|h_{\min}|$ and $|h_{\max}|$ denoting the smallest and the largest Huffman codeword sizes found inside a Huffman tree and $|\tau|$ be the token bit word size; then, the compression ratio for Huffman encoding is:

$$\frac{|h_{\min}|}{|\tau|} \leq \eta_{Huff} \leq \frac{|h_{\max}|}{|\tau|} = \widehat{\eta}_{Huff}$$

If the decompression ratio has to be bounded in such a way that fractions of high entropy data will not completely cut off the throughput, the Huffman code has to be restricted. For example, if the worst case compression ratio $\widehat{\eta}_{Huff}$ has to be restricted to, for example, less than 2, the largest code word size has to be limited to $|h_{\max}| \leq 2 \cdot |\tau|$. In the case $\widehat{\eta}_{Huff}$ is larger than 2, the depth of the Huffman tree has to be reduced, as illustrated in Fig. 3.20.

Reducing a Huffman tree does not necessarily reduce the maximum configuration speed. For example, in the histogram in Fig. 3.21 that depicts the probability of occurrence for the most frequently occurring tokens within all benchmark bitstreams, it can be seen that for a word size of $\tau = |8|$ bit there is one token occurring with a probability of 58%. Consequently, this token will be encoded in a classical Huffman encoding with 1 bit. However, when assuming that the configuration memory interface can deliver 50% of the maximum configuration data rate, up to 4 bits can be spent for the most occurring token (that possesses a word size of 8 bits), before the configuration speed drops down. In other words,

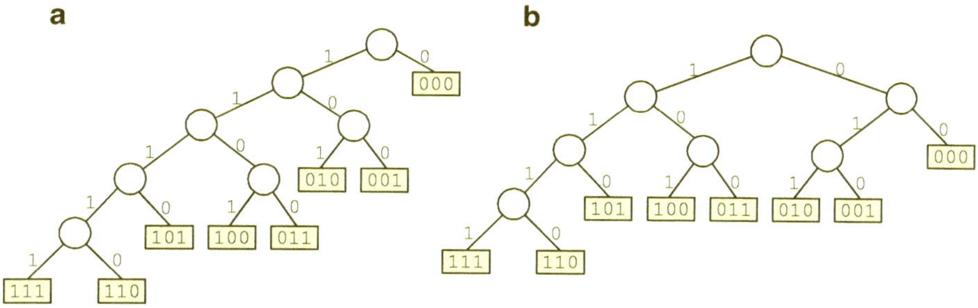

Fig. 3.20 (**a**) Huffman tree and (**b**) a reduced Huffman tree. Both examples are for the same distribution of 3-bit wide tokens (the eight leaves of the tree). The reduced Huffman tree has a better worst case behavior, which means that in the longest encoding of a token requires less bits (compare the encodings for the values `111` and `110` in (**a**) and (**b**)). This reduces the peak input data rate of the decompressor by the cost of an inferior average compression ratio. This can be applied to speed up the configuration process

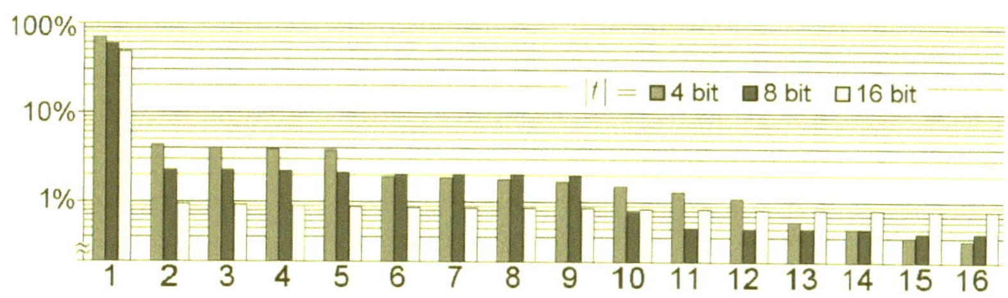

Fig. 3.21 Normalized histogram of the probability of occurrences for the first 16 most frequently appearing tokens found inside all benchmark bitstreams. The abscissa enumerates tokens by their probability of occurrence. Note that in the case of a token bit width of $|\tau| = 4$, the sum over all listed occurrences is 100%; because in this case, only $2^{|\tau|} = 2^4 = 16$ different tokens exist

before this point, the maximum throughput of the configuration interface will limit the configuration speed. By reducing the height of the Huffman tree, the average compression ratio will be decreased. However, this will still enhance the configuration speed, because the longest code word size has been reduced.

3.5.4.7 Huffman Hardware Decompression

For static Huffman trees, a code generator was implemented that can automatically create the hardware for Huffman decompressors. An example of such a decompressor is illustrated in Fig. 3.22. When in this example a new word is decoded upon enabling the `start` input, the values in the so-called *alignment register* (flip-flops $r_0, \ldots, r_2$) will activate exactly one output of the decoder tree. As the words of the original bitstream data are encoded into codewords of different sizes $h(i)$, the alignment register is used to arrange the incoming data d, such that the Huffman tree decoder gets its incoming codeword always at exactly the same

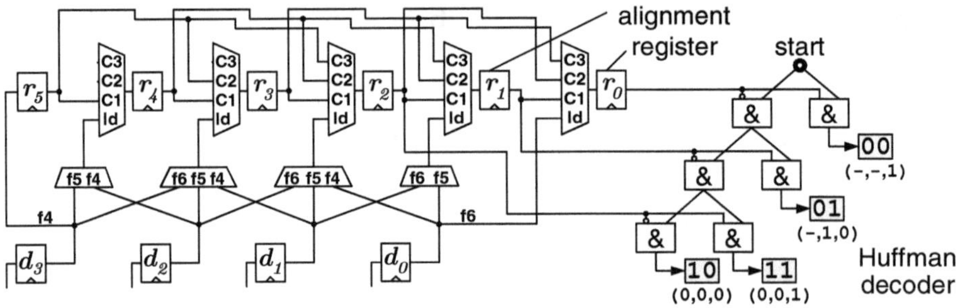

Fig. 3.22 Huffman decompressor. A Huffman tree decoder decodes a new output word of the alphabet {00, 01, 10, 11} ($|t| = 2$) upon start=1. The values in brackets denote the compressed encodings. The multiplexers in front of the *alignment register* $\boldsymbol{r}$ shift the compressed bitstream data according to the size of the current code word $h(i)$ such that the next code word is located again from the right inside r. The bottom multiplexers load new compressed input data to r according to the current fill level

position. The outputs of the Huffman tree decoder indicate the decoded token of the alphabet {00,01,10,11} that is emitted in the following. Concurrently, the alignment register is shifted right by the length $|h(i)|$ of the encoded token. For instance, if the output for token 01 becomes active, the alignment register is shifted by two, which is selected by the multiplexer input C2.

A state machine (not shown in Fig. 3.22) tracks the fill level of the alignment register. When there is sufficient space in the alignment register, the next value from the input $d_0, \ldots, d_3$ will be loaded via the bottom multiplexers to the most right free position. For instance, in a scenario where $r_3, \ldots, r_5$ are empty (not containing a codeword) and where r_0 =0 and r_1 =1, the output 01 of the Huffman tree decoder will become active. Consequently, r_0 and r_1 will be consumed and the last valid fragment of the compressed bitstream stored in r_2 will be shifted into r_0. This will result in five free entries in the alignment register and a new input will be stored in the registers $r_1, \ldots, r_4$ which is selected by the multiplexer input f5.

The size of the alignment register $|r|$ depends on the input data width $|d|$ and the Huffman tree depth which is equal to $|h_{\max}|$. If in the case that the alignment register is missing 1 bit to decode the next token, it must be possible to load the alignment register with a new input word for preventing a deadlock. Thus, the required alignment register size is $|r| \geq |d| + |h_{\max}| - 1$ flip-flops.

The decompressor example in Fig. 3.22 points out that it requires more logic for aligning the input data than for the Huffman decoder itself. The logic overhead for the multiplexers is enormous for larger trees. For each occurrence of a Huffman codeword width, one input for the top multiplexers as well as one input for the bottom ones adjusting the input data is required. In addition, the structure contains long combinatorial paths and is difficult to pipeline, because only when a decode phase has determined the next output token, it is known from which position in the alignment register the next decode phase will start. However, as shown in the following section, static Huffman encoding reaches better compression ratios than the RLE or LZSS variants.

Fig. 3.23 Structure of a configuration bitstream, a software binary, and an ASCII file. Inside the images, a black pixel represents a `'1'`-bit (from [Bec07])

3.5.5 Benchmarking Bitstream Decompression Accelerators

All compression algorithms exploit statistical characteristics within the uncompressed data like non-uniform probability distributions of tokens or some regularities found in the input data. In the case of FPGA configuration bitstreams, such characteristics include sparsely populated configuration data (more `'0'`-bits than `'1'`-bits) and recurrences due to regularities in the mapped circuits. Such regularities can be visualized, as shown in Fig. 3.23.

3.5.5.1 Experimental Setup

For studying bitstream compression algorithms, suitable benchmark bitstreams are required that represent the statistical characteristics for a wide range of different FPGA designs. Most FPGA designs will hold a high logic utilization for allowing smaller FPGAs.[8] This significantly minimizes overheads like monetary cost, static power consumption, or configuration time. Consequently, suitable benchmark bitstreams are requested to hold a logic utilization of at least 90% of the available look-up table resources. In order to detect influences on the compression ratios that are related to the implemented algorithm of a module and not only to the logic utilization, examples have been selected from different application domains.

[8] As the routing fabric of an FPGA requires more area on a die than the logic (mainly the look-up tables), it is advantageous to delimit the routing capabilities. This will result in a lower logic utilization but still allows mapping circuits to FPGAs possessing an on average smaller die [Deh99]. However, this observation does not consider that the place and the route steps are separately performed in the tools of the main FPGA vendors. In addition, customers demand high logic utilization, hence driving FPGA companies to include sufficient routing capabilities in their architectures.

This includes cryptographic accelerators, signal processing cores, communication modules, and an SoC-design. More detailed information on the benchmark can be found in [KBT07].

3.5.5.2 Evaluating Compression Ratios

Figure 3.24 lists measured compression ratios for multiple decompressor variations. As a reference, the figure lists the compression ratios achieved with the program gzip (V 1.3.5 --best) and for two FPGA vendor-specific dedicated configuration memory devices [xcf07, epc07] that both perform bitstream decompression. Beside the different bitstreams of the benchmark modules, the average value over all bitstreams (AV) is quoted. For the toggle run length encoding, a decompressor with fixed length encoding (T-RLE) as well as with a prefix free encoding (T-RLE*) was examined. The two LZSS alternatives differ by their data path size (8/16 bit). The word size of the flag RLE and the Huffman decompressor was selected to 8 bit.

For the LZSS accelerators, it was found that most data inside the compressed stream is spent on *nextTokens* (see also Fig. 3.19). Consequently, hybrid accelerators have been tested that use beside the LZSS decompressor an additional Huffman decompressor for the *nextTokens*. In order to keep the logic overhead in acceptable margins, two variants of Huffman decompressors with 4 bit wide words ($|\tau| = 4$) have been used. The first one (named LZSS+Huff2 in Fig. 3.24) is based on a binary Huffman tree, while the second one (LZSS+Huff4) is based on a quaternary Huffman tree [KBT09b]. The latter one achieves an on average 1.2% inferior compression ratio, but with 436 look-up tables, requires only 63% of the 693 LUTs necessary to implement the accelerator based on binary Huffman decoders.

The best compression ratios have been measured for the Huffman decompressor that in some cases outperforms gzip. Huffman performs good because bitstreams are sparsely populated with `'1'`-bits, hence, leading to a strong non-uniform distribution of the code words. For the Altera Cyclone-II benchmark bitstreams, the flag RLE (F-RLE) ranges always and on average 6% better than the dedicated configuration device EPC16 (available from Altera). For Xilinx Virtex-II and Virtex-5 bitsteams, the toggle RLE using a simple prefix free encoding (T-RLE*) ranges with one exception (the FIR filter on Virtex-II FPGAs) always better than the XCF16P configuration device from Xilinx. The LZSS cannot compete with the RLE algorithms for the Cyclone-II benchmark, but it is the most advantageous one for decompressing configuration data for Xilinx FPGAs, because of its high speed and its low implementation cost as revealed in the following. It is assumed that the configuration data within Cyclone-II FPGAs and the logic tiles are not directly correlated. For example, this happens if the configuration is written row by row to the FPGA while circuits are preferentially mapped in a column-wise manner (e.g., when using carry chains).

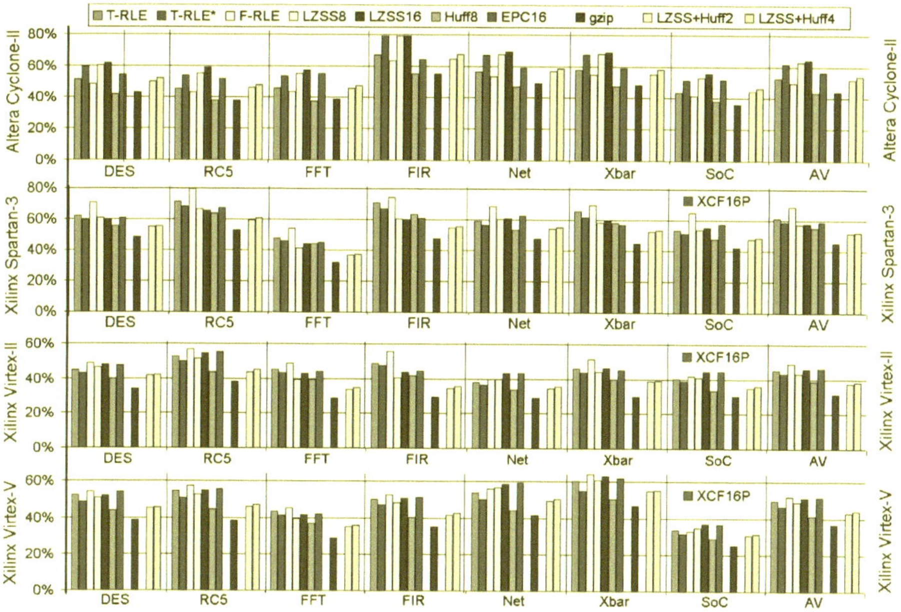

Fig. 3.24 Achieved compression ratios for the benchmark bitstreams

3.5.5.3 Synthesis Results for the Bitstream Decompressors

The different decompression accelerators have been implemented and the synthesis results are listed in Table 3.1. The table reveals that less than a 100 look-up tables are sufficient to achieve average compression ratios of 50% on an ESM [MTAB07] (see also Fig. 3.24). For instance, the LZSS8 accelerator costs only 83 LUTs (0.14% of the XC2V6000 main FPGA) but may decompress data with double the peak data rate of that of a Virtex-II configuration interface. By doubling the size of the data path, the LZSS16 is capable to decompress configuration data at 400 MB/s, which is in turn the peak configuration speed of Xilinx Virtex-4 / 5 FPGAs.

The best hardware compression ratios have been reached with Huffman encoding that indeed turns out to become much more costly than the other accelerators. According to the data sheets [xcf07, epc07], both dedicated configuration devices, the EPC16 and the XCF16P, provide a maximum output data rate of up to 40 MB/s, which is an order of magnitude below the LZSS16. Assuming that the input stream arrives fast enough at the decompression modules, all accelerators are capable to emit one token of $|\tau|$ bit per clock cycle.

3.5.5.4 Configuration Time Improvements

As discussed in Sect. 3.5.3.1, the maximum clock frequency and the achieved compression ratio will not automatically deliver the minimum reconfiguration time because the decompressor may be stalled on missing input data. This impact of

Table 3.1 Synthesis results of the RLE and LZSS decompression accelerators

	T-RLE$_{FPGA}$[a]	T-RLE$_{CPLD}$[b]	F-RLE[a]	LZSS8[a]	LZSS16[a]
LUTs/MCs	111	57	85	83	120
F_{max}[MHz]	165	67	193	198	200

Implemented on:
[a] Xilinx XC2V40-5 FPGA,
[b] Altera EPM7064-10 CPLD

Synthesis results for the static Huffman decompressors

Huffman	Cyclone	Spartan-III	Virtex-II	Virtex-5
LUTs	4,451	4,223	5,286	4,114
F_{max}[MHz]	50	60	45	66

All timing results are based on the slowest available speed-grade.

the memory bandwidth on the configuration speed has been measured for the LZSS decompression accelerator that is best suitable for Xilinx FPGAs. The exact decompression time was determined with the help of RTL simulation. Furthermore, the impact of a FIFO between the memory and the decompressor has been investigated for different FIFO sizes. During simulation, accesses to the memory was restricted by allowing a read operation only every γ cycles. For instance, $\gamma = 3$ denotes that the peak throughput rate of the decompressor is three times higher than the memory data rate. The optimal time is bounded by the achieved compression ratio or limited by the configuration interface throughput in the case when $\eta \leq \frac{1}{\gamma}$.

As can be seen in Table 3.2, the optimal reconfiguration times cannot be reached, but still the configuration speed was significantly enhanced by balancing compression ratio and data rates between the configuration memory and configuration interface. Furthermore, the table points out that the FIFO size does not have much impact on the final configuration time. Experiments with larger FIFOs have not demonstrated a significant further improvement of the configuration speed.

3.5.6 *High-speed Defragmentation*

As shown throughout the last sections, runtime reconfiguration can be accomplished fast at high configuration data rates. However, if modules are loaded and replaced at runtime, the reconfigurable area may get fragmented, which means that not all resource slots form a continuous area. This is known as *external fragmentation*. If assuming that a module requires resource slots as one complete consecutive area, the number of free slots alone is not sufficient to decide whether a module can be placed. By rearranging the module layout among the resource slots, the area with free consecutive resource slots can be increased, as illustrated in Fig. 3.25. The example shows that module m_2 has to be moved before placing m_4. As the defragmentation arranges the module layout into larger blocks of continuous free slots and continuous occupied slots, this process is also called *compaction* [FKS+08]. Dependent on

Table 3.2 Summarized decompression times for various FIFO sizes and diverse memory bandwidths $\gamma = \frac{d_{FPGA}}{d_{MEM}}$. The table lists the relative time compared to a non compressed case. Italic values denote that the optimal speed-up is limited by the maximum decompressor output data rate and not by the compression rate

			FIFO size in byte				
	γ	opt.	1	4	8	16	128
Altera Cyclone-II	2	63.1	78.1	74.1	69.6	67.1	65.0
	3	63.1	71.5	67.9	65.3	64.4	63.7
	4	63.1	67.5	65.3	64.1	63.7	63.2
	5	63.1	65.7	64.1	63.6	63.4	63.1
Xilinx Virtex-II	2	*50.0*	67.5	63.7	62.3	61.5	59.5
	3	43.5	60.5	54.7	53.7	53.1	51.9
	4	43.5	51.8	50.8	50.2	49.8	49.0
	5	43.5	49.2	48.6	48.3	48.1	47.4
Xilinx Spartan-3	2	59.1	75.8	72.9	71.6	70.9	69.2
	3	59.1	69.4	66.5	65.7	65.3	64.2
	4	59.1	64.6	63.7	63.2	62.9	62.3
	5	59.1	62.7	62.2	61.8	61.6	61.3
Xilinx Virtex-5	2	*50.0*	69.4	65.6	64.2	63.5	62.1
	3	49.3	58.8	58.2	57.5	57.0	56.3
	4	49.3	56.1	55.1	54.7	54.4	53.9
	5	49.3	54.0	53.5	53.1	53.0	52.6

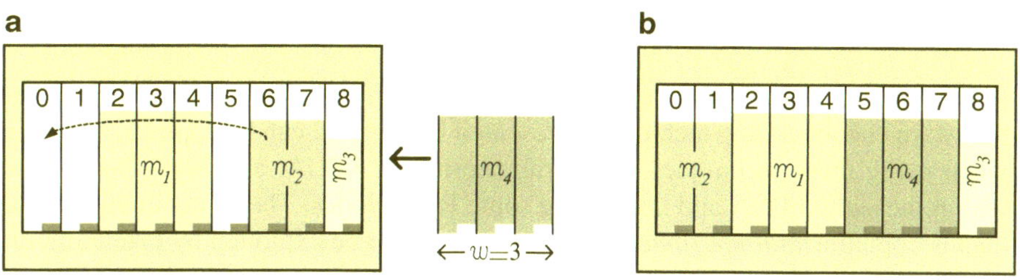

Fig. 3.25 (**a**) Scenario where module m_4 cannot be placed despite that there are sufficient free resource slots available because of the free slots are not continuous (external fragmentation); (**b**) By moving module m_2 to slot 0, the reconfigurable area is defragmented and m_4 can be placed

the entire task of a moved module, there are different scenarios possible of how the defragmentation influences the execution of a particular module:

1. Restarting with a new iteration on an additional instance of m_2 allows to perform the defragmentation without interrupting the task execution. This is only applicable if there are no data dependencies between different iterations of m_2. The lane detection in the ReCoNets demonstrator, for example, as shown in Fig. 4.17, can be restarted at another placement position after accomplishing the segmentation within a full frame of the input video stream.

2. Transferring the internal state to a new instance of m_2 before its restart is required when m_2 demands some results from preceding iterations. This means that m_2 contains an iteration dependent internal state. Note that the module reconfiguration can be carried out before the state transfer for continuing execution in parallel when the new instance of m_2 is reconfigured at the new placement position.
3. If the module to be moved cannot be fully placed to the new position, one or more modules (1) have to be stopped, (2) their internal state has to be captured, then, (3) the modules can be placed to the new position, and finally, (4) can be restarted after restoring their internal state. For example, instead of moving m_2 in the example in Fig. 3.25 a, it is alternatively possible to move m_1 either one slot to the right or two slots to the left for placing m_4. In both cases, module m_1 has to stop execution as it must be partially overwritten with its displaced instance.

In the first scenario, no internal state of a moved module is considered. This restricted model has been examined by Platzner [WP02] for two-dimensional module placement where modules of any arbitrary shape (not necessary rectangular) are replaced only after task completion. In the last two scenarios, a module is preempted at one position and resumed somewhere else for enlarging the continuous free resource area. The scenarios point out that defragmenting an FPGA module layout can comprise either an overhead in terms of free resource slots (*space*), in terms of stopping module execution (*time*), or a mixture of space and time.

3.5.6.1 Hardware Supported FPGA Defragmentaion

If a hardware module has to be stopped, state extracted, and resumed later, the hardware checkpointing techniques presented in Sect. 4.2 can be applied. However, when using these techniques for defragmenting an FPGA, a module is migrated within the same FPGA and not among some ReCoNodes. Therefore, configuration data is copied *within the same FPGA* only. This can be exploited by extending the FPGA configuration logic as it is possible to transfer data on-chip at rates that are orders of magnitudes beyond of-chip communication via a configuration port.

Brebner and Diessel proposed in [BD01] for a one-dimensional slot-based reconfigurable system to arrange all configuration SRAM cells as shift registers such that a full configuration can be shifted further by one logic column within a single clock cycle. The shifting can be individually masked for each logic column, such that in the case that a module is terminated all succeeding slots can be shifted towards the next used preceding slot. This allows compacting the reconfigurable area within w^{BE} clock cycles after terminating a module being w^{BE} logic columns wide (in terms of basic elements; see also Sect. 2.3). The main restriction of this approach is that placed modules are obstacles for the shift process, hence, not allowing inhomogeneous resources such as memory columns to be located within the reconfigurable area.

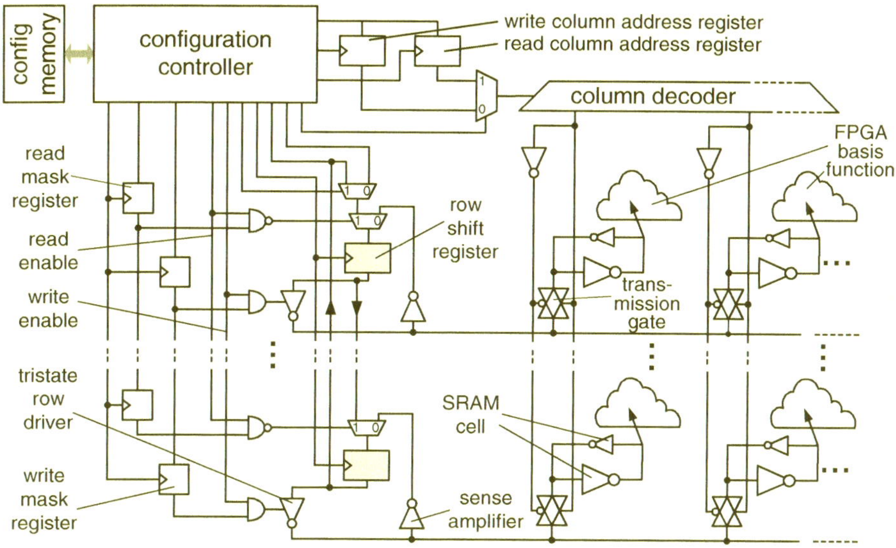

Fig. 3.26 Extended configuration logic for hardware supported module defragmentation. A maskable subcolumn can be selectively copied into the row shift register, then ring-shifted, and selectively written to another maskable subcolumn. Note that one mask register pair can control access to all configuration SRAM cells within a row of logic basic elements

For allowing obstacles when compacting the module layout, Compton et al. suggested to move modules via a shadow register capable to store one full logic column [CLC$^+$02]. In this approach, a module is moved by reading the first logic column of a module from its original position to a shadow register and writing it from there to the new position. This process takes two clock cycles per logic column and is repeated for all other logic columns until the complete module is moved. This technique requires only small changes to the current configuration logic of Xilinx FPGAs that already contains shadow registers. However, in Xilinx FPGAs there exist two independent registers, one for configuring the fabric and a separate one for reading back configuration data. Furthermore, the shadow registers can only hold a frame that contains a subset of the configuration bits ($\frac{1}{21}$ in the case of a Virtex-II device) required to configure a full logic column. Consequently, moving a module would take the amount of frames per logic column times longer on a Xilinx FPGA, if the two shadow registers will be merged.

Both the module shifting [BD01] and the shadow register approach [CLC$^+$02] are limited to compact a one-dimensional module layout with slots spanning the entire height of an FPGA. For supporting two-dimensional grid style reconfigurable systems, an according enhancement was developed that also uses a shadow register. But, as compared to [CLC$^+$02] access to the shadow register can be masked and the register can be ring-shifted as shown in Fig. 3.26.

The mask registers allow to control which particular part of a logic column will be accessed. By reserving one pair of mask flip-flops per row of logic elements for masking read and write operations, modules of any size can be placed freely on the FPGA fabric. In Xilinx Virtex-4 and Virtex-5 FPGAs, the device can also be configured in subcolumns rather than in full logic columns, but the granularity is coarser grained with a height of 16 CLBs (10 CLBs) in the case of Virtex-4 (Virtex-5) devices. This leads to internal fragmentation when the subcolumn-wise reconfiguration scheme is exploited for two-dimensional grid based systems, because in general, a system may not perfectly fit this coarser granularity.

For configuring the FPGA fabric with a new module, the write mask register chain is updated with the according row write mask and the configuration data is shifted through the row shift register to its target position. The configuration bitstream contains only data for the masked region and the configuration of the different subcolumns (or subframes) is concatenated to one module string. By activating all mask bits, the device configuration would follow the manner provided in Xilinx Virtex-II FPGAs.

Figure 3.27 exemplarily reveals how a maskable row shift register chain is used for defragmenting an FPGA. By firstly loading a particular module column to the register and then shifting this data to the target row position, a module can be moved completely two-dimensional within the FPGA. Because of the high on-chip bandwidth, this process can be carried out quickly. With Δ_{AB} denoting the vertical displacement and w^{BE} and h^{BE} stating the width and height of a module in basic elements, the amount of clock cycles for moving this module is:

$$t_{defrag} = \begin{cases} (2 + \Delta_{AB}) \cdot w^{BE} & if\ \Delta_{AB} < h^{BE} \\ 2 + w^{BE} \cdot h^{BE} + (\Delta_{AB} - h^{BE}) + w^{BE} & else \end{cases} \qquad (3.7)$$

When applying this technique to the column wise reconfigurable Virtex-II-6000 FPGA of the ESM, the width w^{BE} has to be multiplied with the number of frames per logic column which is 21 for this device. For a quantitative example related to this platform, a module of 3,000 look-up tables is assumed, what corresponds to the logic of an average 32-bit soft-core CPU. When such a module is located in the top half of the device, it will occupy a block of $w^{BE} \times h^{BE} = 8 \times 48$ consecutive configurable logic blocks (CLB). According to (3.7), moving this module to another position in the bottom half of the device ($\Delta = 48$) requires $2+8\cdot21\cdot48+(48-48)+8\cdot21 = 8{,}234$ clock cycles. At 100 MHz, this corresponds to 82.34 μs for the complete module move. The total amount of configuration data transferred in this example is 80,640 byte. For the assumed 100 MHz this results in a data rate of 979 MB/s. For this high transfer rate, it requires 80 flip-flops within the row shift register per CLB row and according 80 vertical wires downwards and upwards for implementing the ring-shifting.

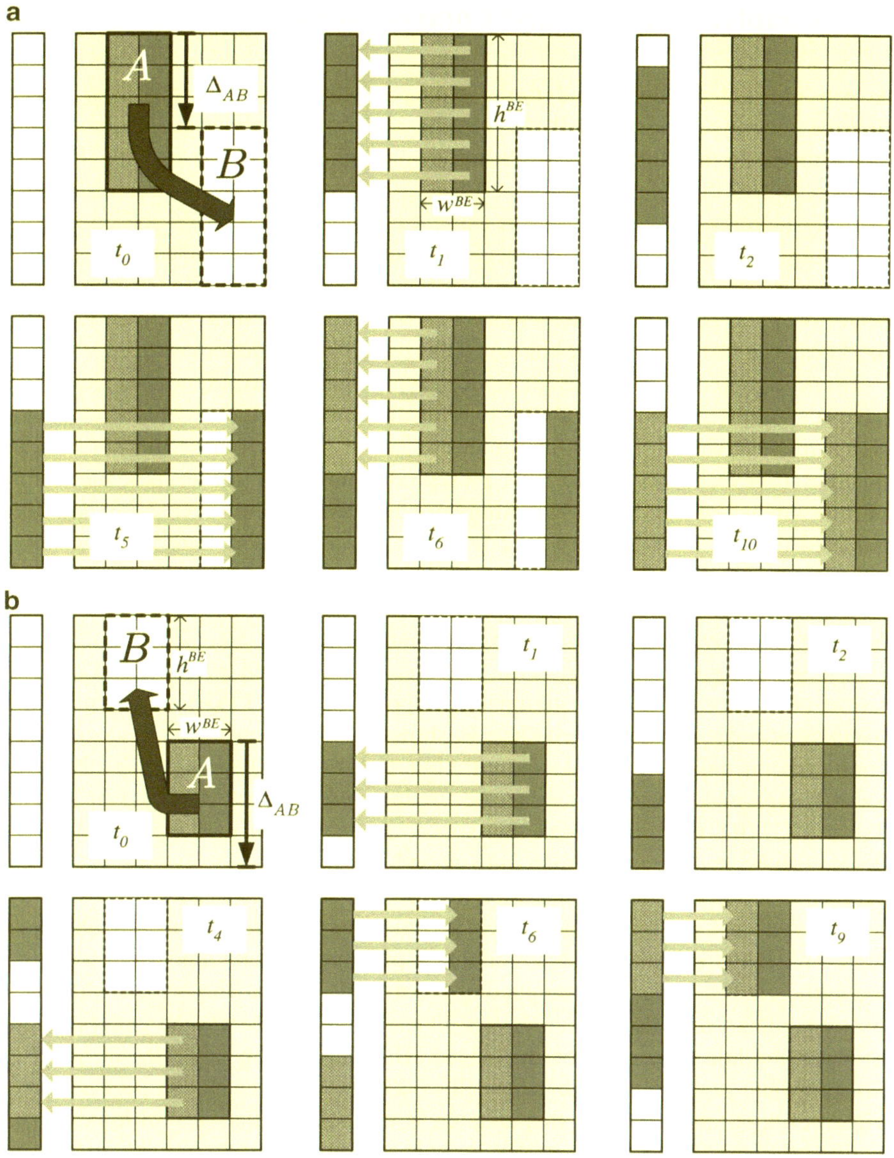

Fig. 3.27 Shift register-based defragmentation. (**a**) If some rows overlap ($\Delta_{AB} < h^{BE}$), it requires $(2 + \Delta_{AB}) \cdot w^{BE}$ clock cycles to move the complete module to the new position, because reading and writing the row shift register cannot be pipelined as the column decoder is only allowed to activate one column at a point of time (see also Fig. 3.26). (**b**) In the case that the source and the destination have no overlapping rows, a module can be moved within $2 + w^{BE} \cdot h^{BE} + (\Delta_{AB} - h^{BE})$ clock cycles. This holds only true if there is no simultaneous read and write operation to the row shift register required. In (**a**), the read mask is set to `11111000` and the write mask to `00011111`. In (**b**), the according settings are `00001110` for the read mask and `11100000` for write mask register

3.6 Chapter Summary and Contributions

A gap in methodologies and tools is one of the major reasons why partial runtime reconfiguration is still very exotic and only rarely used in commercial applications. In this chapter, a novel framework called ReCoBus-Builder [KBT08] was presented that closes this gap. Based on new and efficient realizations of partially reconfigurable bus-systems using the proposed ReCoBus and connection bar communication architecture, the framework assists in specifying the system, simulating reconfiguration processes, partitioning of the FPGA resources and performing the physical implementation of the static system as well as all partial modules. The last step includes in particular the bitstream generation.

In addition to the offline design aspects, sophisticated reconfiguration techniques have been implemented. This allows to relocate modules arbitrarily in a two-dimensional manner within a reconfigurable area on an FPGA. Furthermore, bitstream manipulation techniques have been presented. With these techniques, the setting of the switches of the circuit-switching network, presented in Sect. 2.7.5, had been adjusted directly within the routing fabric of an FPGA. The here presented two-dimensional circuit-switching approach is the first one that is fully carried out within the switch matrices and that is embedded in the modules without constituting any obstacle that may restrict the module placement.

For speeding-up reconfiguration process, configuration prefetching, configuration port overclocking, bitstream decompression, and FPGA defragmentation techniques have been revealed. It was shown that it is possible to decompress configuration data at the full configuration speed of 400 MB/s, achievable with today's high-end FPGAs, whereas the implementation cost is only 120 look-up Tables [KBT07, KBT09b]. With this speed, a one million gate equivalent FPGA design can be exchanged within a millisecond.

When partially reconfigurable modules are loaded and terminated, the reconfigurable area may get fragmented, such that the free resources form no continuous area. This can prevent to place a module despite that the total amount of resources are sufficient. However, by compacting the module layout (the present placement position of the modules), a larger free consecutive space may be generated and a larger module may now being placed. But performing compaction takes an inappropriate amount of time when all configuration data has to be transferred via an configuration port. In this chapter, it was shown that very little additional logic within the FPGA configuration logic is sufficient for substantually enhancing defragmentation speed by moving modules directly by the configuration logic within the FPGA [KBAT07].

Chapter 4
Self-adaptive Reconfigurable Networks

Connecting several control units via a network is becoming important for a rising number of complex embedded systems. In such systems, each control unit is specialized to execute certain functionality. With the introduction of reconfigurable hardware (commonly FPGAs) as a centric part of the control units, a new dimension of flexibility can be implemented with the help of *runtime reconfiguration*. On a *node level*, this includes reconfiguring hardware and software tasks on a single control unit, called *ReCoNode* in the following, while the reconfiguration may also involve a *network level* where tasks are *migrated* among different ReCoNodes of a so-called *ReCoNet*. An example of a ReCoNet is depicted in Fig. 4.1a. The capability of reconfiguring a ReCoNode or a ReCoNet allows the system to automatically adapt to environmental changes, including faults, network topology changes, or varying workload scenarios, as illustrated by the different examples in Fig. 4.1a–e.

This comprises many novel theoretical aspects like autonomously taking decisions which task is executed *where*, hence answering the question on which node (network level) and at which placement position on a particular reconfigurable device (node level) a task is executed. Additionally, reconfiguration permits to decide *how* – in software or in hardware – a task is executed. Consequently, reconfiguration performs typical offline design problems like hardware/software-partitioning and the task binding dynamically at runtime.

With a rising amount of control units in distributed embedded systems, like for example, 60 and more ECUs (electronic control unit) in a car, reliability becomes a major concern. The *self-adaptive* capabilities of a ReCoNet/ReCoNode provides novel strategies for improving fault tolerance in distributed embedded control systems by *self-healing* techniques and a better resource utilization by *self-optimization*. For instance, instead of using *Triple Modular Redundancy* (TMR) with a fixed binding of the redundant task, a ReCoNet may adapt communication, task binding, or even the implementation style (hardware or software) according to the currently available resources or load of the network components at runtime.

The goal of this chapter is to propose novel methodologies for exploiting the – so far mostly unused – hardware runtime reconfiguration of FPGAs to fulfill the

D. Koch, *Partial Reconfiguration on FPGAs: Architectures, Tools and Applications*, Lecture Notes in Electrical Engineering 153, DOI 10.1007/978-1-4614-1225-0_4,

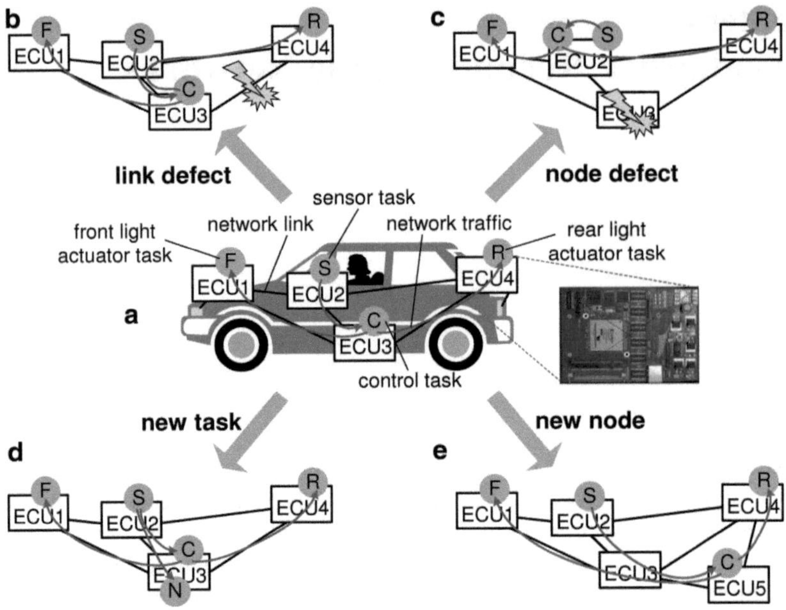

Fig. 4.1 Automotive example of a self-adaptive and self-repairing distributed reconfigurable ReCoNet. (**a**) The initial system consists of four electronic control units (ECU1, ..., ECU4) that are linked together to a small network. The network provides a control application where a sensor task S gathers the state of the cockpit switches and sends it further to a controller task C who is in charge of activating the front and rear lights via the actuator tasks F and R. Each ECU consists of reconfigurable hardware (an FPGA) and a CPU, such that hardware and software tasks can be arbitrarily executed or exchanged at runtime. For this reason, such ECUs are also called *ReCoNodes*. In addition to this reconfiguration on a *node level*, the system can also reconfigure on a *network level* (e.g., by *migrating a task* from one ECU to another one). In case of a link failure, the system reroutes traffic over alternative paths. (**c**) When a complete node drops out, the system will *detect the node failure* and automatically *activate a redundant task* on another node. (**d**) The system is further capable to integrate *new tasks* (see task N) into the system. (**e**) A self-adaptive system may also *integrate new nodes* (see node ECU5). Due to the *self-optimizing* capabilities of the system, tasks may autonomously migrate to this node without affecting the currently running application

requirements, which, in particular, includes the reliability, of future distributed embedded systems. The proposed and examined methods include the following services:

Hardware Checkpointing: For recovering from task failures or complete node crashes, internal states of hardware modules will be sent to redundant nodes. The state information can be used by a replica task to seamlessly continuing operation.

Hardware/Software Morphing: The possibility of reconfiguring both, the software and the hardware of the system allows to swap the execution between these two implementation variants. This capability provides additional freedom for self-optimization and self-healing the system.

With these services, the binding of tasks and communication demands to resources, as well as the partitioning between providing functionality in hardware and software, can be modified and optimized at runtime. This will be termed *dynamic hardware/software partitioning*. Dynamic hardware/software partitioning has to consider several constraints and objectives. This includes resource limitations like the available CPU performance, the free reconfigurable hardware resources (in case of an FPGA, this is the free available area of the device), and the available capacity on the network links. Furthermore, reliability aspects of the dynamic hardware/software partitioning will typically be incorporated, as well as real-time aspects. These topics are beyond the topic of this bock and this chapter was limited to the runtime reconfiguration aspects of a ReCoNet. Further information about the ReCoNets project can be found in [KRS$^+$10].

The central issues of this chapter are the technical aspects behind *online hardware/software-partitioning* which describes the procedure of binding functionality to resources in a ReCoNet at runtime. This chapter will continue with the introduction of *hardware checkpointing* which includes concepts for saving and restoring internal states of hardware tasks (i.e. modules). Checkpointing is needed for recovering from, for example, node failures in a ReCoNet. Furthermore, it is used to seamlessly *migrating* hardware tasks within the network. This includes transferring configuration bitstreams as well as internal state data. Note that we omit a discussion about the task-to-task communication that has to be able to support migration of tasks within a ReCoNet.

Checkpointing is further used for *task morphing* which allows to swap the execution of a task between either hardware implementation on an FPGA or software implementation on a CPU, as presented in Sect. 4.3. Here, the theoretical aspects of swapping the execution between hardware and software execution without losing internal states will be modeled and analyzed.

The novel techniques for self-adaptivity have been implemented and integrated into a ReCoNet that demonstrates self-adaptive and self-healing in a distributed automotive driver assistance application.

4.1 Hardware Checkpointing

Dynamic hardware/software partitioning comprises the migration of hardware tasks among different nodes in a ReCoNet. This permits, for example, that a ReCoNode delegates one or more hardware tasks to other nodes in a ReCoNet if, for example, the local hardware resources (FPGA area) are heavily loaded. Consequently, methods are required for integrating modules into a ReCoNode with the help of partial reconfiguration at runtime. This aspect is deeply examined in Chaps. 2 and 3.

In general, a task contains an internal state that has to be sent together with the configuration data in case of a task migration. In addition, tasks can exchange checkpoints for seamless continuation of operation, for example, in case of a node failure. In the latter case, the state is not transferred once but periodically from one ReCoNode to another. Consequently, techniques are demanded that permit to read a *consistent state* from a module as well as to initialize a module with this state. The state of a hardware module is a collection of the individual states of all memory elements (registers and flip-flops) used in processing, communication, or data storage parts of the module. A state is consistent, if it represents the global state off all these memory elements of the whole module as it may occur during failure free, correct execution. This state must be observable at a dedicated point of time, like for example at a certain clock cycle.[1]

In this section, it is investigated how *checkpointing* mechanisms that are well known for software systems [EAWJ02], can be utilized in hardware modules running on reconfigurable devices. From a theoretical perspective, a hardware task will be modeled as a finite state machine that has to be transformed such that its internal state can be accessed via the input and output for taking and restoring a checkpoint. With a hardware task, the state is represented by the content of its memory elements. Therefore, for the implementation of hardware checkpointing, all flip-flops and memories must be made accessible within a hardware module.

Besides discussing the theoretical fundamentals on hardware checkpointing, a further important aspect, examined throughout this section, is the question of how the state of a hardware module can be extracted and restored with (a) low latency penalty and (b) with low resource overhead. In addition, these penalties and resource overheads have to be estimated as soon as possible (e.g., when information about the state space and the data types used to implement a hardware module is available) in order to take the optimal checkpointing technique for a particular use case.

After an overview of related publications on accessing the state from modules implemented on FPGAs in the next section, Sect. 4.1.2 reveals the model and theoretical background of hardware checkpointing. Next, in Sect. 4.1.3, the states in the interfaces will be considered. Later in this chapter, multiple techniques for taking and restoring checkpoints in hardware tasks are investigated (Sect. 4.2).

Implementation details on automatically integrating hardware checkpointing into given modules, as well as an experimental evaluation, will be presented later within this chapter in Sect. 4.2.4.

4.1.1 Related Work on Hardware Checkpointing

State extraction and restoration of hardware modules using only the FPGA configuration interface has been proposed in systems where FPGAs are used for preemptive task execution [Bre97, SLM00, LWH02, KP05]. In the approaches, multiple hardware tasks share the same FPGA resources over time. One main

---

[1] There might exist situations where a register snapshot, taken at a certain clock cycle, is not representing a consistent state. For example, if the snapshot is taken during the evaluation of a multy cycle path.

difference between hardware checkpointing and preemptive hardware task execution is that checkpoints are rapidly taken in order to decrease the loss of computation in case of a fault, while a hardware task preemption will be typically triggered rarely because of the time consuming reconfiguration process. Consequently, the time overhead to access the state of a hardware module is more critical when the state extraction is used for hardware checkpointing. Another difference is that the extracted state data will be used on the same device in case of hardware task preemption, while in the case of hardware checkpointing, the state data should be transferable between different sized devices or even different FPGA types. In other words, it can be beneficial to abstract away from the target FPGA in order to share state data in a heterogeneous FPGA environment.

Scalera and Trimberger proposed multi-context FPGAs for swapping between multiple configuration planes in a single cycle [TCJW97, SV98]. As these planes are assumed to be read/writable, it would be possible to use multi-context FPGAs for hardware checkpointing. In this case, taking and restoring a checkpoint may be performed on an inactive plane. However, such FPGAs would be very costly and are not available, as described in Sect. 1.5.2.

Asadi and Tahoori proposed to use hardware checkpointing for recovering FPGAs from single event upsets [AT05]. In their system, the configuration data is continuously read back and compared with the original configuration by computing CRC checksums. The authors aim to use checkpointing to restart execution with an error-free state without presenting any further details.

Reading back or initiating the present register state of a module is widely used in ASIC design for test and debug purpose. Here, the flip-flops of an ASIC are included into a *scan chain* for providing access to the state of the hardware. For the same reason, scan chains have been implemented on FPGAs [WGNH01, TT03].

Besides these general approaches, hardware checkpointing on FPGAs has been addressed rarely. Huang and McCluskey proposed a checkpoint scheme for a fault-tolerant LZ-compressor [HM01]. The approach is highly optimized but no automatic design flow or tool for checkpointing was presented.

4.1.2 FSM-based Task Model

Each hardware module is modeled by a Finite State Machine (FSM). An FSM is a sextuple $(\mathcal{I}, \mathcal{O}, \mathcal{S}, \delta, \omega, s_0)$ with a set of inputs $\mathcal{I}$, a set of outputs $\mathcal{O}$, a set of internal states $\mathcal{S}$, a state transition function δ, an output function ω, and an initial state s_0 (see also Fig. 4.2a). In order to allow restoring a module from a checkpoint, the input values $i \in \mathcal{I}$ and the output values $o \in \mathcal{O}$ have to be restored together with the internal state $s \in \mathcal{S}$ that are all captured during the checkpointing process. This feature has to be supplied by the interface of the hardware module. An FSM providing the capability to save and to restore checkpoints will be called a *checkpoint FSM* (CFSM). The most general case of a CFSM is presented in Fig. 4.2b.

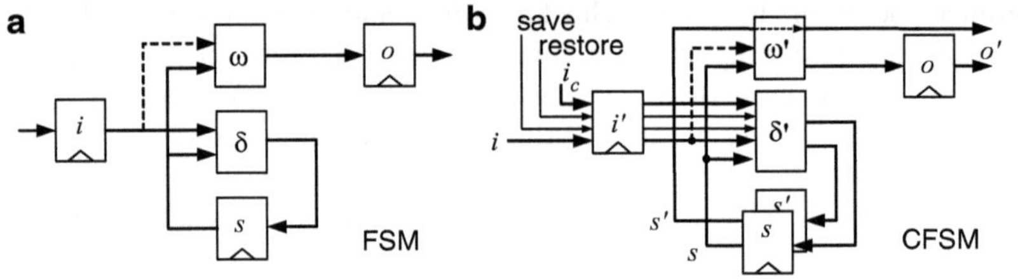

Fig. 4.2 (**a**) Finite State Machine (FSM) and (**b**) the corresponding Checkpoint-FSM (CFSM). The original module will be transformed, such that its state can be read or set from outside. The read and write process is controlled by the external signals save and restore

For supporting to take and to restore checkpoints in a hardware module, a given original FSM specification has to be extended. Given a conventional FSM $a = (\mathcal{I}, \mathcal{O}, \mathcal{S}, \delta, \omega, s_0)$ and a set of checkpoints $\mathcal{S}_c \subseteq \mathcal{S}$, the corresponding CFSM $a_c = (\mathcal{I}', \mathcal{O}', \mathcal{S}', \delta', \omega', s_0')$ can be derived as follows:

$$\mathcal{I}' = \mathcal{I} \times \mathcal{S}_c \times \mathcal{I}_{save} \times \mathcal{I}_{restore} \;; \mathcal{I}_{save} = \mathcal{I}_{restore} = \{0, 1\},$$
$$\mathcal{O}' = \mathcal{O} \times \mathcal{S}_c,$$
$$\mathcal{S}' = \mathcal{S} \times \mathcal{S}_c.$$

A given original FSM is transformed into a CFSM by extending the original input set $\mathcal{I}$ by two control signals I_{save}, and $I_{restore}$, and a checkpoint[2] $i_c \in \mathcal{S}_c$. With the current state being $(s, s') \in \mathcal{S}'$, the state transition function $\delta' : \mathcal{S}' \times I' \rightarrow \mathcal{S}'$ is given by:

$$\delta' = \begin{cases} (\delta(s,i), s') \text{ if } i' = (i, -, 0, 0) & \text{run mode, } s' \text{ remains checkpoint} \\ (\delta(s,i), s) \text{ if } i' = (i, -, 1, 0) \wedge s \in \mathcal{S}_c & \text{save checkpoint} \\ (\delta(s,i), s') \text{ if } i' = (i, -, 1, 0) \wedge s \notin \mathcal{S}_c & \text{continue to save checkpoint} \\ (i_c, s') \quad\quad \text{ if } i' = (i, i_c, 0, 1) & \text{restore checkpoint} \\ (i_c, s) \quad\quad \text{ if } i' = (i, i_c, 1, 1) \wedge s \in \mathcal{S}_c & \text{swap state } s \text{ with checkpoint } s_c \\ (\delta(s,i), s') \text{ if } i' = (i, i_c, 1, 1) \wedge s \notin \mathcal{S}_c & \text{continue to swap checkpoint} \end{cases}$$

Similarly, the output is expanded to support checkpoint extraction by linking the present state to O'. The output function ω' is defined as: $\omega' = (\omega(s, i), s')$; and the initial state is $s_0' = (s_0, s_0)$.

This system model provides full control over saving checkpoints or initiating a rollback by restoring the last saved checkpoint. If a checkpoint store operation is requested in case of $s \notin S_c$, the CFSM will continue its operation in the run mode until the next possible checkpoint has been reached. Note that the CFSM can alternatively swap between the present working state and the checkpoint state. This can be used, for example, for sharing the same hardware resources by multiple

[2]In general, depending on the checkpoint protocol, multiple checkpoints can exist,including the state at the input and output. For clearness, only one checkpoint is assumed throughout this chapter.

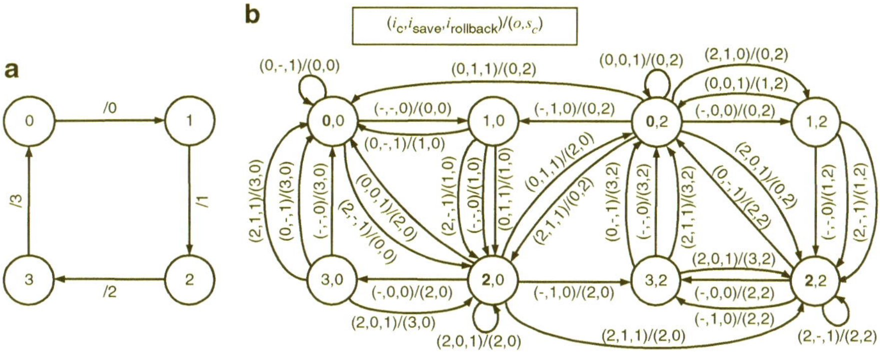

Fig. 4.3 (**a**) Given FSM of a modulo-4-counter and (**b**) corresponding CFSM for $S_c = \{0, 2\}$, i.e., only in states 0 and 2 saving of a checkpoint is permitted. The state space S' is given by $S \times S_c$, which is $\{0, \ldots, 3\} \times \{0, 2\}$. When exemplary starting from state $(0, 0)$ in run mode, which is set by the input assignment $(-, 0, 0)$, the counter will change to state $(1, 0)$. If now a checkpoint is requested to be stored with $i' = (-, 1, 0)$, the CFSM will count further to state $(2, 0)$, because the counter state 1 is not a checkpoint state in this example ($s \notin S_c$). After reaching the state $(2, 0)$, the counter is incremented and the stored checkpoint is overwritten with the present counter state 2. Therefore, the next state will be $(3, 2)$, if the input is still $i' = (-, 1, 0)$

hardware tasks (e.g., multiple identical cryptographic modules for multiple secured connections running in parallel). In case $S_c = S$, it is possible to save a checkpoint in each state.

However, it can be advantageous to omit some states from the set of possible checkpoints. If, for example, a hardware module contains a lot of intermediate register values during execution, it is possible to reduce the amount of memory required to store a checkpoint by taking checkpoints only when this module is in an idle state, thus, allowing to omit saving the intermediate register values. In this case, only the idle states are in the set of checkpoints (S_c). Figure 4.3 gives an example of a modulo-4-counter module with the states $S = \{0, 1, 2, 3\}$ and the derived CFSM with the set of checkpoint states $S_c = \{0, 2\}$. In this example, the state space of the checkpoint is reduced which allows to reduce the memory footprint of the checkpoint. In general, it is possible to omit a register from the checkpoint if this register is written before it will be read (*write before read*) under all possible state sequences started from a checkpoint state s_c. It is further possible to initialize saving a checkpoint and restoring another checkpoint at the same time. In this case, s is exchanged with s' at the next possible checkpoint.

4.1.3 Checkpointing Input and Output States

In most cases, not only the internal state s has to be restored in case of a rollback but also the input and output states. Furthermore, a rollback of one task is not transparent to its environment and all tasks in the entire sensor-controller-actuator chain have

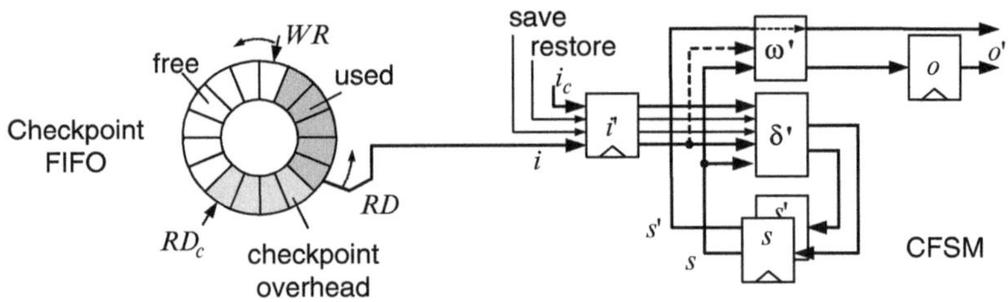

Fig. 4.4 Checkpoint FIFO for fault-tolerant storing input events in front of a checkpoint FSM. Whenever a checkpoint is taken, the present read pointer position RD is saved as part of the checkpoint in RD_c. The value RD_c might be saved together with the state of the hardware module. When a rollback is performed, e.g., in case of a fault, the read pointer RD will be set back to the stored RD_c value. As a consequence, all incoming data since the last stored checkpoint is preserved and can be sent again to the attached hardware

to restore their last checkpoint when a coordinated checkpoint scheme is used. However, external events between the checkpoint and the fault will still get lost, for example, input values from a sensor task. For systems where losing external events is prohibited, a special memory, called *checkpoint FIFO* can be used. An example of this FIFO is sketched in Fig. 4.4, where the checkpoint FIFO is illustrated as a ring buffer with a write pointer WR, and a read pointer RD. A specific capability of a checkpoint FIFO is its ability to reset the read pointer to an entry of the FIFO that has been read already before. The idea of a checkpoint FIFO, as opposed to a commonly used FIFO, is to keep data that has to be read from the FIFO for allowing a rollback of the checkpoint FIFO. Whenever a checkpoint is saved in the attached CFSM, the present read pointer position RD is included into the checkpoint of the CFSM. The read pointer, as part of the checkpoint, is labeled RD_c in Fig. 4.4. FIFO data between RD_c and RD is not allowed to be overwritten. In the rollback case, the read pointer RD is set to the position of the stored read pointer at the last taken checkpoint (RD_c). Consequently, the output values since the last checkpoint will be supplied again to the CFSM. Note that RD and WR play the role of typical FIFO pointers with a corresponding read and write port each, while RD_c is a marker for the last stored checkpoint. The checkpoint FIFO concept ensures to maintain the order of all events but not the exact chronology. Keeping also the chronology was demonstrated for software checkpointing by Fidge [Fid88] with the help of time stamps that are annotated to each event.

Figure 4.4 illustrates that the checkpoint FIFO is separated from the CFSM. Therefore, the checkpoint FIFO is not necessarily bound to the same FPGA as the CFSM. If the checkpoint FIFO cannot be assumed to be robust to potential node failures, it has to be implemented redundantly.

4.2 Accessing Module States

In this section, different techniques suitable for automatically transforming a module for accessing its internal state will be discussed. This is of interest for a context switching of hardware tasks as well as for load balancing and checkpointing in a ReCoNet, as described in the introduction to this chapter. We will discuss the state access from a fault-tolerant control applications' point of view where hardware checkpointing is used to recover from faults. Here the objectives are the checkpoint latency and the hardware overhead. This will be examined for three different methods: (1) *memory-mapped state access* (MM), (2) *scan chain based state access* (SC), and (3) *shadow based scan chain state access* (SHC). All these methodologies differ in terms of the latency t_c to extract a checkpoint and the additional hardware overhead L required for saving and restoring the checkpoints. It is presumed that all hardware modules are implemented fully synchronous without any latches. Then, the state of a hardware module is represented by the value of all memory elements inside the module. Consequently, the different approaches provide techniques to read and write all or a subset of the flip-flops inside a module.

Assume that the original hardware module has the following parameters: (1) K_{LUT}, K states the number of look-up tables (LUTs) and flip-flops of the module, and (2) $F_{\max}$ denotes the modules maximal operating frequency. Additionally, the corresponding variables of the checkpointable hardware module after modification are K^*_{LUT}, K^*, and $F^*_{\max}$. Then, the different methodologies can be compared using the following indicators:

- The *Checkpoint Hardware Overhead* specifies the amount of additional resources required by a specific checkpoint technique. It will be distinguished between $L_{LUT} = K^*_{LUT} - K_{LUT}$ and $L_{FF} = K^* - K$, denoting the number of required extra look-up tables and flip-flops, respectively.
- The *Checkpoint Frequency Reduction* f_r denotes the reduction of the maximal achievable clock frequency due to the increased module complexity ($f_r = F^*_{\max} - F_{\max}$).
- The *Checkpoint Overhead* t_c states the time, a module must be interrupted when saving a checkpoint. This interrupt leads to an increase in the execution time.
- The *Checkpoint Latency* t_l specifies the amount of time required until the complete checkpoint data arrives at the device that will take over the task execution in case of a fault.

Based on the overhead t_c and the latency t_l, further indicators can be derived for a quantitative comparison of hardware checkpointing techniques. The relationship between t_c and t_l is depicted in Fig. 4.5. The example shows a task T sending a checkpoint to another task T' with the checkpoint period t_p. Due to the communication to the replica task, the latency t_l is larger than the overhead t_c. The *checkpoint efficiency*, denoting the relative amount of time a module spends on its execution, is defined as $\eta = 1 - \frac{t_c}{t_p}$. Figure 4.5 shows an example where a fault of

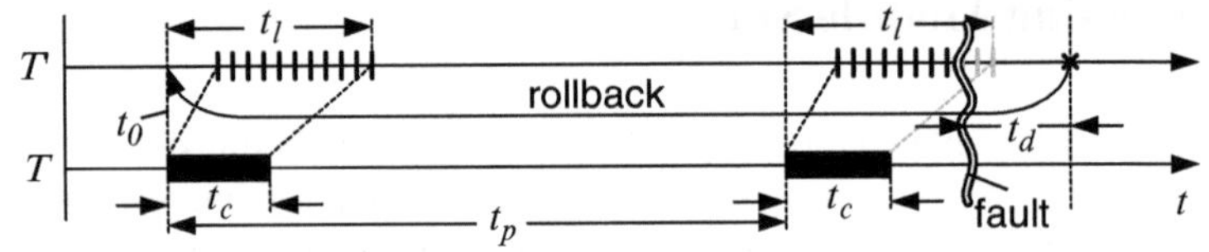

Fig. 4.5 Relationship between the checkpoint overhead t_c and the checkpoint latency t_l. Checkpoints are stored continuously with the period t_p. In order to read a consistent state, the hardware task has to stop its operation for the time t_c. The value t_l denotes the time required to save the checkpoint into a secure memory (typically located on another ReCoNode). After a fault occurs, it takes the time t_d to recognize this event. In this case, the redundant CFSM performs a rollback to the last completely stored checkpoint (Here, the state stored at time t_0)

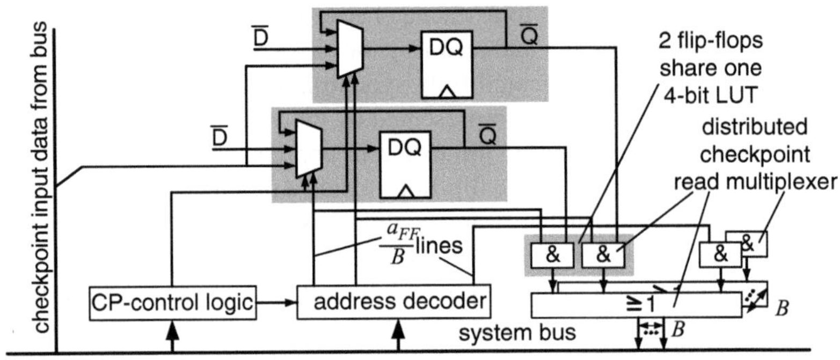

Fig. 4.6 Hardware checkpointing by memory-mapped state access. Flip-flops are clustered together according to the word size of the bus (B), thus, reducing the complexity of the address decoder to $\frac{K}{B}$ select lines for K flip-flops

task T is recognized after a fault detection latency t_d. After fault detection, task T' takes over the execution starting with the last completely stored state (here at t_0). The maximal *loss of computation* is $t_{loss} = t_p + t_l + t_d$.

4.2.1 Memory-Mapped State Access

In the memory-mapped methodology (MM), a processor accesses the module's state by integrating the flip-flops storing the checkpoint into a read/writable memory space of a system. In order to keep the state consistent during read or write operations, the module must not operate until a state extraction or a rollback process has been completed. This is achieved by a multiplexer that feeds back the present value of a flip-flop while saving a checkpoint. As depicted in Fig. 4.6, this is implemented by extending each individual flip-flop of the module. For reducing the checkpoint overhead t_c when storing a checkpoint, flip-flops will be grouped together according to the data-bus size B.

The extra amount of flip-flops L_{FF} is mainly related to the module interface registers that are marginal as compared to the amount of flip-flops inside the module

The logic overhead for providing access to K module flip-flops using memory mapping to a B-bit wide bus using 4-bit LUTs can be estimated as:

$$L_{LUT}^{MM} \approx \underbrace{\frac{K}{B} + \sum_{n=0}^{\lceil log_8(\frac{K}{B}) \rceil - 1} 8^n}_{\text{address decoder}} + \underbrace{\frac{K}{2} + B \left\lceil \frac{\frac{K}{2B} - 1}{3} \right\rceil}_{\text{read mux}} + \underbrace{K}_{\text{feedback mux}} \tag{4.1}$$

It must be noted for the address decoder circuit needs a LUT for each output select signal. In case of 4-bit LUTs, it is possible to evaluate three address bits in addition to the enable signal generated by the evaluation of the higher address lines. Thus, the decoder can be modeled as a B-tree with eight children (selected by three address bits) for each select output wire. The read multiplexer consists of an OR-gate for every bit connected to the read data bus. Wide input OR-gates are mapped as a B-tree with k children when using look-up tables with k inputs.

Equation 4.1 states a worst case estimation for the logic overhead resulting from the address decoder, the feedback multiplexers, and the restore logic. However, current architectures and tools support to translate the feedback multiplexer to a flip-flop enable signal, and the restore multiplexer is mapped together with the original circuit sharing resources. Therefore, L_{LUT} will be less than the number of flip-flops K in practice.

The checkpoint overhead t_c can be predicted by counting the bus read/write operations necessary to save or restore a checkpoint. With τ_{bus} denoting the time for a bus cycle, the overhead is about:

$$t_c^{MM} \approx \frac{K}{B} \cdot \tau_{bus}$$

4.2.2 *State Access via Scan Chains*

Instead of mapping flip-flop values into the memory address space, the flip-flops can be chained together to a long shift register chain (SC). This is known from established ASIC design techniques. A multiplexer in front of every flip-flop is used to switch between normal operation and the shift register operation where all flip-flops are connected to a long shift register. In general, it is not possible to read the complete chain atomically in one bus cycle. For keeping the state of the chain consistent without additional logic for each flip-flop of the module, the chain ends are connected to a ring shift during scan mode (t_c). An additional state machine keeps track of the present shift position of the state data inside the chain.

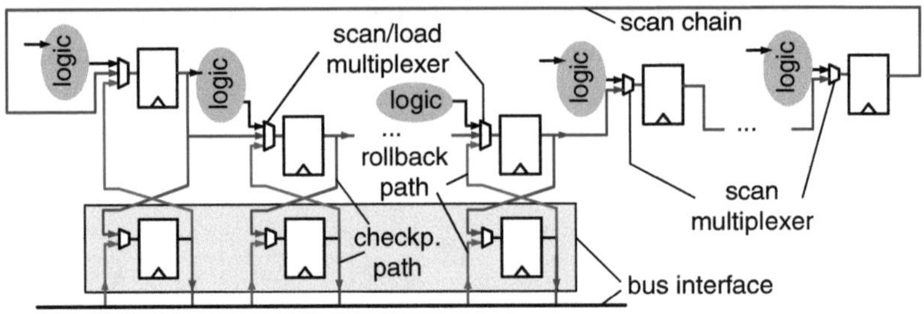

Fig. 4.7 Hardware checkpointing using a scan chan. Scan multiplexers allow to connect all flip-flops to one large ring shift register. The bus interface contains B extra flip-flops for taking a checkpoint or for performing rollback. A state machine (not shown) keeps track of the current shift position within the scan chain. Furthermore, this automate coordinates the data exchange between the scan chain and the bus interface

Furthermore, the additional state machine is used to automatically and selectively read and write portions of the flip-flop values through an extra register for accessing the state at the full data bus width B, as illustrated in Fig. 4.7.

Again, the hardware overhead L_{FF} depends on the module interface Z. In addition, some memory is required for the state machine controlling the scan chain access. Beside a constant value Z_{const} (for the control logic), it needs $\lceil log_2 K \rceil$ flip-flops for a scan shift position counter and $\lceil log_2 K/K \rceil$ flip-flops for the word position register for a module with K flip-flops connected to a B-bit wide bus. Thus, the flip-flop overhead is:

$$H_{FF}^{SC} \approx Z_{const} + Z + \lceil log_2 K \rceil + \left\lceil log_2 \frac{K}{B} \right\rceil + B$$

In the worst case, the additional amount of look-up tables L_{LUT} is given by the logic for the scan chain interface plus additional LUTs for the scan multiplexer in front of each flip-flop of the module. This overhead will be less than K in practice, because of the scan multiplexer may be often integrated into the module logic without additional logic cost. This holds in particular for recent 6-input LUT devices. The advantage of the ring-shift is that state consistency can be guaranteed with a low resource overhead by the cost of an increased transfer time for the CPU subsystem. When a specific fraction of the ring shift register is addressed, it takes on average $K/2$ cycles until the state data has passed the extra register for the state access. $\frac{K}{B}$ times a value has to be transferred between the extra register and the CPU subsystem. Therefore, the checkpoint overhead t_c scales with $O(K^2)$ leading to a poor checkpoint efficiency η for large checkpoints. Instead of using one ring shift register of length K, it is alternatively possible to generate K' parallel running shift-register chains each of length $\frac{K}{K'}$. Note that K' must be chosen to be less or equal than B. Then, the checkpoint overhead t_c is about a factor of $\left(\frac{1}{2}\frac{K}{K'}\right) \cdot \frac{K}{B}$ larger as compared to the memory-mapped approach.

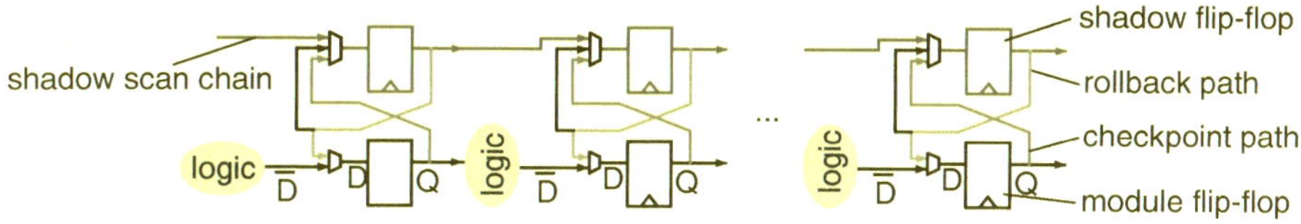

Fig. 4.8 Hardware checkpointing using shadow scan chains. Buy duplicating all flip-flops and arranging them into a chain, checkpoint data transfer can be performed concurrently with the module execution

4.2.3 State Access via Shadow Scan Chains

The checkpoint efficiency η decreases dramatically in systems where checkpoints are frequently taken. In this case, it is advantageous to duplicate all flip-flops of the original module to allow copying all values in one single clock cycle, as depicted in Fig. 4.8. This permits to continue the execution of a module while moving the state data transfer in the background through the shadow register chain. With respect to the CFSM model, the original module registers store the state s and the shadow registers the checkpoint state s'. The multiplexers implement the state transition function δ.

For the shadow scan chain approach, the checkpoint overhead is zero and the efficiency is $\eta = 1$. However, t_c^{SHC} is used to denote the time required to read the shadow chain in the background while the module keeps running. In case of a rollback, the state information is shifted into the shadow chain, then loaded into the working registers and operation will start immediately. If the checkpoint data is stored inside a redundant CFSM, the rollback latency can be reduced to a minimum.

If an FSM refinement with K flip-flops is transferred to a CFSM, the overhead for the additional flip-flops and LUTs is obviously: $L_{FF}^{SHC} \approx K$ and $L_{LUT}^{SHC} \approx 2K$ (without the interface).

The shadow scan chain approach fits well to recent 6-input look-up table devices, such as Xilinx Spartan-6 or Virtex-6 FPGAs. On these devices, the 6-input LUTs can be split into two 5-input LUTs with shared LUT input. In this case, both LUTs can provide an individual state flip-flop. Consequently, the state flip-flop extension that is shown in Fig. 4.8 can be mapped into a single 6-input LUT and might not increase the logic cost at all on these devices.

4.2.4 Automatic Hardware Transformation with StateAccess

This section covers the implementation of hardware checkpointing with the tool *StateAccess* [KHT07]. StateAccess performs all required transformations to automatically extend a module with the capability of checkpointing. It supports all presented state extraction techniques (see Sect. 4.2) with multiple options. For

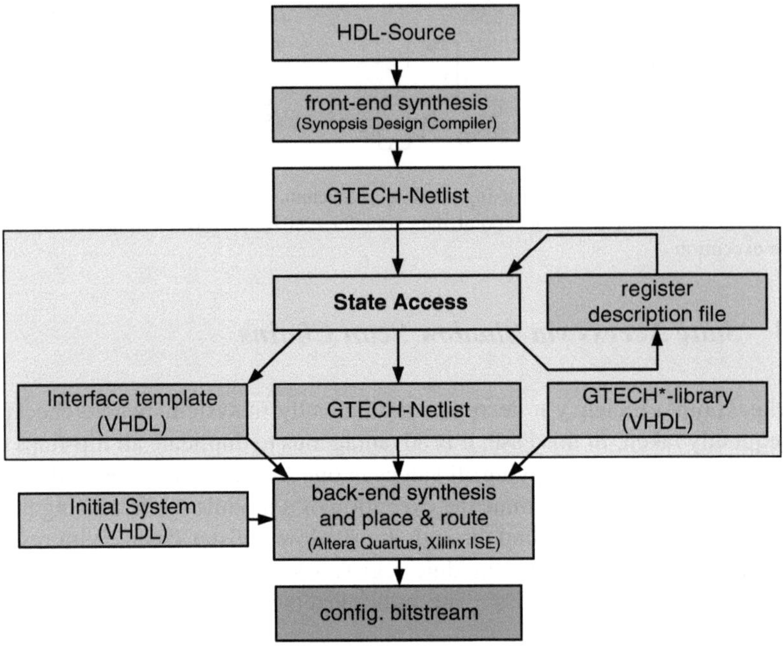

Fig. 4.9 Modified design flow to enable hardware checkpointing with StateAccess

example, instead of implementing one scan chain, it is possible to generate multiple shorter ones that are used in parallel. The tool works at the netlist level where all types of flip-flops of a module can be directly identified. The original flip-flops will be replaced with *checkpoint flip-flops* according to the specified state extraction technique. Next, the state extraction interface logic to the system bus will be created and all connections will be automatically generated. The result of this process is again a netlist that will be synthesized together with the remaining system.

The design flow for implementing hardware checkpointing is depicted in Fig. 4.9. Starting from an (verified) HDL specification, a netlist is generated in a front-end synthesis step. This uses the Synopsys Design Compiler for generating an EDIF netlist based on GTECH-primitives (Generic TECHnology) [Syn08].

In addition to identifying all flip-flops in the netlist, StateAccess executes design rule checks to detect unsupported constructs like combinatorial feedback paths or latches. The identified flip-flops are listed in a human readable form in a description file. By editing this list, the set of flip-flops and their order within the checkpoint data can be adjusted for guiding the transformation process.

The back-end synthesis step comprises an extended GTECH library with dedicated flip-flop replacement primitives and a library providing templates to interface the state access logic with the rest of the system. The last step does not demand a special synthesis tool. Hence, it is possible to use the checkpoints in heterogeneous

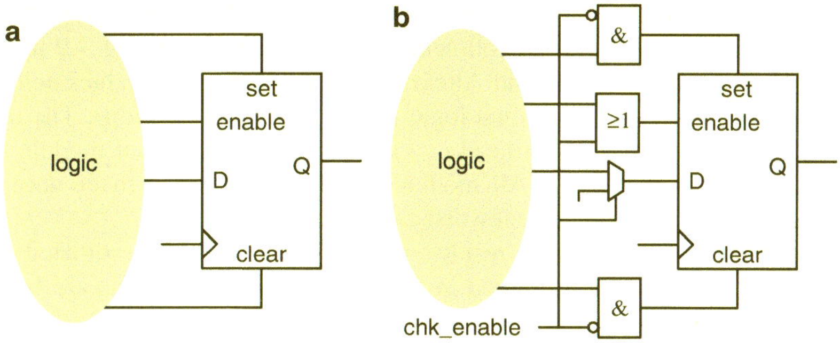

Fig. 4.10 (**a**) Flip-flop providing set, clear, and enable capability. (**b**) adaptations with additional logic to allow checkpoint read or restore operations. These transformations are automatically performed by StateAccess

FPGA environments. For example, an Altera FPGA-based ReCoNode can directly send a checkpoint to a Xilinx counterpart without any incompatibility or further transformations. This is exemplary implemented in Sect. 4.2.8.

4.2.5 Further Module Transformations and Limitations

Besides the presented hardware extensions applied at the netlist level, it has to be ensured that during state extraction and state restoration all set or clear signals of the flip-flops are disabled. Furthermore, if flip-flops use enable signals, these have to be forced active. StateAccess supports these signals by adding additional control logic around the flip-flops, as depicted in Fig. 4.10. This logic overhead is not considered in the estimates presented above because the amount of additional logic is normally very small and the control signals of the flip-flops can be often shared for complete register banks.

If a hardware module instantiates on-chip memory blocks, a wrapper similar to the memory-mapped approach is automatically generated around the memory. This allows the CPU to randomly access the state stored in on-chip memories. However, StateAccess supports this only for synchronous single port memories that must be directly located inside the module netlist.

4.2.6 Experimental Evaluation

The capability of checkpointing is bought with additional hardware resources and a decreased operating frequency. This section presents quantitative trends on these overheads for each proposed checkpointing mechanism. In addition, the results will be quantitatively compared with checkpointing using FPGA configuration readback.

Table 4.1 lists synthesis results for a set of benchmark modules [KBT07]. The checkpointing was implemented following the flow depicted in Fig. 4.9 using the Altera Quartus-II design suite and Altera Cyclone FPGAs. The values denote the cost originating from the additional logic for accessing all flip-flops. The relative logic (L_{LUT}) and flip-flop overhead (L_{FF}) refers to the original module netlist (named "blank" in the table). All module flip-flops have been made accessible. Consequently, the values represent worst case results.

In addition to the synthesis results, the table also reveals estimated values according to the formulas presented in Sect. 4.2. These estimations are based on the number of flip-flops found in the unmodified module (e.g., for the DES56 module, $K = 862$). The data bus width is set to $B = 32$ bits for all experiments. The results point out that the estimates match the measured overheads well. In the worst case, the difference between the measured and the estimated logic overhead is 28% (found for the look-up table overhead of the DES56 module when applying the memory-mapped approach). The amount of additional LUTs for the FIR16 module is 1.44 times larger than the number of flip-flops found in the unmodified module when using a simple scan chain to access the state. This has two reasons: Firstly, the unmodified module has a high degree of fully used 4-bit LUTs (82%), hence, preventing to fit a larger amount of additional logic into the existing LUTs. Secondly, the FIR16 module possesses a very complex control structure with a huge amount of individual flip-flop enable signals.

The impact of inserting scan chains into FPGA designs on the netlist level for testing hardware modules was examined in [WGNH01]. That work focuses on just one approach similar to the simple scan chain presented in Sect. 4.2.2. But instead of using a ring shifter for keeping the state consistent, flip-flop enable logic was used. The reported results state on average look-up table overhead of 130% that is more than the worst case result for all examined checkpointing techniques in this chapter. Unfortunately, the paper does not list overheads on the achievable clock frequency $F_{\max}$ or the flip-flops. Tiwari and Tomko reported an overhead of 46% in average for their benchmark circuits that have been extended with simple scan chains [TT03] for debugging purposes. The scan process in that work was controlled by gating the clock on Xilinx Virtex-II FPGAs. On these devices, clock gating requires a dedicated global clock tree and additional care must be taken because of the skew among different clock trees. Opposed to this, all presented techniques in this chapter are not touching any clock net and the overhead of the simple scan chain technique is only 33% on average.

In addition to the logic overhead, the impact on the maximal achievable clock frequency was examined for the benchmark. As stated in Table 4.2, no strong influence was found due to hardware checkpointing.

4.2.6.1 Checkpoint Overhead

For determining the checkpoint overhead t_c, and consequently, the time to take or to restore a checkpoint, the benchmark modules have been integrated into an SoC containing a CPU (Altera NIOS-II), a memory controller, and some further

Table 4.1 Synthesis results on flip-flop (L_{FF}), look-up table (L_{LUT}) overheads for the proposed hardware checkpointing methodologies

		flip-flops			4-bit LUTs		
		Synth. results	Overhead Measured	Overhead Estimated	Synth. results	Overhead Measured	Overhead Estimated
DES56	Blank	862			1,899		
	MM	903	$L_{FF}^{MM^*} = 41\ (5\%)$	Negligible	2,794	$L_{LUT}^{MM^*} = 895\ (47\%)$	$L_{LUT}^{MM} \approx 1{,}425\ (75\%)$
	SC	928	$L_{FF}^{SC^*} = 66\ (8\%)$	$L_{FF}^{SC} \approx 47\ (5\%)$	2,438	$L_{LUT}^{SC^*} = 539\ (28\%)$	$L_{LUT}^{SC} \leq 862$
	SHC	1,831	$L_{FF}^{SHC^*} = 969\ (112\%)$	$L_{FF}^{SHC} \approx 894\ (103\%)$	3,712	$L_{LUT}^{SHC^*} = 1{,}813\ (96\%)$	$L_{LUT}^{SHC} \approx 1{,}756\ (93\%)$
FIR16	Blank	563			1,223		
	MM	576	$L_{FF}^{MM^*} = 13\ (2\%)$	Negligible	2,243	$L_{LUT}^{MM^*} = 1{,}020\ (83\%)$	$L_{LUT}^{MM} \approx 967\ (80\%)$
	SC	583	$L_{FF}^{SC^*} = 20\ (4\%)$	$L_{FF}^{SC} \approx 47\ (8\%)$	2,034	$L_{LUT}^{SC^*} = 811\ (66\%)$	$L_{LUT}^{SC} \leq 563$
	SHC	1142	$L_{FF}^{SHC^*} = 579\ (103\%)$	$L_{FF}^{SHC} \approx 595\ (106\%)$	2,707	$L_{LUT}^{SHC^*} = 1{,}484\ (121\%)$	$L_{LUT}^{SHC} \approx 1{,}158\ (95\%)$
FFT1024	Blank	5057			17,626		
	MM	5632	$L_{FF}^{MM^*} = 575\ (11\%)$	Negligible	23,320	$L_{LUT}^{MM^*} = 5{,}694\ (32\%)$	$L_{LUT}^{MM} \approx 7{,}945\ (45\%)$
	SC	5651	$L_{FF}^{SC^*} = 594\ (12\%)$	$L_{FF}^{SC} \approx 53\ (1\%)$	18,493	$L_{LUT}^{SC^*} = 867\ (5\%)$	$L_{LUT}^{SC} \leq 17{,}626$
	SHC	11242	$L_{FF}^{SHC^*} = 6{,}185\ (122\%)$	$L_{FF}^{SHC} \approx 5{,}089\ (101\%)$	30,322	$L_{LUT}^{SHC^*} = 12{,}696\ (72\%)$	$L_{LUT}^{SHC} \approx 10{,}146\ (58\%)$

Table 4.2 Reduction of the achievable clock frequency for the proposed hardware checkpointing methodologies

	DES56	FIR16	FFT 1024
Blank	113	22.7	27.7
MM	104 (−8%)	21.0 (−7%)	25.6 (−8%)
SC	111 (−2%)	21.7 (−4%)	26.6 (−4%)
SHC	94 (−17%)	21.5 (−5%)	24.8 (−10%)

peripherals. Most commonly, FPGAs are logic bounded and typically not restricted to the amount of flip-flops in a circuit. This observation is confirmed by the benchmark circuits that require more than double the amount of LUTs than of flip-flops. Consequently, focus has to be put on L_{LUT}, when evaluating the different checkpointing techniques.

The measured checkpoint overheads are plotted together with the LUT overheads L_{LUT} from Table 4.1 in Fig. 4.11. For each test module, all results for the different techniques represent a Pareto point, hence, all techniques have their justification depending on the entire requirements (low checkpoint overhead or low logic overhead). The shadow scan chain technique is listed with two different t_c values for the overhead for each module. The first value is zero, because of taking a checkpoint can be performed without stopping the module, while the second value denotes the time for transferring the state (average of checkpoint read and restore).

4.2.7 Checkpointing via FPGA Readback Techniques

As a reference, checkpointing with the help of configuration readback and bitstream manipulation can be quantitatively compared with the results in Fig. 4.11. Assuming a Xilinx Virtex-II FPGA, the time required to read back configuration data is related to the number of configuration columns utilized by a module and, hence, to the module width. In the best case, when a module is constrained to a rectangle shape of minimal size and maximal height, the readback data contains 20 times more data than the data required to store just the flip-flop values (refer to Sect. 3.4 for further details on the bitstream format). This holds, if all flip-flops will be used, and if only the configuration columns containing register values are selectively read. This overhead in the readback data will be pretty much the same for recent Xilinx FPGAs as it is still not required to read the configuration frames that configure the routing, when reading back state values.

In the case of the DES56 core, for example, this overhead ratio is in the best case $\frac{1{,}899}{862} \cdot 20 = 44$ because not all flip-flops provided by the FPGA were used. This means that instead of reading just the 862 flip-flop values of the module, it is necessary to read about $862 \cdot 44 = 37$ kB of configuration data which takes roughly 10k cycles at the 32-bit configuration interface. In addition, we have to consider

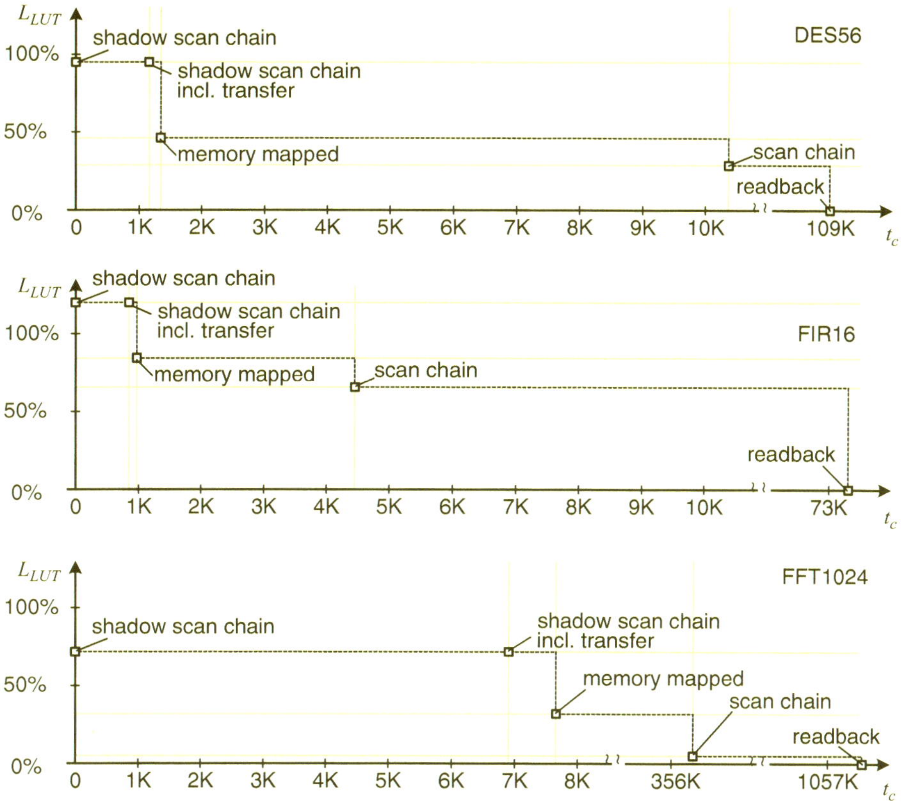

Fig. 4.11 Hardware checkpointing logic overhead L_{LUT} over the checkpoint overhead t_c in terms of clock cycles for different benchmark modules

the clock cycles to filter out the flip-flop states in order to store and transfer the checkpoint efficiently. Consequently, it will requires at least 100k cycles to extract the state. Note that the readback configuration data needs to be manipulated before it can be used to reconfigure the device again, and that there is no trivial method on Xilinx FPGAs to preset just a submodule with an initial state during partial reconfiguration. Furthermore, the configuration readback approach demands to gate the module clock in order to keep the state consistent when a checkpoint rollback occurs.

In [LWH02] readback techniques are used to implement multitasking on a Virtex FPGA. It is reported that it takes 407 ms to read back the configuration (766 kB), 465 ms to extract the state information and another 365 ms to configure the device. These values confirm that the state access via the configuration port is relatively slow. It can be concluded that it is sometimes worth or even required to spend some additional logic for reducing the latency overhead by at least one order of magnitude.

For a sake of completeness, the state capturing and state initialization process for Xilinx FPGAs will be sketched in the following. For capturing the sate of all state flip-flops of a module, Xilinx provides a "*capture*" primitive. This primitive has an enable and a clock input and it can be instantiated in the HDL used design. By activating the enable input, a *global snapshot* of all state flip-flops will be taken synchronously and stored at the position of the state flip-flop init values. These values can then be accessed by normal configuration readback. This feature is preliminary intended to be used for debug purpose and Xilinx allows the generation of a mapping file that describes which relative bit-offset corresponds to a specific user signal bit. The mapping file can be generated using the `bitgen -l` option. If we freeze the clock to the module during readback, we can also capture the look-up table state (when using distributed memory). For efficient state storage, we can filter the part of the readback data that stores the state. This could be done globally (with some modulo counters for picking the state information bits from the stream) or selectively in order to only store the state of used state flip-flops or distributed memory LUTs rather than all flip-flop values.

Initializing a module with a certain state is unfortunately a bit more complex. It is possible to start an FPGA with the init values; but however, this can only be initialized globally for the whole fabric. This would be basically a device reset. However, there are a few strategies that can be used to set initial values. First of all, we have to disable the clock during reconfiguration. Then we will activate for all used state flip-flops the asynchronous reset option. The requested init values (i.e., the module state), will be loaded into the part of the bitstream that is used for the reset values. Consequently, we can initialize the state flip-flops by applying a reset. This is done by two times toggling the reset enable multiplexer that, in each slice, selects if a reset is active high or active low. This requires corresponding configuration bitstreams. As we are in the asynchronous reset mode, the state flip-flops will be updated without a clock and regardless to the present level of the reset enable input. Before releasing the clock and therefore enabling (= resuming) the module, we have to set the original reset values (just consider that we use reset for a modulo counter) and the original synchronous flip-flop reset mode.

As an alternative, we can keep the clock active during reconfiguration. Instead of using the reset values, we can initialize the LUTs with the state. We will set all LUT table values to the same value, with respect to the state we have to initialize. Then, regardless to the current LUT input, the flip-flop will be initialized with the requested value. Note that we have to temporarily clear the flip-flop enable input routing multiplexers in order to enable the flip-flops. This ensures that flip-flops will sample the state stored in the LUT values. For starting the module, we have to gate the clock, set the original LUT table entries, and set the original flip-flop enable routing multiplexer configuration, before activating the clock again. With these two strategies, preemptive multitasking using read configuration readback is possible.

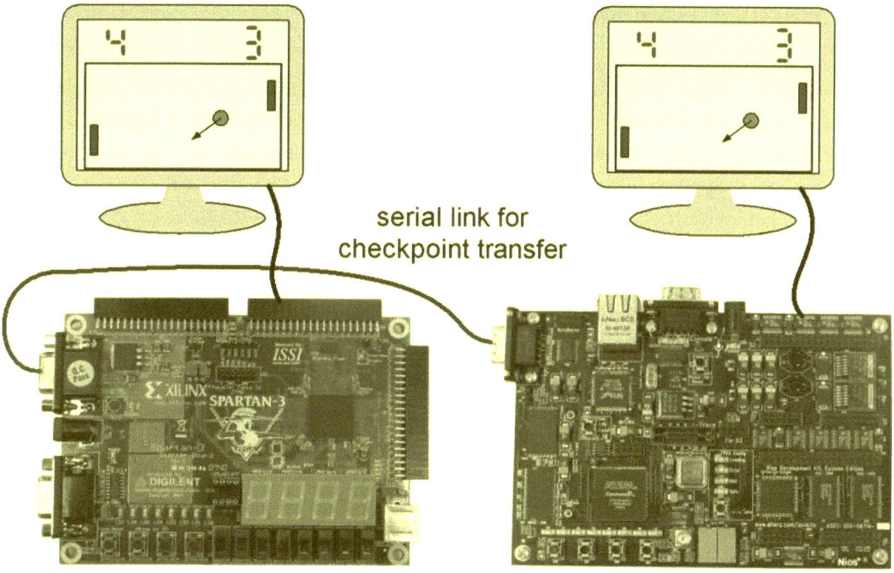

Fig. 4.12 Two FPGA boards linked for remotely playing Pong The synchronization between the boards is completely carried out by sending checkpoints between the boards. Note that this is possible despite of using completely different FPGAs

4.2.8 Hardware Checkpointing in Heterogeneous FPGA Environments

For demonstrating the capability of hardware checkpointing in heterogeneous FPGA environments, an existing implementation of the video game Pong (a table tennis like game) was automatically extended with a simple scan chain using StateAccess. In this example, checkpointing is used to synchronize two gamers playing remotely on different nodes against each other, as illustrated in Fig. 4.12. If the ball of the right player hits the right border of the screen or the racket, a checkpoint is sent to the FPGA board of the left player. The left side performs a rollback with this state that contains all necessary information to continue the game (e.g., score, ball direction). The analogous process happens for the left player at the left hand side of the screen.

By omitting all flip-flops from the register description that are used by the video output module of the game, the checkpoint data shrinks from 162 down to 57 flip-flops. The only modification required to the original Pong specification is to generate a signal that indicates whenever a ball hits the left or right border. This signal triggers the actual checkpoint saving process. The back end synthesis step (see Fig. 4.9) was performed for FPGAs from different vendors. The synthesis results differ with respect to the amount of LUTs and flip-flops instantiated for the game. However, the checkpoints are still compatible for both FPGA boards. By using an additional RS232 compatible UART at 115,200 baud, the checkpoint latency is below 5 ms.

4.3 Hardware/Software Morphing

The decision which particular task is executed by dedicated hardware or as a software program running on an ISP (Instruction Set Processor) is traditionally taken at design time. This partitioning has massive impact to the overall performance of a system, and methods have been proposed that allow to take the final decision, which function should be implemented in hardware or in software, as late as possible [VSV$^+$01] within the design flow. With the introduction of reconfigurable hardware, the hardware/software partitioning problem can be solved online. Consequently an FPGA for running dedicated tasks in hardware, provides an ideal platform for allowing *dynamic hardware/software partitioning*. This permits to adapt the system to environmental changes, such as the current computational demands or the allowed power consumption, at runtime. While some tasks in a system have high throughput requirements that can only be fulfilled with dedicated hardware modules (e.g., video preprocessing), other tasks might be highly control flow dominated (e.g. a graphical user interface), and hence, are more appropriate for software implementation. However, for a wide range of applications, both implementation alternatives are feasible and are consequently suitable for applying *hardware/software morphing*. This allows for swapping arbitrary between hardware or software execution of a task at runtime. Morphing is especially useful if the throughput demand of an algorithm varies at runtime.

Note that morphing is not limited to change between hardware and software execution. It may also be used to switch among different implementation alternatives within the same domain (hardware or software). This states a further option of using runtime reconfiguration for self-adapting a systems. For example, if data is sent at a variable data rate via a secured connection, the corresponding cryptographic task implementation may be swapped between different hardware modules (for trading off throughput versus area) or even software execution, depending on the present data rate.

The remainder of this section is organized as follows. After revealing related work on hardware/software morphing, a formal model for morphing will be sketched in Sect. 4.3.2. After this, technical aspects required to translate states between the hardware and software domain will be discussed in Sect. 4.3.3. Later in this chapter, in Sect. 4.3.5, a case study will be presented.

4.3.1 Related Work on Hardware/Software Morphing

Adapting systems to environmental changes is proposed by Stitt et al. by profiling the software that is currently running on an ISP (MIPS). With the acquired data, hardware modules have been generated, synthesized, and mapped to FPGAs at runtime [SLV03] for accelerating software kernels. However, the method was only demonstrated for simple Boolean functions and the authors omit information about

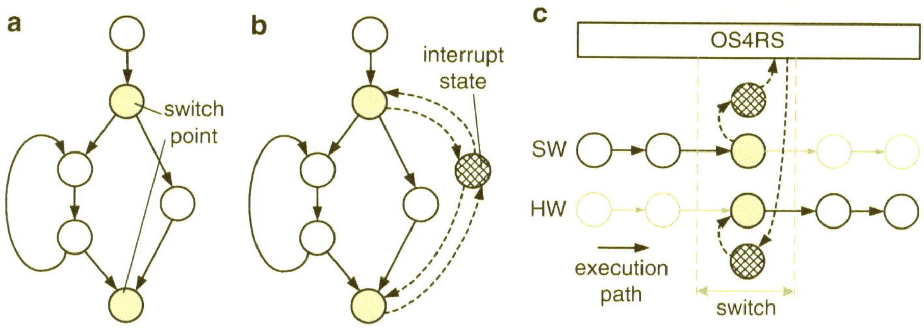

Fig. 4.13 (**a**) Task with two *switch points* at which the implementation style can be changed between hardware and software. (**b**) For performing the switching, the task is brought from a switch point into an interrupt state from where the OS can access the state of the hardware or software implementation. (**c**) Example of switching from software to hardware execution (The figures are taken from [MNM+04])

how this approach could be properly implemented in real-world systems. Gathering information about the current state of a system and adapting it at runtime was also proposed by Xu et al. [XRT03]. In that work, information about FPGA defects are transferred via a network to a remote reconfiguration server that is responsible to generate a functional system specific configuration to be used by the client located in the field. This technique was not used to automatically generate hardware accelerators for software binaries, but it demonstrates that it is in principle feasible that an embedded system initiates the synthesis and generation of new configurations at runtime, which may be combined with the approach from Stitt.

Instead of running the synthesis at runtime, it is more realistic to implement design alternatives offline and allowing to switch between the variants depending on the current demands or available resources of the system. Such a system was developed at the Interuniversity Microelectronics Centre (IMAC) [MNC+03, MNM+04] and is based on a state machine model, as depicted in Fig. 4.13. The idea of this approach is to identify certain states at which the execution can be switched between hardware and software. When the systems takes the decision to change the implementation style, the task will continue operation up to the next change point and then invoke the OS for controlling the switching between the implementation alternatives. It is assumed that the OS can also migrate the internal state between the different implementations. However, the authors consider FPGA readback and bitstream manipulations for this purpose but restrict their system to tasks that run to completion and where no internal state has to be transferred from a software to a hardware implementation or vice versa. This model is considerable limiting the algorithms that are suitable for implementing hardware/software morphing and the technique cannot be applied if the next iteration of an algorithm depends on a prior run (e.g., when implementing a FIR-filter).

4.3.2 Modeling Hardware/Software Morphing

The checkpoint FSM model, which is based on finite state machines and which has been introduced for modeling the checkpointing of hardware modules in Sect. 4.1.2, is the base for modeling hardware/software morphing. Regardless of the final implementation in hardware or in software, a checkpoint FSM allows (1) to control and (2) to perform accessing the internal state of a ReCoNet task. But that model has to be refined for explicitly allowing *state-dependable hardware/software morphing*, as revealed in the following section. While the proposed software and hardware checkpointing techniques are the base to read or to write the state of a task, focus will now be put on how and when the state can be translated between a hardware and a software refinement of a task specification. Such a specification (that is independent with respect to its refinement) is assumed to be the base for implementing both, the hardware task refinement, as well as the software counterpart.

In a system providing the morphing of tasks, the set of task implementations $\mathcal{T}$ is bipartitioned in hardware $\mathcal{T}^{HW}$ and software tasks $\mathcal{T}^{SW}$, where $\mathcal{T} = \mathcal{T}^{HW} \cup \mathcal{T}^{SW}$ with $\mathcal{T}^{HW} \cap \mathcal{T}^{SW} = \emptyset$. When performing hardware/software morphing, the implementation style is changed from hardware to software, or vice versa. However, changing the implementation style of a task also affects its binding and, for example, a hardware task, that was bound to an FPGA will have to be bound to an ISP after morphing. Consequently, this will affect the communication binding and the task interfaces, and hence the interface implementation as well.[3]

Considering the FSM-based task model, introduced in Sect. 4.1.2 on Page 203, the state $\mathcal{S}$ of an implementation independent task specification will be transformed into a software state $F_{SW} : \mathcal{S} \rightarrow \mathcal{S}^{SW}$, when compiling the specification to a program binary or into a hardware state $F_{HW} : \mathcal{S} \rightarrow \mathcal{S}^{HW}$, when synthesizing the specification to a netlist for generating the FPGA configuration data. The two functions F_{SW} and F_{HW} denote a mapping of states specified by an abstract specification to states encoded in the entire hardware or software refinements. $\mathcal{S}^{HW}$ is an abstraction of the state of a hardware module and $\mathcal{S}^{SW}$ an abstraction of the state of a software process (represented, e.g., by CPU registers or variables), respectively. In general, $\mathcal{S}^{HW}$ and $\mathcal{S}^{SW}$ are not equal to $\mathcal{S}$, because some elements in $\mathcal{S}$ may be trimmed during optimization. On the other hand, the refinements will typically increase the state space, thus, $|\mathcal{S}_{SW}| > |\mathcal{S}|$ and $|\mathcal{S}_{HW}| > |\mathcal{S}|$. For instance, when utilizing a library function, this may result in additional internal states not specified in $\mathcal{S}$. Furthermore, the hardware and software refinements are in general not identical: $\mathcal{S}_{HW} \neq \mathcal{S}_{SW}$. For example, assuming the following function:

```
boolean example function{A, B, C, D : boolean}
  {    tmp1 = A AND B;
       tmp2 = C AND D;
       return{tmp1 OR tmp2};   }
```

[3]Refer to [JW05] for a comprehensive overview on hardware/software interface techniques.

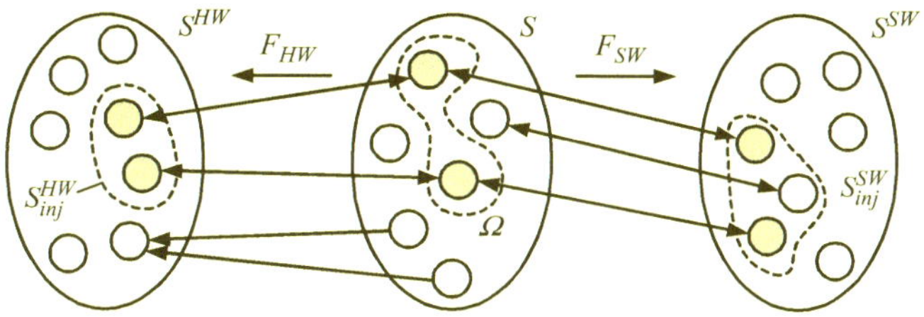

Fig. 4.14 Starting from an implementation independent specification with the states $\mathcal{S}$, compiling will transform $\mathcal{S}$ to software states $\mathcal{S}^{SW}$ and synthesizing will transform $\mathcal{S}$ to hardware states $\mathcal{S}^{HW}$. Due to possible optimizations, not all states of $\mathcal{S}$ will be mapped and the number of states typically increases after compilation or synthesis. Only from the subsets $\mathcal{S}^{HW}_{inj}$ and $\mathcal{S}^{SW}_{inj}$ that denote the mapping of the injective fractions of the functions F_{SW} and F_{HW}, it is possible to uniquely identify a state in the original state set $\mathcal{S}$. Morphing is only possible in states $\Omega \subseteq \mathcal{S}$ that can be uniquely identified from both, the software and the hardware domain.

then a software implementation of this function will execute the instructions in line 2–4 sequentially on an ISP, while it is possible to evaluate the complete function in parallel by a single look-up table, if it is mapped to an FPGA. In the latter case, all input operands are directly connected to the table inputs and the states of the signals tmp_1 and tmp_2 will be eliminated.

The example points out that morphing is not possible in all states of $\mathcal{S}^{HW}$ and $\mathcal{S}^{SW}$. In general the mapping from $\mathcal{S}$ to $\mathcal{S}^{HW}$ and $\mathcal{S}^{SW}$ by the functions F_{HW} and F_{SW} is not injective. Consequently, identifying a state from the hardware or software refinements that has been mapped from $\mathcal{S}$ (i.e. the original specification) is only feasible within the subsets $\mathcal{S}^{HW}_{inj} \in \mathcal{S}^{HW}$ and $\mathcal{S}^{SW}_{inj} \in \mathcal{S}^{SW}$, as illustrated in Fig. 4.14. Then, the mapping $F^{HW}_{inj} : \mathcal{S}^{HW}_{inj} \rightarrow \mathcal{S}$ and $F^{HW}_{inj} : \mathcal{S}^{HW}_{inj} \rightarrow \mathcal{S}$ denote injective functions from the refinements back to the states defined in the original specification. Note that such states $\mathcal{S}^{HW}_{inj}$ and $\mathcal{S}^{SW}_{inj}$ can be forced,[4] such that morphing is possible in a definable subset of the states that will be called *morph points* $\Omega \subseteq \mathcal{S}$ (see also Fig. 4.14).

Definition: Morphpoint A *morphpoint* represents a state $s \in \Omega \subseteq \mathcal{S}$ for which equivalent counterparts exist in both refinements, the hardware and the software domain:

$$\Omega = F^{HW}_{inj}(\mathcal{S}^{HW}_{inj}) \cap F^{SW}_{inj}(\mathcal{S}^{SW}_{inj}).$$

For identifying such morph points, the state $\mathcal{S}$ has to be separated into two sets: (1) a state that is related to the control path $\mathcal{S}_{ctrl}$ and (2) to the data path $\mathcal{S}_{data}$, with $\mathcal{S} = \mathcal{S}_{ctrl} \times \mathcal{S}_{data}$. Morph points are defined in $\Omega \subseteq \mathcal{S}_{ctrl}$ and the state of the data path has to be transformed into the entire other domain (hardware or software) by translating between the state encoding of $\mathcal{S}^{SW}_{data}$ and $\mathcal{S}^{HW}_{data}$.

[4]This can be achieved by preventing possible optimizations. For example, synthesis tools typically provide *keep pragmas* for maintaining specific signals even if optimizations could remove these signals (e.g., the signals tmp_1 and tmp_2 in the Boolean example function).

The checkpoint FSM model, presented in Sect. 4.1.2 on page 203, includes explicitly that saving a checkpoint can be restricted to be performed only within a set of definable states (called checkpoint states $\mathcal{S}_c$). Then the checkpoint states are a subset of the states and the morphpoints are a subset of the checkpoint states: $\Omega \subseteq \mathcal{S}_c \subseteq \mathcal{S}$. Note that $\mathcal{S}_c$ can be defined different for hardware and software tasks. Additionally, it is possible to take checkpoints that are not morphable, if $\mathcal{S}_c \setminus \Omega \neq \emptyset$. This may be used, for example, if the software implementation of a task is much slower than the hardware counterpart, thus requiring much more time to rollback and recompute a task in case of a fault. Then, taking software checkpoints more frequently than just at the morphpoints may reduce the worst case time to recover from a fault.

4.3.3 State Translation

For providing hardware/software morphing without losing internal states, the hardware and software state encodings must be translated among these domains. Each implementation style may use individual encodings for data items, like enumerated states, variables, or constants. The encodings will commonly determined during software compilation or hardware synthesis. However, there exist corresponding data types in the hardware and the software domain and some of them can be directly exchanged. For instance, integer types exist in hardware as well as in software. Consequently, it might be possible to use an identical encoding in both, the hardware and the software refinements. In general three strategies for translating a state between hardware and software have to be considered:

Direct state exchange allows to use the same identical state for the software and the hardware implementation of a task. For copying, checkpointing techniques can be used. Note that this may require to change the order of the individual data items that are stored in a checkpoint.

State translation via inverse functions. Here, the idea is to firstly translate the state back into the original state space $\mathcal{S}$, and then, to use the mapping functions (F_{SW} and F_{HW}, depending on the morph direction) to translate the state into the target domain. This requires that both, the mapping functions and the inverse mapping functions are known.

Direct state translation aims on directly translating the state encoding without the intermediate step via the original state space $\mathcal{S}$. This might be possible, as data types are defined in the original specification and when their encoding in the hardware and software refinements are known.

These strategies may be mixed in a hybrid fashion for different parts of the state. The following paragraphs will reveal the state translation of various data types more detailed.

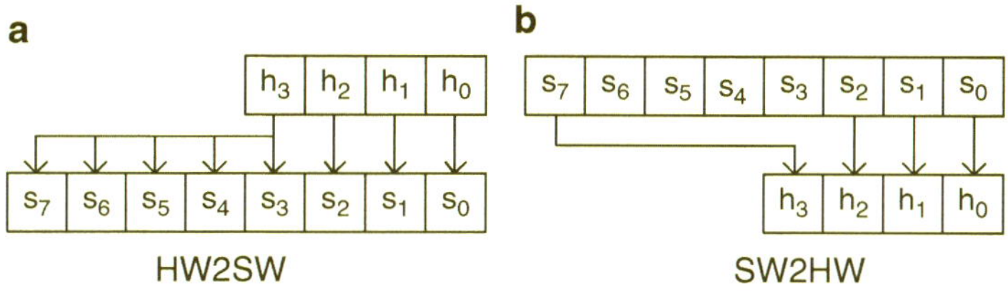

Fig. 4.15 State translation between a byte value in software and a 4-bit bitvector in hardware

4.3.3.1 Integers

While in the software case, integer types are coded in fixed sizes (e.g., as a byte or as a 32-bit signed/unsigned), the hardware counterparts are more flexible as they are based on *bitvectors* that are optimized individually for specific number ranges. The different integer number encodings can be converted when respecting the particular number ranges. Assuming the bit vector representations of the two integer values $Int^{SW} = \{Int_m^{SW}, \ldots, Int_0^{SW}\}$ and $Int^{HW} = \{Int_n^{HW}, \ldots, Int_0^{HW}\}$, then their translation between hardware and software encoding is as follows (see also Fig. 4.15):

$$translate\_Int_{HW2SW} = \begin{cases} Int_i^{SW} = Int_i^{HW} \ \forall i \leq n-1 \\ Int_i^{SW} = Int_n^{HW} \ \forall i \geq n \end{cases}$$

$$translate\_Int_{SW2HW} = \begin{cases} Int_i^{HW} = Int_i^{SW} \ \forall i \leq n-1 \\ Int_n^{HW} = Int_m^{SW} \end{cases}$$

For integer representations in hardware modules, further optimizations behind the truncation to the required size may apply. This comprises replacing unchanged bits of a register with constant values and a substitution of multiple bits that always possess an identical state to a single bit reference. Such optimizations for constant values or redundancy can be handled by considering report files that are generated by the synthesis tools and allow to reconstruct integer representations for software execution or to compose compatible bitvectors to be used within the hardware. Alternatively, synthesis tools can be forced to omit such optimizations. This may be set with the help of dedicated pragmas, special synthesis options, or by putting such integer registers into the interface of the module.

Boolean values are a special case of an unsigned integer with the number range from 0 to 1. Consequently, Boolean values are represented with a single bit signal in hardware. In software, this type is usually mapped to an integer.[5]

[5]The size of the used integer number depends on language and compiler version. For example, Visual C++ 4.2 equates bool with an 32-bit integer inside the standard C++ header files that contain the type definition, while all later versions implement Boolean variables with a size of 1 byte.

4.3.3.2 Enumerated Type

Besides integer types, it is possible to define enumerated types that are in particular used to specify states in state machines. While in the software domain enumerates are mapped per default to consecutive integer values, this is unusual in hardware. One well known format for state encoding in hardware is *one-hot encoding*. In this format, an enumerated type with n states is mapped to an n-bit wide bitvector and each state is encoded by an individual 1-out-of-n code. With this encoding, it is possible to determine a specific state by evaluating a single bit value.

The principally different state representations of enumerated types in software and hardware can be translated using the following techniques: (1) forcing the encoding of enumerates in both domains to predefined values (e.g., one hot), (2) forcing the hardware domain to use the same state encoding as determined during software compilation,[6] (3) forcing the software domain to use the same state encoding as determined during hardware synthesis, and (4) generate a translate function that performs a mapping between states in the hardware or software domain (e.g., by a state look-up table). As the state encoding has massive impact to the hardware module with respect to the performance (achievable clock frequency) and cost (logic resources), the latter two techniques should be used.

4.3.3.3 Further Discussion on State Translation

With the capability of translating integer types and enumerated types, hardware/software-morphing can be applied to the majority of tasks that are suitable for an implementation in both domains. Floating point numbers, as a further common data type, are costly to be implemented in hardware on FPGAs and various optimization techniques are typically applied for this data type, like dedicated number representations that differ from the standard IEEE 754 format or special circuitry like operator chaining [DL09]. Consequently, translating the state of floating point can be very complex and might prevent to apply morphing.

A further issue, when translating states between hardware and software, is how tasks represent *hierarchies* and how the hierarchy level is controlled. In the software domain, it is common to use a stack that represents hierarchies and functions can be reentered, thus, allowing recursion. In contrast to this, a task implemented in hardware may typically express hierarchies by communicating state machines and recursion is not appropriate to be implemented in hardware. Furthermore, multiple dedicated instances of the same or different functionality may run in parallel in hardware tasks. These issues can be solved by selecting morphpoints at states that will not result in hierarchies, like, for example, idle states.

---

[6]Note that in the VHDL language, the state encoding of enumerated tapes is the only specified synthesizable attribute in this language [Ash08].

The state translation can either be done in the hardware or in the software domain of the system, depending on allowed latencies and on the rate how often a state translation process is required.

With these restrictions, a state translation that is required for stateful hardware/software morphing is possible and the execution style of a task can be transparently exchanged at runtime. In Sect. 4.3.4, a tool with a corresponding design flow will demonstrate that morphing capabilities can be automatically added to a task specification in order to implement both, the morphable hardware and the corresponding software task that are based on the same specification. This design flow will be used to implement a case study that is evaluated in Sect. 4.3.5.

4.3.4 Design Flow for Implementing Morphable Tasks

For implementing morphable tasks for a ReCoNet, a design flow was developed around the tool StateAccess, which is introduced in Sect. 4.2.4. With this tool, hardware modules can be automatically extended with logic for accessing all or a definable part of their internal flip-flops. The tool includes also a FSM into the module to control the checkpointing, which, in particular, allows to trigger the checkpointing with signals from the module itself, which will be used to take checkpoints only if the hardware module is in a morphpoint.

The design flow for implementing morphable tasks is illustrated in Fig. 4.16. Based on a C++ specification of the task functionality, the software binary is compiled. In addition, an RTL model is synthesized with the help of an behavioral compiler that uses the same C++ specification. Note that the behavioral compiler is capable to build design alternatives that may be used for morphing between multiple hardware variants. With multiple hardware implementation alternatives, the system may adapt more flexible to environmental changes with hardware modules that are optimized for different scenarios (e.g., throughput demands).

For each RTL model, StateAccess includes additional logic into the module for accessing its internal state. This can be controlled in multiple ways. Besides selecting between various options for accessing the state, the logic generation can be further controlled with a register description file. By modifying this file, StateAccess can be guided to only include a subset of the flip-flops into hardware checkpoint. Furthermore, it is possible to sort the bitvectors such that there is no bit shuffling required during morphing at runtime. Note that the identifiers of the original C++ specification will be conserved within the RTL model as well as in the register description file. The result of the transformation by StateAccess is a modified netlist that is used together with templates (for interfacing the hardware checkpointing logic with the CPU) for generating the final module configuration bitstream.

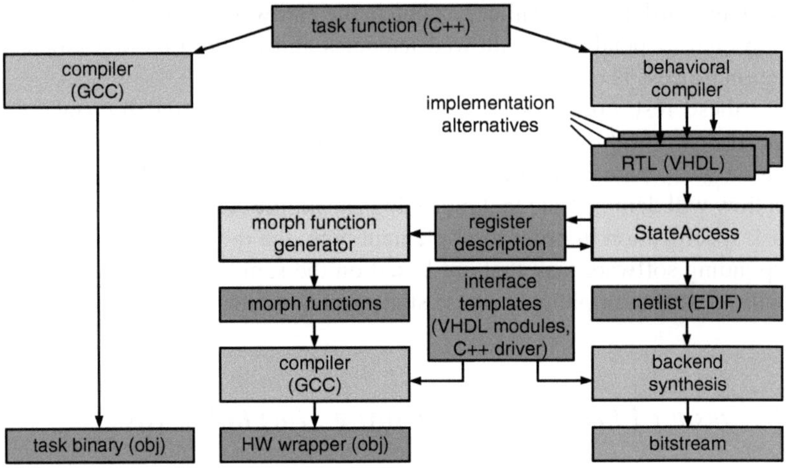

Fig. 4.16 Design flow for implementing hardware/software morphable tasks. Starting from a task function, the task binary (software implementation) as well as an RTL model (hardware implementation) is generated. The RTL model is automatically transformed for taking and restoring hardware checkpoints with the help of StateAccess. StateAccess further provides a register description file that is indicating the flip-flop mapping within the checkpoint. An example of such a register description is given in the next section. With this information, morph functions for translating the state encoding between hardware and software execution are generated, Together with some predefined templates for carrying out the interprocess communication, the binary for the hardware task will be generated

4.3.4.1 Automatic Hardware/Software State-Translation

The final register description file is used to automatically generate morph functions. This is obtained by firstly scanning the file for figuring out the number, names, and sizes of all bitvectors occurring inside the hardware checkpoint (`HW_CP`). With this information, a struct storing the software checkpoint (`SW_CP`) is generated that combines all different bitvectors such that they are represented by the smallest fitting (unsigned) integer type. The following register description file denotes an example for a hardware checkpoint consisting of two registers (`Reg_A` and `Reg_B`) that are each 2 bit wide (`reg_0` to `reg_1`):

```
HDL_hierarchy_Reg_A_reg_1    // LSB of HW_CP
HDL_hierarchy_Reg_B_reg_0
HDL_hierarchy_Reg_B_reg_1
HDL_hierarchy_Reg_A_reg_0    // MSB of HW_CP
```

The generated struct representing the corresponding software checkpoint is:

```
struct SW_checkpoint {
    unsigned char Reg_A;
    unsigned char Reg_B;
} SW_CP;
```

After this, the actual state translation functions are generated:

```
SW_CP translate_to_SW(HW_CP) {
  SW_CP.Reg_A = 0;
  SW_CP.Reg_B = 0;
  SW_CP.Reg_A = SW_CP.Reg_A | ((HW_CP & 0x01) << 0x01);
  SW_CP.Reg_B = SW_CP.Reg_B | ((HW_CP & 0x06) >> 0x01);
  SW_CP.Reg_A = SW_CP.Reg_A | ((HW_CP & 0x08) >> 0x03);}

HW_CP translate_to_HW(SW_CP) {
  HW_CP = 0;
  HW_CP = HW_CP | ((SW_CP.Reg_A & 0x02) >> 0x01);
  HW_CP = HW_CP | ((SW_CP.Reg_B & 0x03) << 0x01);
  HW_CP = HW_CP | ((SW_CP.Reg_A & 0x01) << 0x03); }
```

The state translation is carried out with mask and shift operations completely in software. The entire bit-masks are adapted to the maximal amount of correctly ordered consecutive bits of a subvector within the hardware checkpoint (HW_CP). This helps, for example, to translate the state of Reg_B in just one line. By moving the first line of the register description file to the end (before synthesizing the hardware module) also Reg_A can be translated in a single line.

The integration of the translate functions into the software task has currently to be completed manually. A further manual step is required if a bitvector represents an enumerated type. In this case, a state encoding look-up table has to be extracted from the report files, generated by the hardware synthesis tool, and integrated as well. Following these steps, tasks for hardware/software morphing can be implemented almost fully automated.

4.3.5 Morphing Case Study

In this section, the implementation and experimental results of applying hardware/-software-morphing to a FIR-filter module will be presented. It will be exemplary demonstrated that morphable tasks can be automatically implemented starting from the same C++ specification of a function that contains the functionality of a task. A FIR-filter has been chosen, because it contains an internal state that is changed with every invocation of the filter. Furthermore, it can be easily compiled to different implementation alternatives with the behavioral compiler (Catapult C). For testing hardware/software morphing, a 16-tap FIR-filter task with constant coefficients and a 32-bit wide data path was implemented. Upon receiving a new input value via the message passing system, the task computes the next output value either in software or in hardware. For the latter one, a module is used that is implemented following the flow presented in the last section.

After the behavioral synthesis of the C++ specification, the resulting RTL model uses 563 flip-flops and 1,223 look-up tables to implement the filter. After applying a constraint for keeping all tap registers that represent the actual state of the filter and after modifying the netlist to allow hardware checkpointing with StateAccess, the module requires 571 flip-flops (+4%) and 1,890 LUTs (+54%). For the experiment, the flip-flops of the module have been made accessible using the simple scan-chain technique (see also Page 209). The achieved clock frequency is in both cases 20 MHz when implemented on an Altera EP2C20-7 FPGA. The test system provides a NIOS-II softcore CPU that requires 1,559 flip-flops (not including the register file that is implemented in block-RAMs) and 2,718 LUTs. The CPU and bus clock frequency in this system is 50 MHz while the FIR-filter module runs with a separate clock of 20 MHz. Synchronization between the two clock domains is achieved with a simple handshake protocol.

The measured execution time of processing the filter as a software task is $t_{SW} = 60\,\mu s$. When the task is morphed to hardware the execution time decreases to $t_{HW} = 2,7\,\mu s$, which results in a speed-up of 22. The values are acquired with a hardware timer for precisely measuring the processing time for one message. Similarly, the time was measured that is spent for morphing between software and hardware execution, which is $t_{morph} = 104\,\mu s$. Most of this time is required to copy the state between the two implementation alternatives. Note that different implementation alternatives of the FIR-filter module would result in roughly the same time, because the number of flip-flops and the position within the checkpoint would be identical.

In this example, it will take two software task activations before the CPU load will be reduced by the accelerator. Note that all variables have been identically specified as 32-bit integer values for the hardware as well as for the software implementation, thus, no translation of data items between the hardware and software domain is required. When omitting any other latencies caused by the operating system, the morphing is performed within $L = t_{morph} + max\{t_{HW}, t_{SW}\} = 164\,\mu s$.

4.4 Chapter Summary

In this chapter, it was demonstrated that the unique capability of partial runtime reconfiguration provided by certain FPGAs can be used for self-adapting and self-repairing a ReCoNet at runtime. This was shown at the network level by migrating hardware tasks among multiple nodes as well as at the node level by providing the ability of dynamically integrating hardware tasks on a single ReCoNode [HKT03, SSK+07, SKHT06]. Additionally, with the introduction of hardware/software morphing, the flexibility was further enhanced by allowing to change the execution style of a task between different implementation alternatives at runtime [KHST07]. These novel techniques have been introduced based on a formal model, that, in particular, provides state dependent migration of hardware tasks, as well as a state dependent swapping of the execution between hardware

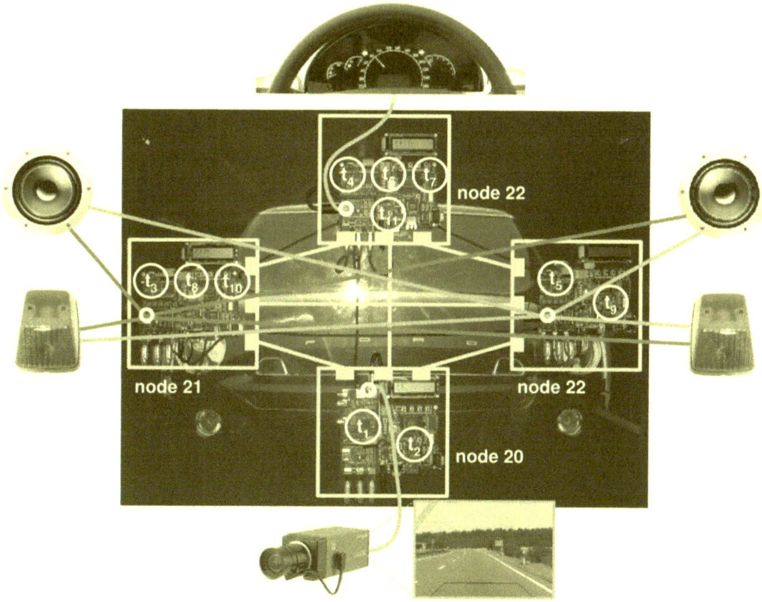

Fig. 4.17 The ReCoNets demonstrator. Four control nodes, each equipped with an FPGA, host multiple hardware and software tasks from the automotive domain in a ReCoNet. Among other things, this system features a driver assistance application for acoustically notifying the driver in case of an unintended lane change. For this reason, the current lane on the street is detected by evaluating a video stream after segmentation and by combining the result with the state of the direction-indicator control. The application is partitioned in multiple sensor, controller, and actuator tasks that are distributed to the ReCoNodes. Note that some actors can be controlled redundantly from two nodes, e.g. the speakers can be controlled by node 21 and node 22. In this system, link defects can be automatically compensated by rerouting traffic over alternative paths. Additionally task or node defects can be self-repaired with the help of task migration and hardware/software morphing. While sensor and actuator tasks are bound to the corresponding connected nodes, all control tasks can be arbitrarily migrated within this ReCoNet

and software implementations of a task. All these techniques can be transparently applied to the application tasks that may be migrated or morphed without losing internal states, hence seamlessly providing their service even when balancing load between different nodes or optimizing the partition between hardware and software load (used FPGA area or CPU time).

All these techniques have been integrated in the ReCoNets demonstrator for proving the feasibility of the self-adaptive and self-repair methodologies based on an automotive application example [SKHT06, CS1]. As revealed in Fig. 4.17, each ReCoNode of the demonstrator provides an Altera Cyclone FPGA. These devices do not support partial runtime reconfiguration and reconfiguration was emulated by implementing multiple design alternatives in parallel that are activated according to the present task binding in this ReCoNet.

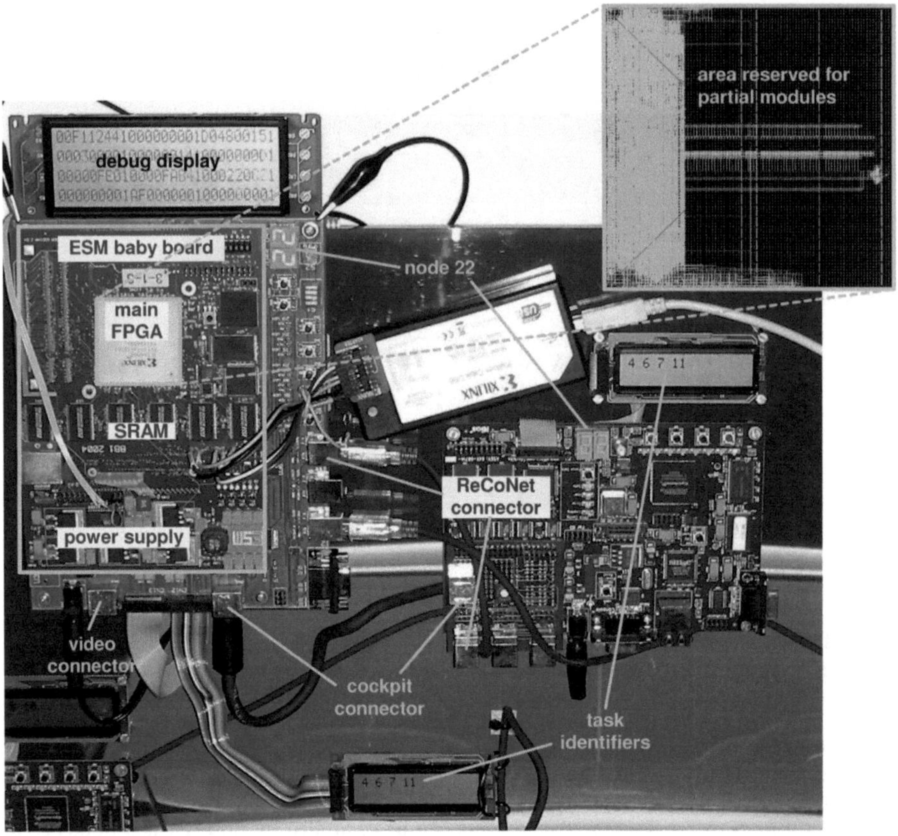

Fig. 4.18 ReCoNets demonstrator with an ESM. The ESM consists of a mother boards that was especially designed for this demonstrator. The board provides additional 8 MB SRAM, a compact flash card slot for storing the firmware and configuration data of partial modules, a video capture device, and connectors for the ReCoNet links and for the sensors or actuators. This hardware platform provides the same functionality as the original ReCoNodes but provides in addition the capability of partial runtime reconfiguration. The top right corner depicts the partition of the main FPGA (a Xilinx Virtex-II 6000) into a static region and an area for hosting reconfigurable modules. The system provides a ReCoBus and I/O connections bars to integrate partial hardware tasks into a ReCoNode at runtime. The system was implemented using the ReCoBus-Builder from Sect. 3.2

For providing partial self-reconfiguration at runtime, the *Erlangen Slot Machine* (ESM) [MTAB07] has been integrated into the demonstrator, as depicted in Fig. 4.18. The ESM platform is a modularized design consisting of a baby-board – providing a large FPGA, six banks of fast SRAM, and configuration logic – and a main-board containing the periphery. The large main FPGA is a Xilinx Virtex-II 6,000 device featuring 6 million system gates [Inc05b]. The Virtex-II family supports dynamic partial self-reconfiguration and multiple configuration options are available on the ESM including serial JTAG, parallel select map via a dedicated configuration controller and the internal configuration access port (ICAP).

The resulting ReCoNet demonstrates that state dependent hardware task migration and checkpointing can be applied even to heterogeneous FPGA environments with ReCoNodes based of FPGAs from different vendors.

With this ESM, the integration of hardware tasks as well as task morphing between hardware and software execution was implemented supported by the tool ReCoBus-Builder (see Sect. 3.2). By using the ReCoBus and I/O connection bar communication architecture, various modules can be dynamically integrated into the system or morphed without interfering other parts of the system.

Chapter 5
Reconfigurable CPU Instruction Set Extensions

Swapping just small fractions of the configuration of an FPGA can be very beneficial in many applications. This is in particular useful for reconfiguring the instruction set of embedded soft core processors. This is highly relevant for software driven design flows. Here, the system is initially implemented as far as possible in software (which is faster to accomplish than hardware development). By profiling the application, hot spots will be identified and kernals will be implemented for the FPGA for acceleration until performance requirements are met. There are several methodologies to integrate such accelerator modules. This ranges from small CPU instruction set extensions to large and fully autonomous modules that work concurrently with the CPU.

In this chapter, we will investigate how CPU instruction set extensions can be used efficiently with the help of partial runtime reconfiguration. The base idea of extending a CPU with exchangeable instructions is sketched in Fig. 5.1. Custom instructions access the register file in the same way as the ALU. By decoding unused instruction in the CPU ISA (instruction set architecture) a multiplexer may select between normal ALU operation or one or more user defined instructions.

Softcore CPUs with statically implemented custom instructions are well supported. For example the NIOS-II CPU from Altera can be easily extended with custom instructions when using the SOPC builder wizard of the Quartus design tools. Similarly, Xilinx allows to add custom hardware to thier Microblaze softcore CPU using FSL ports. These ports provide basically a streaming port interface between the Microblaze core and the custom hardware. However, for implementing *runtime reconfigurable* custom instructions, the support is weak, hence omitting this powerful opportunity.

In the following section, we will firstly demonstrate that commonly used techniques, like the Xilinx bus macro approach or the recent proxy logic technique is not well suited for integrating custom instructions. After this, in Sect. 5.2, we will demonstrate for a reconfigurable soft core processor that instructions can be integrated into the system without causing any additional logic overhead for the communication. In Sect. 5.3, we reveal how such systems can be easily implemented with the tool *ReCoBus-Builder*. Rather than providing reconfigurable islands, we

D. Koch, *Partial Reconfiguration on FPGAs: Architectures, Tools and Applications*, Lecture Notes in Electrical Engineering 153, DOI 10.1007/978-1-4614-1225-0_5,

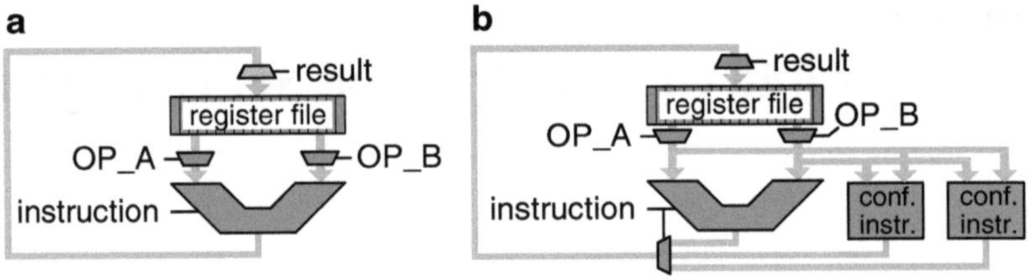

Fig. 5.1 (**a**) Example of a CPU data path. (**b**) Extension of the ALU with multiple configurable instructions. The modules are connected to the operands from the register file and a multiplexer selects between the different results

will integrate multiple custom instructions in a slot-based fashion. Finally, in Sect. 5.4, an experimental evaluation of a system providing a MIPS CPU extended to support reconfigurable custom instructions will be presented.

5.1 On-FPGA Communication for Custom Instructions

One basic problem to be solved in the design of partially reconfigurable systems is to constrain the routing of the interface signals for a partial module during its physical implementation. As introduced in Chap. 2, there are several ways to accomplish this. However, when considering a custom instruction set extension, as shown in Fig. 5.1, we have to consider that a relatively large wire count is required for connecting a relatively small reconfigurable area. For example, if we consider a bit permutation function, we have to connect 32 wires towards an island hosting the permutation accelerator and an additional 32 wires for the result. Note that this example would not demand any logic on the FPGA as a permutation is basically wiring on the FPGA. But still, as compared to a software implementation of a permute function, we can easily save a hundred or more assembly instructions, even if fully unroling the function. Or, if we consider a 64-bit XOR gate (over both operands) to be hosted in the same reconfigurable island, it requires two times 32 wires for connecting the input operands. But also in this case, we need only 21 4-input LUTs (e.g., on a Xilinx Virtex-II FPGA) or 13 or 6-input LUTs on a (Xilinx Spartan-6 FPGA) for implementing the 64 bit XOR gate. Again, this instruction would save about a hundred instructions per call of the function. In other words, for some programs, we could gain a substantial speed-up by just adding little additional logic. And by making this configurable, we could host virtually an infinite amount of different accelerators for supporting various software tasks.

When implementing such custom instructions with slice-based bus macros, as illustratrated in Fig. 5.2a, it takes two LUTs per signal wire only for providing the accelerator connection. If we consider in total 100 wires for linking two times a 32-bit operand, a 32-bit result vector and a few additional signals, The overhead is 200 LUTs. This is roughly 10 × more than actually needed for the XOR gate! Moreover,

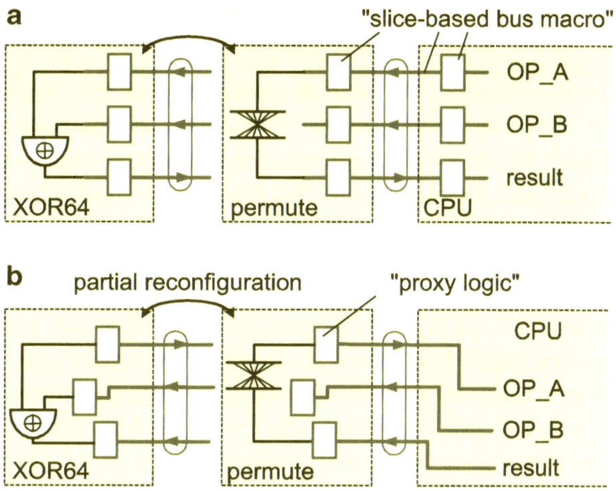

Fig. 5.2 Integrating custom instructions using (**a**) Xilinx bus macros, (**b**) proxy logic

the look-up tables constitute not only a logic overhead, but also a latency overhead which is roughly 0.4 ns on a Virtex-II FPGA per LUT. Finally, adding LUTs for the communication can negatively impact the placement of both, the static system and the partial modules. For example, a placed bus macro LUT interrupts carry chains and it can further force to spread a module over more area.

With the recent proxy-logic approach, the situation has improved, as shown in Fig. 5.2b. However, it still needs 100 LUTs for the communication. Again this is pure overhead in terms of resources and latency. And as explained in Sect. 2.4.5 on Page 71, the proxy logic approach is not well suited to implement systems with many different reconfigurable modules.

At this point, someone might think to use static only implementations instead, if custom instructions are that small. This is probably the better option for very few instructions. With a rising number of instructions, the CPU gets larger and consequently slower. When assuming the simplified diagram of a CPU datapath in Fig. 5.1a, the ALU contains a multiplexer for selecting between the different sets of instructions of the ALU (e.g., Boolean logic, simple arithmetic, shifter, etc.). This multiplexer is in the critical path and unlikely to be pipelined [Met04], and despite that an FPGA fabric is mainly based on multiplexers, it is poor in implementing wide input multiplexers (see also Sect. 2.6 on Page 104). If carefully applied, runtime reconfiguration allows to integrate more instructions while providing higher performance than a static system. Note that this is in many cases still valid even when considering the configuration overhead. Moreover, partial reconfiguration adds a flexibility to the system that allows to integrate hardware accelerators dynamically to a system like known from the software world.

5.2 Zero Logic Overhead Integration

In this section, we will demonstrate how the Xilinx vendor tools can be used to integrate reconfigurable instructions without any logic overhead. As shown in Fig. 5.3a, we are only interested in binding the signals between the static system (the CPU) and the partial modules (the custom instructions) to a preceisly defined wire of the fabric, called a "PR link", in the following. In order to occupy a wire segment (i.e., use a PR link), we need a path that will use this wire. In other words, there must by somewhere a primitive source (e.g., a LUT output) and another primitive destination (e.g., a LUT input) in our netlist with a requested connection from the source to the destination. However, this creates a path in our netlist but we have still not constrained the routing. This is done by generating blocker macros, that occupy a user specified set of routing resources such that the Xilinx vendor router cannot use these wires for further implementation steps. The blocker concept is introduced in Sect. 3.2.4. Note that we cannot constrain the routing directly in a way that we say "*use* $wire_x$ *for* $signal_y$". We are basically defining a *wire allocation* in a way that we define "*do not use* $wire_z$". However, if we ensure by our allocation, that there is only one possible path remaining, we can actually achieve our goal to bind a signal path to a wire.

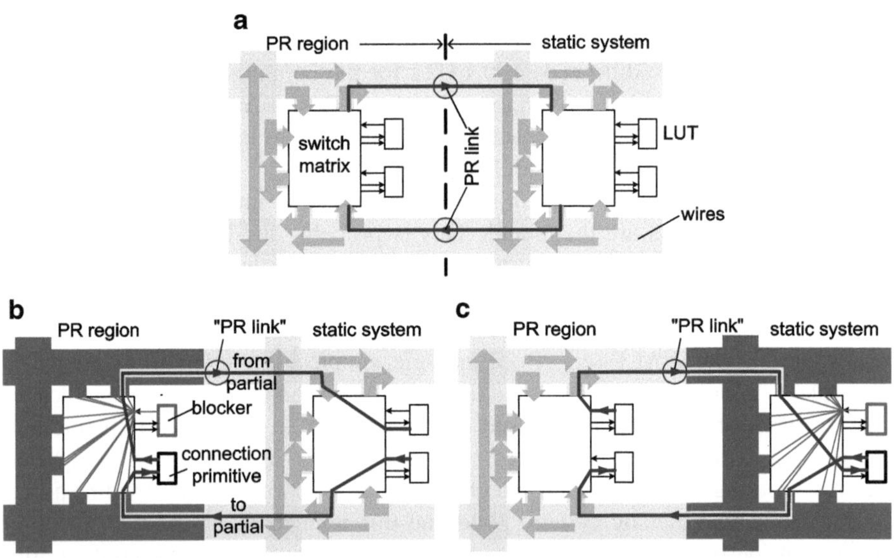

Fig. 5.3 Constraining the routing to use PR links. (**a**) The allocated wires to be used as PR links. These wires carry out the static-partial communication. (**b**) During the implementation of the static system, we place connection primitives into the reconfigurable region that act as a placeholders for the partial modules (here the custom instructions). In addition, we place a blocker *into* the reconfigurable region that conguests all wires except some tunnels for the PR link. (**c**) During the implementation of the partial modules, we place connection primitives into the static region that act as a placeholder fot the static system (here the CPU). In addition, we place a blocker *around* the reconfigurable region that conguests all wires except some tunnels for the PR link

5.2.1 Static System Constraints

With the knowledge of how to create a path and how to constrain this path to certain wire resources of the FPGA fabric, we can implement the static system. The static system contains the CPU and a reconfigurable region. In order to create paths into this region for connecting the operands `OP_A` and `OP_B` (see Fig. 5.1), we place a connection primitive into the reconfigurable region (PR region), as depicted in Fig. 5.3b. This primitive acts as a *placeholder* for the partial module and is the destination for the operand routing. Similarly, for creating a path for the result vector back to the CPU in the static part, we place a placeholder acting as the source for the path. Note that the same LUT primitive (or, to be more preceise, a slice) might be used as a placeholder for multiple input and output signals at the same time. So far, this seams to be pretty much identical to the proxy logic approach. However, we will now add a blocker into the reconfigurable region that blocks all routing resources in this region, except the wires to be used as PR links. Note, that the placement of the placeholders and the blocking is not random and has to support the intended PR link. If we now start the router, we will create the routing of the static system including paths to and from the partial region that are routed using the requested PR links. There are two things to remember: (1) we have not added any logic overhead to the static system, and (2) we only blocked wire resources *inside* the reconfigurable region.

5.2.2 Partial Module Constraints

The partial module implementation (here the custom instructions) is very similar to what we did for the static system. However, all signals directions are now changed and with respect to a custom instruction, the operands are no inputs and the result vector is an output. Consequently, we place a source placeholder as the start for the operands outside the reconfigurable region (i.e., the static region). Respectively, we add also placeholders acting as the destinations for the result vector. Again, placeholders for inputs and outputs can share the same FPGA primitive, as shown in Fig. 5.3. We will now add a blocker around the partial module that conguests all routing resources, except the ones needed to route the operands and results over the PR links. Here it is important, that the blocker releases PR links that are compatible to the PR links used in the static design. Again, there are two things to remember: (1) we have added no logic overhead to the static system, and (2) we only blocked wires *outside* the reconfigurable region. Consequently, when loading a reconfigurable instruction into an reconfigurable island that was created as described for the static system in the last section, there will be no placeholder module visible. The placeholders are only temporarily required to create a path over the PR link.

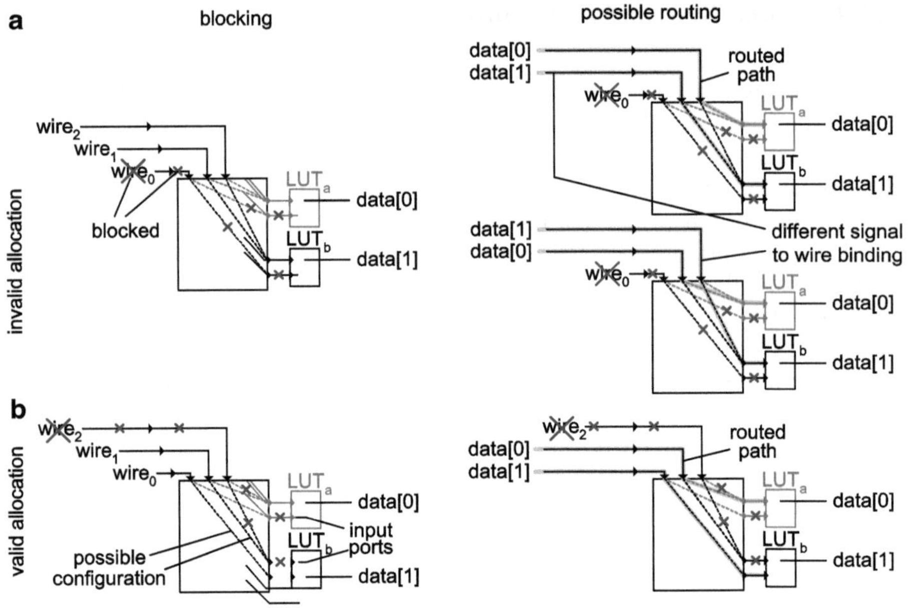

Fig. 5.4 Wire allocation for the zero overhead integration technique. (**a**) Allocating wire_1 and wire_2 results in an unpredictable routing, while (**b**) allocating wire_0 and wire_1 results in an unique routing

5.2.3 Communication Binding by Wire Allocation

The zero logic overhead technique has to follow some rules. Again, by blocking, we can only select the set of wires that are allowed for routing (i.e., wire allocation) but this does not necessary ensure a particular binding of a logical signal to a physical wire. However, the binding is achieved by allocating wires such that *only one unique routing path* can be used to reach the connection macro (see Fig. 5.4). As a consequence, not all wires within a CLB can be used at the same time to implement the routing between the static part and the partial part of a system. This is because in the case that multiple wires are routed from one configurable logic block (CLB) to another, wires must be allocated that cannot be swapped. A possible swapping of wires would allow the router to decide between more than one option for a PR link, which cannot be accepted. A situation of allocating swappable wire resources is shown in Fig. 5.4a. Here, the problem is that both allocated wires can be arbitrary used to connect to both placeholders that used for the data signals `data[0]` and `data[1]`. Consequently, the router has two possibilities to chose from and we cannot gurantee a signal binding to a specific PR link. However, by allocating a different wire set, we leave only one possible path per data signal and we achieve an exact binding to wires, as shown in Fig. 5.4b. Note that designing PR link paths needs deep knowledge about the FPGA routing fabric including wire resources and possible switch matrix settings. This information is provided by Xilinx individually for each FPGA in a language called XDL [BKT11].

5.3 Implementing Reconfigurable Instructions with the ReCoBus-Builder

The ReCoBus-Builder is originally designed for implementing bus-based systems consisting of many small resource slots that are integrated with the help of macros, as revealed in Sect. 3.2. At this point, we focus only on macros implementing the connection bar architecture (Sect. 2.6.1). For implementing the zero logic overhead approach, we follow the original ReCoBus-Builder flow and perform resource budgeting and define a floorplan that fulfills the resource requirements. Then, we create our communication architecture that will provide connection primitives in the static part of the system as well as in each resource slot. Let us consider the simple case of a connection bar to connect only a single resource slot. We would then basically generate a Xilinx bus macro for an island reconfiguration style. When following the default ReCoBus-Builder flow, we will generate two blocker macros, one for the static design and one for the reconfigurable modules. We will use these blockers for implementing the PR link approach shown in Fig. 5.3. As the blockers generated by the ReCoBus-Builder will not block the wires that are already used for the connection bar macro, the blocker will contain a tunnel for a PR link. The only thing that is now missing are the placeholder primitives. These primitives are taken directly from the generated connection bar macro. Consequently, we can generate compatible placeholder/blocker pairs for both the static system and the partial modules. If we assume a connection bar with one internal wire towards east and another wire towards westwards direction, the resulting primitives and blockers would match the example in Fig. 5.3. The ReCoBus-Builder has a wire database for each supported device. This is used by the tool to check if a wire allocation can ensure PR links without possible swaps as discussed in the last paragraph. With this approach, we can provide four double wire PR links per CLB on a Xilinx Virtex-II FPGA.

As a case study, we consider to integrate up to five different instructions into the system at the same time. Instead of using five individual islands for hosting the instruction modules (as it would be necessary following the Xilinx PR flow), the system uses a more flexible approach with one reconfigurable area that is tiled into five resource slots, as depicted in Fig. 5.5. This has the advantage that modules of different size can be more efficiently integrated into the system by taking a variable amount of slots. The communication architecture has to link the two operands to each slot and the result vector back individually for each slot to an instruction multiplexer. By using different wire resources for the operands and the result vectors that route over different distances, both requirements can

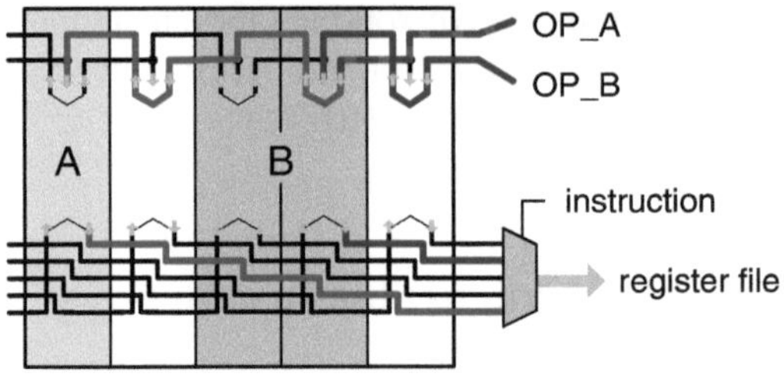

Fig. 5.5 Partial region of the reconfigurable ALU part. The slots can host different sized modules. Due to the interleaved routing of the operands with double lines and due to the available middle access in the middle of a double line, both operators can be accessed in each slot

be properly implemented. By taking advantage of the regular FPGA fabric, the slots can be arranged completely identically, hence allowing free placement of instructions into the reconfigurable ALU. Figure 5.5 reveals a detail of the routing architecture of Xilinx Virtex-II FPGAs that was used to provide slots that are smaller than the routing distance of a wire. In the example, it is assumed that one resource slot is only one CLB wide and that the operands are routed using *double lines* that route two CLBs wide. However, by using a connection in the middle of the wire, which is provided by the routing fabric after a distance of one CLB, and by displacing the start points of the regular routing structure of the two operands by one CLB in horizontal direction, both operands can be accessed in any slot. This is possible by routing the signals in an interleaved manner. Note that it is also possible to route paths by cascading multiple different wires, which would allow to widen the slots (in terms of CLB columns) and to extend the total amount of slots for hosting modules (see Sect. 2.5.2 on Page 81 for more details). The interleaving results in swapping the operands with respect to the placement position (odd or even start slot). However, for instructions that are not commutative, we can use two physical implementations in order to omit the alignment multiplexing. See Sect. 2.5.3 on Page 92 for more details on interleaving.

5.4 Case Study on Custom Instructions

The case study has been implemented with the ReCo-Bus-Builder on a Xilinx Virtex-II XC2V500-5 FPGA. The tool generates regular structured macros together with the surrounding blocker macros that constrain the routing. The implementation follows directly the methodology revealed in Sect. 5.2. The communication macros provide the connection primitives and fix the wire resources. The ReCoBus-Builer generates the all macros (including the blocker) in the Xilinx design language (XDL). While communication macros are instantiated using the HDL flow, the blockers are integrated into the design just before the final route step. A floor-planning view on the system is depicted in Fig. 5.6. The area reserved for hosting

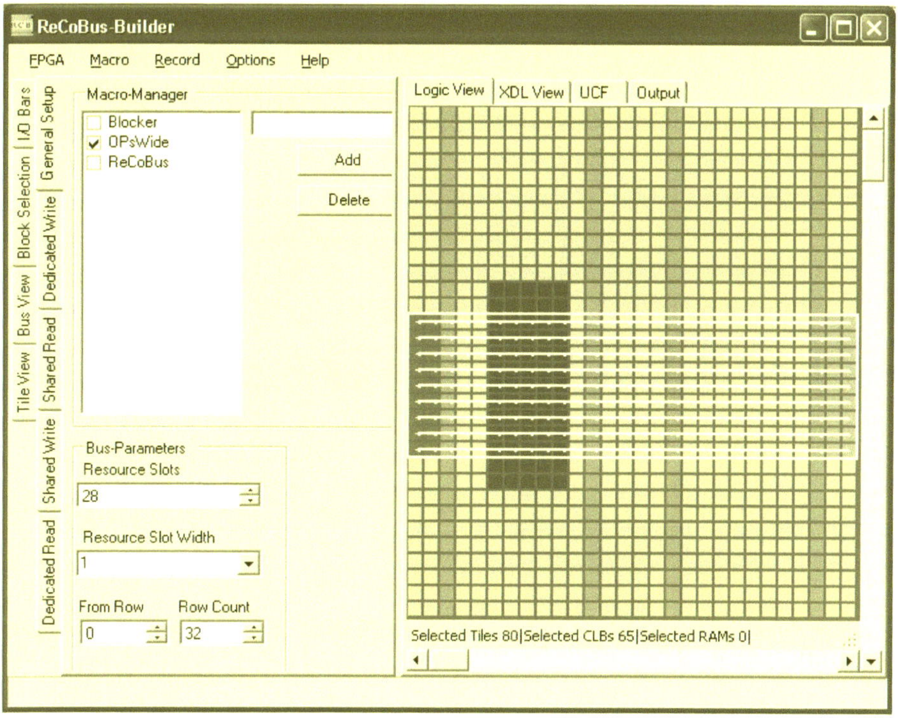

Fig. 5.6 Floorplanning view of the case study. Each gray square represents a CLB that provides eight 4-input LUTs. The five highlighted columns are reserved for hosting up to five different 32-bit instructions

reconfigurable instructions is 8% of the total amount of CLBs that are available on the used device. With five times 48 slices, the PR region provides roughly 15–20% the amount of logic that would be required by an optimized 32 bit soft core processor, such as the Xilinx Microblaze. For the experiments, we used our own MIPS processor implementation that has not been optimized for speed or area, but which can be easily adapted to include reconfigurable instructions.

5.4.1 Static System Implementation

During implementation of the static system, connection primitives that are placed inside the reconfigurable region and that are surrounded with blocker macros have been used to constrain all signals required to integrate the instructions. A screenshot with the static system is shown in Fig. 5.7. The amount of wires that are connected from the static part of the system to the PR region is 2×32 for the operands plus additional eight wires of control signals. In reverse direction, each one out of the five slot delivers a 32 bit result plus additional four flags. This results in a total amount of $64 + 8 + 5 \times (4 + 32) = 252$ wires.

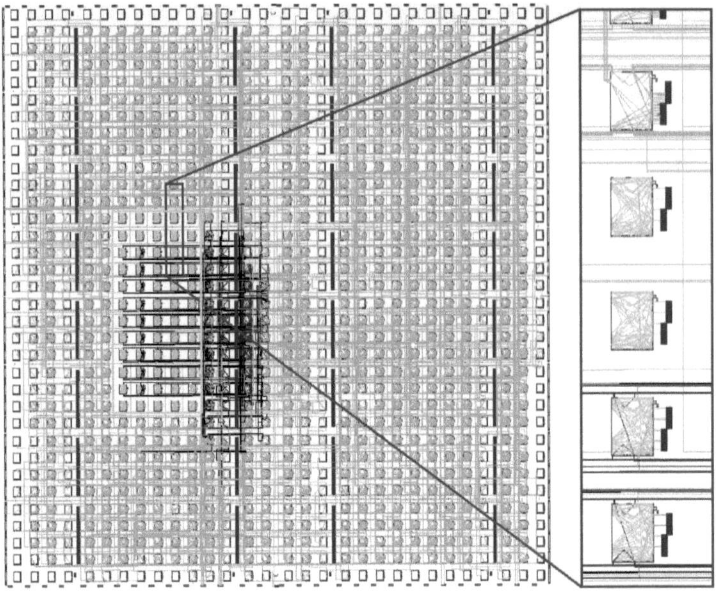

Fig. 5.7 Xilinx FPGA editor view of the static system. Blockers in the PR region prohibit routes of the static system

According to the partial design flow provided by Xilinx, the number of operand bits and control signals has to be multiplied by the number of slots, as that flow does not consider multicast routing to multiple slots without additional connection primitives. Then the slice based macro approach would cost $2 \times 5 \times (72+36) = 1{,}080$ LUTs only for the communication. This is 18% of the available LUTs on the target device and roughly one third of the logic a fully featured 32 bit Microblaze soft core processor would take. Even using the new flow that is based on proxy logic, would still result in a remarkable unnecessary overhead.

When floorplanning a reconfigurable system, it is recommended to consider the underlying FPGA architecture. For example, Xilinx FPGAs are column-wise reconfigured, which should be taken into account by designing the slots vertically. This optimizes the reconfiguration time. A restriction derived from the full column reconfiguration scheme is that no distributed memory can be used directly above or below the PR region as this would corrupt the state of these primitives. Following this rule, partial reconfiguration can be carried out while continuing the system to operate.

FPGAs provide carry chain logic, which are used for different kinds of arithmetic operations. On Xilinx FPGAs, the carry chains include four LUTs per CLB and the chains are arranged in upwards direction. Consequently, we built the system such that exactly two times four operand signal bits and four bits of the result vector are connected in a CLB. Furthermore, the signal vector bits are connected bottom-up (LSBs in the bottom) to follow the carry chain. Without this physical port

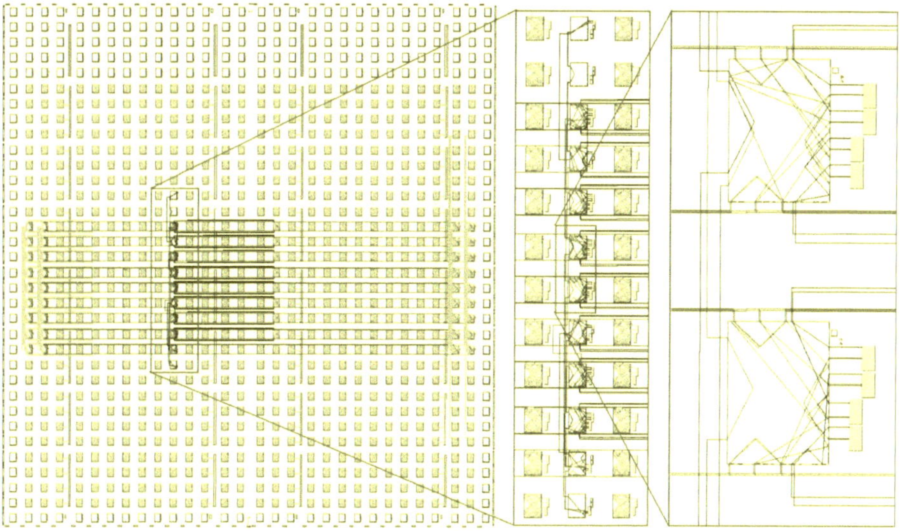

Fig. 5.8 View on the implementation of a CCITT CRC checker instruction

mapping, routing will get very conguested for the modules. In [CPF09], a tool using a simulated annealing heuristic was used to place communication macros around a reconfigurable region that was also used for reconfigurable CPU extensions. Such tools have an excessive runtime as they require a place and route step for each annealing step. It can be assumed, that the final result would be very similar to the here proposed *rule based port mapping* that needs only one place and route run.

5.4.2 Reconfigurable Instructions

For implementing the reconfigurable modules, the complete static system was substituted with a connection bar macro, as depicted in Fig. 5.8. This permits to implement reconfigurable modules in absence of the static system. As can be seen in Fig. 5.8 for a CRC checksum function, a module is surrounded with a blocker macro for restricting modules into strict bounding boxes. This design has no connections to external pins. The timing was constrained with the Xilinx TPSYNC parameter.

5.4.3 Results and Overhead Analysis

Swapping instructions comprises a significant time for writing the corresponding partial bit stream to the right target position. In addition, extra time might be required for computing a placement position or performing some bitstream

Table 5.1 Implementation and performance details

Instruction	# Slices	Slots	Latency (max/av)	b.stream (KB)	t_{conf} (ms)	SW (cycles)	k @ 50 MHz
64-bit XOR	19 (40%)	1	7.04 / 5.95 ns	2.64	0.6	61	492
CCITT CRC	33 (34%)	2*	5.32 / 3.98 ns	5.28	1.2	215/257*	279/233
sat. add/sub	70 (73%)	2	9.89 / 7.81 ns	5.28	1.2	12	5,000
Barrel shifter	90 (94%)	2	11.07 / 7.88 ns	5.28	1.2	143	420
'1' bit counter	214 (89%)	5	11.37 / 8.25 ns	13.2	3	102	1,471
Mask & permute	16 (33%)	1	5.94 / 4.05 ns	2.64	0.6	98	306

*Additional slot to accomplish routing

manipulations. This extra time overhead is implementation dependent and not further considered in the following. However, due to the small size of the systems, most work could be precomputed offline (e.g., a table for the placement position). When taking the decision to use reconfigurable instructions, it is important to know the latency that has to be considered for the reconfiguration process (response time) and the time the processor will require when executing the instructions alternatively as simple software function calls. This determines the breakeven factor k and the system has to trigger a reconfigurable instruction at least k times before gaining a benefit in the total execution time of the system. Note that we use function calls and no traps, as traps are very specific for emulating CPU instructions in software and because traps have a tiny additional overhead that would not occur in case of normal function calls. The configuration times and the execution times for software implementations of the custom instructions (determined in a simulator) are listed in Table 5.1.

The reconfiguration process is relatively slow and would consequently prevent using custom instructions in time critical parts of the software (e.g., interrupts). However this is not problematic as critical software parts should typically not perform complex computations. The breakeven factor k is the number of possible invocations of a particular instruction during the time to configure this instruction. As can be seen, for complex operations, such as the CRC instruction, less than 300 calls of this reconfigurable instruction would pay of the configuration overhead; and even if an instruction can save only a few cycles, this can pay of after just a few thousand cycles. Considering that the saturation addition/subtraction module is used in an image processing application, it can be assumed that it is very likely to trigger this function an sufficient amount of times. It must be mentioned that the listed values are theoretical and the breakeven points will probably be likely higher. This is because the configuration data transfer is in our system in conflict with the CPU (shared memory buses); and even having only a few KB of configuration data results in a burst affecting the CPU. However, reconfigurable instructions are still an interesting option for both saving FPGA resources and gaining performance.

The values in brackets denote the utilization within the occupied slots. Despite that the CRC logic would easily fit into one slot, an additional slot was required to fully route the module. The bitstream size states only the fraction of the partial module and no static parts. The reconfiguration time is mainly related to the amount of slots that have to be written to the device. A single slot configuration is 11.6 KB on this device which results in 0.6 ms configuration time, when assuming a configuration speed of 20 MB/s. The latency was determined using the FPGA editor. The values are measured between the operand fetching pipeline register through the combinatory path of the instruction and further towards the output of the instruction select multiplexer. The max value denotes the critical path delay and the average delay over all paths.

The examples point out that small FPGA areas are sufficient to include very valuable instructions into a CPU with the help of partial runtime reconfiguration. Despite the small slots, a high number of signals can be interfaced to partial modules.

Chapter 6
Concluding Remarks

In this book, methods architectures and tools have been proposed for runtime reconfigurable systems based on FPGAs. This results in substantial improvements in the efficiency of integrating reconfigurable modules into an FPGA, the tools to build such systems, and the capabilities to reconfigure a system at runtime. The key contributions are as follows:

Communication architecture for module-based on-FPGA communication: It has been proven that existing methodologies and communication architectures provide a tiling of the reconfigurable area that is too coarse-grained for most practical applications. With the novel ReCoBus and connection bar architecture, the size of the tiles is reduced by more than an order of magnitude while reducing the cost to implement the complete infrastructure and to increase throughput at the same time. This is the key technology improvement required to efficiently implement runtime reconfigurable systems.

Design flow and tools: For implementing the ReCoBus and connection bar communication architecture, a powerful tool has been developed that is capable to map the communication infrastructure in a regular structured manner on an FPGA. Furthermore, this tool provides a floorplanner capable to generate all location and resource constraints required to build the static system or the configurable modules.

Techniques for flexible high-speed runtime reconfiguration: For partial runtime reconfiguration, techniques have been proposed that allow to place modules in a two-dimensional fashion at high speed. In addition it was demonstrated how modules can be parameterized by configuration bitstream manipulation at runtime (e.g., setting a module base address). For a further acceleration of the configuration, additional techniques, including prefetching, configuration port overclocking, configuration decompression, and defragmenting the module layout have been examined.

With these architectures, methods, and tools, a technical basics for achieving a considerable benefit over completely statically implemented systems on FPGAs has been provided and demonstrated in different applications. This will help design

D. Koch, *Partial Reconfiguration on FPGAs: Architectures, Tools and Applications*, Lecture Notes in Electrical Engineering 153, DOI 10.1007/978-1-4614-1225-0_6,

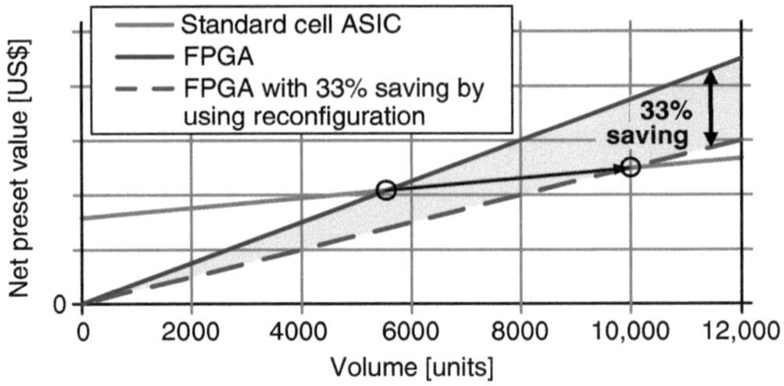

Fig. 6.1 Production cost of a hypothetical system over the units for an assumed ASIC or FPGA implementation (The values are taken from [Kot]). The FPGA has zero initial cost, but a much higher price per unit than compared with the ASIC solution. By utilizing partial runtime reconfiguration, the FPGA domain is shifted remarkable into higher volume regions, hence, squeezing the ASIC out of the mid range market

engineers in the field of reconfigurable computing to exploit the advantages of adapting systems at runtime. Additionally, the results are relevant for commercial applications and electronic design automation in general. With partial runtime reconfiguration, systems can be implemented cheaper and with lower power consumption by using smaller devices. Furthermore, this is an option to add additional functionality to existing systems. Even if only a relatively small part of the worldwide multi billion US$ FPGA market [EW0] can substantial benefit from runtime reconfiguration, the possible savings are many tens of millions US$ per year. However these savings will not decrease the total FPGA market. As illustrated in Fig. 6.1, the reduction of the FPGA unit cost – due to runtime reconfiguration – yields in a much wider range of designs that dominate an ASIC solution from an economical point of view. And the FPGA has further advantages, including faster time to market, possible after sales business by updates, lower economical risk due to the lower initial investment – just to mention a few of them. Moreover, runtime reconfiguration provides new opportunities in fast system start and system modularity. Consequently, partial runtime reconfiguration is advantageous for both, FPGA users and vendors.

A further important aspect is the fully encapsulated component-based design methodology that is advantageous for FPGA design and hardware development in general. With the here presented design flow, implemented and foremost fully placed and routed modules are composed together to complex systems based on interface specifications that also include the timing behavior. This enormously simplifies the system integration phase and addresses the demands of an easy design reuse to close the design productivity gap, as sketched in Fig. 6.2. With introducing design reuse, more functionality can be designed in less time and fully implemented modules will in particular ensure timing and reduce the module test expenses. Furthermore, the encapsulated design methodology prevents that changes in one part of the system will influence the rest of the system. Also the synthesis, and place and route itself will be accelerated as they can be carried out fully parallel for each module and

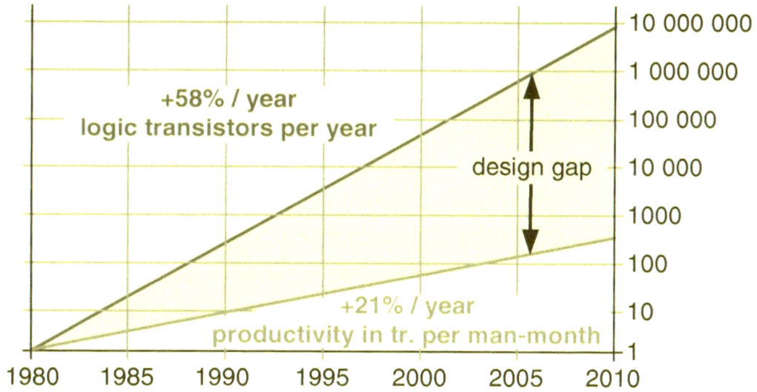

Fig. 6.2 Illustration of the design productivity gap. The progress in silicon process technology rises much stronger than the progress in electronic design automation (The values are taken from [Fly05])

because of the reduced problem size by partition the implementation steps in smaller units. In other words, this book provides the technical basics required for performing place and route in the cloud.

6.1 Future Directions

The range of applications and the field of research on runtime reconfigurable systems is huge; and in this book, focus was put on technical aspects required for efficient implementations of according FPGA-based systems. In this context, *efficiency* is used for both, the implementation process (supported by tools) as well as for the resulting systems (supported by architectures and methods). However, there remain challenging open problems for future research and development.

Most embedded systems integrate various hardware and software components. So far, the hardware/software partitioning (i.e., the decision, which part of the system will be implemented in hardware and which in software) is performed completely at design-time. As presented in Chap. 4, runtime reconfiguration – together with hardware/software morphing – provides interesting opportunities for self-adapting a system according to various objectives, including variable throughput demands, power consumption, or the currently available resources. However, the different hardware and software components have to communicate among each other. This demands compatible hardware/software interfaces that allow to link components that are arbitrary implemented in hardware or software (and that might change, even at runtime). This might also include a network layer with a dynamic binding of task among multiple nodes, as revealed for a ReCoNet. Here exists already good work that targets efficient [LK10, HHK10] interfaces or even complete message passing systems [GSS09, SPLC12, LP08]. It would be interesting to combine these approaches with partial reconfiguration and the dynamic hardware/software partitioning.

This could be further combined with modern design languages and behavioral compilers, such as Catapult-C from Mentor-Graphics or Impulse C from Impulse Accelerated Technologies. Such tools are not only helping in rising design productivity, they can automatically provide a range of implementation alternatives that are optimized for different objectives (typically resources versus throughput /latency). Consequently, various design alternatives for software as well as different hardware implementations may be automatically generated from the same specification. And these implementation might then be integrated on demand at runtime. This would combine the advantages of behavioral synthesis with the advanced system integration capabilities provided by the ReCoBus and connection bar communication architecture. Behavioral compilers allow communication constructs that may be automatically mapped to a customized or standardized communication architecture. For instance, pointers may result in a bus master interface to a ReCoBus and streaming ports may be mapped to corresponding connection bars.

Throughout this book, many issues that arise when implementing more advanced runtime reconfigurable systems have been presented. With tools like the RoCoBus-Builder, or the upcoming GOAHEAD (introduced in the appendix) most of these issues have been solved and automated such that a designer has not to care much about low-level details of the FPGA. In particular, this provides an easy migration path from one FPGA family to another one. However, there is still further room for automating the design of runtime reconfigurable systems. Here, it requires language extensions to specify specific properties of a reconfigurable system. Present RTL HDL languages are made for completely static ASIC designs and do not include constructs to express dynamic aspects, as they might arise for FPGAs. For example there are no statements to express that modules are executed mutually exclusive to each other (which means that they can share FPGA resources), or that a module might be preempted. There are also no nice ways to describe reconfiguration sequences of modules. Moreover, simple data flow models describe mainly an order of modules rather than a precise communication behavior. For instance, it is a big difference if a module is pipelined in the sense that it consumes an input value and produces an output value in, lets say, every clock cycle (e.g., a FIR filter) or if a module produces only a single result after some complex processing (e.g., a CRC checker). Consequently, the temporal module placement example of a DFG in Fig. 1.20 on Page 37 has to be refined to automatically derive an optimized module packing and communication synthesis. In this field exist already good entry points. Besides the system level initiatives, there are very promising approaches using polymorphism in object oriented languages to express reconfiguration [SNH$^+$10, Abe11].

We can conclude, that runtime reconfiguration is ready to use and that it will be used in many upcoming systems order to meat performance, cost, or power requirements.

Appendix A

When writing this book, our new tool GoAhead becomes ready to be released. This tool will replace the ReCoBus-Builder that is limited to Xilinx Virtex-II(-Pro) and Spartan-3 families. GoAhead supports all recent Xilinx FPGAs (V4,V5,V6,V7[1] and S6) and is designed for portability in order to also provide device support beyond the Virtex-7 series. The tool provides a user friendly GUI, similar as the ReCoBus-Builder, but also a script interface which allows to run the whole implementation process from a VHDL input description and constraint file all the way down to the final bitstream as a single batch job. This simplifies the integration of GoAhead into higher level tools. For example, this can be used for automatically exploring different design parameters, such as different module shapes, different design alternatives (i.e., area versus throughput alternatives), or for an automatic floorplanning and partitioning of a system into a static part and a set of reconfigurable regions.

GoAhead supports all communication techniques for integrating reconfigurable modules that have been introduced in this book, including bus macros, proxy logic, and the zero logic overhead technique. It supports the generation of homogeneous communication architectures such as the connection bar approach or the ReCoBus architecture (the ReCoBus wizard is currently under development). The following sections introduce the tool and go once through all the implementation steps of a tiny reconfigurable system.

A.1 The Tool GoAhead

The tool GoAhead provides floorplanning capabilities, communication infrastructure generation (required to integrate reconfigurable modules), and constraints

---

[1]Support for Virtex-4 and Virtex-7 is at present only tested by tools, but not on an FPGA itself. This will be done with the availability of corresponding boards.

D. Koch, *Partial Reconfiguration on FPGAs: Architectures, Tools and Applications*, Lecture Notes in Electrical Engineering 153, DOI 10.1007/978-1-4614-1225-0,

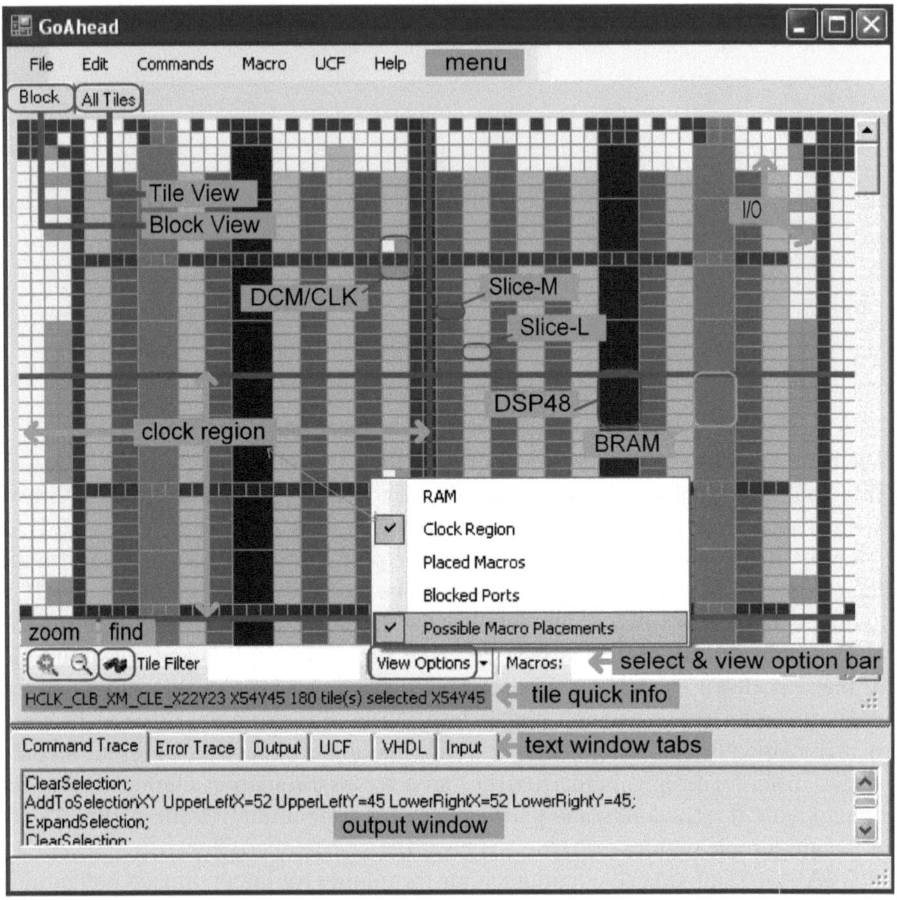

Fig. A.1 GoAhead GUI showing a Spartan-6 LX16 FPGA. All GUI actions will be recorded in the `Command Trace` output window

generation required for the place and route implementation phase. GoAhead provides an intuitive graphical user interface as well as a scripting interface. A screen shot of the tool is shown in Fig. A.1. All GUI actions are recorded by corresponding script commands. Consequently, there is no need to learn the GoAhead scripting language. However, it is recommended to use the GUI to create scripts. This removes error sources and ensures reproducible results. When calling GoAhead without parameters, it will start with the GUI. As an alternative, it is possible to call the tool in batch mode:

```
goahead my_script.goa
```

This allows to run the whole implementation process including the generation of all bitstreams as a single shot batch job. GoAhead is not restricted to work together with a particular synthesis tool. However, throughout this appendix, we will demonstrate how GoAhead can interact with the Xilinx XST synthesis tool.

GoAhead uses the Xilinx vendor place and route tools and GoAhead creates constraints to guide the vendor tools. As the different versions of the vendor tools behave slightly different, the constraint generation backend can be retargeted to different versions of the Xilinx ISE tools.

GoAhead has been written in C# and runs under Windows. Linux users can use GoAhead with the help of MONO. GoAhead needs FPGA device descriptions that can be easily generated from the GUI or by a batch script from the command line. This will involve the Xilinx vendor tools to generate device descriptions in XDL [BKT11] that are then parsed in by GoAhead to build up its own device data base.

A.2 A Reconfigurable `Hello World` System

The following sections describe the implementation of a tiny system using simple island style reconfiguration. The system will replace a module with `AND/NAND` gates by a corresponding one with `OR/NOR` gates. We will firstly implement the static system and then, completely independent from this, the two reconfigurable modules.

As shown in Fig. A.2, the static system comprises a reconfigurable island for hosting two different reconfigurable modules. In order to guarantee that always the same wires are used for the communication between the static system and the reconfigurable island, a *bus-macro* will be used. Such a macro is nothing else as two lock-up tables (in this example placed left and right in two consecutive CLB columns) and a set of wires between them. To be more precise, the bus macro consists of two slices (with four LUTs) each for bridging four signals from left-to-right as well as for bridging four signals from right-to-left. Except any clock signals (which are routed via dedicated clock networks), all signals connecting a partial module (i.e., the top level port signals of the reconfigurable modules) have to be bridged using a bus macro (or with any other provided option). Partial module integration using bus macros was chosen as this was the most widely used technique. A design using the more efficient zero logic overhead technique (as demonstrated in Sect. 5.2 on Page 238), would look pretty much the same and a corresponding tutorial will be released together with the GoAhead tool.

A.3 Implementation of the Static System

We will start the design process with floorplanning the FPGA. This includes the definition of a reconfigurable island and the placement of the bus macro to the static design. We have to ensure that no static logic and static routing is used inside

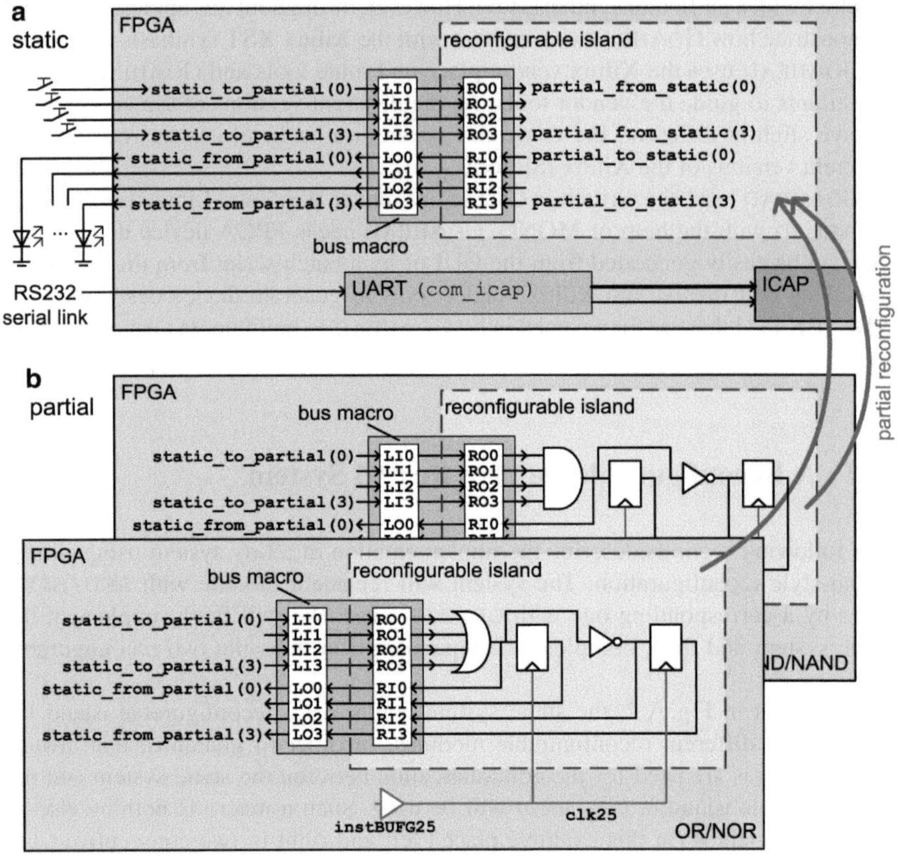

Fig. A.2 Reconfigurable ``Hello World'' system. The system consists of a static system (**a**) which comprises a reconfigurable island to host modules that are implemented independently in (**b**)

the reconfigurable island.[2] This avoids resource conflicts with the partial modules hosted in the reconfigurable island at runtime.

For the following steps, we have to start GoAhead and open the device description of the used device target FPGA. For this example on a Xilinx Spartan-6 LX16 FPGA, we use the command `FPGA -> Open Raw` and select the file `xc6slx16csg324 -3.binFPGA` from the devices library.

Defining the Reconfigurable Island After loading the device description in GoAhead , we see an empty device in the `Block` view. We select the reconfigurable

[2]Especially for Spartan-6 FPGAs, there are good reasons to relax this constraint (e.g., for crossing a static signal through the reconfigurable island).

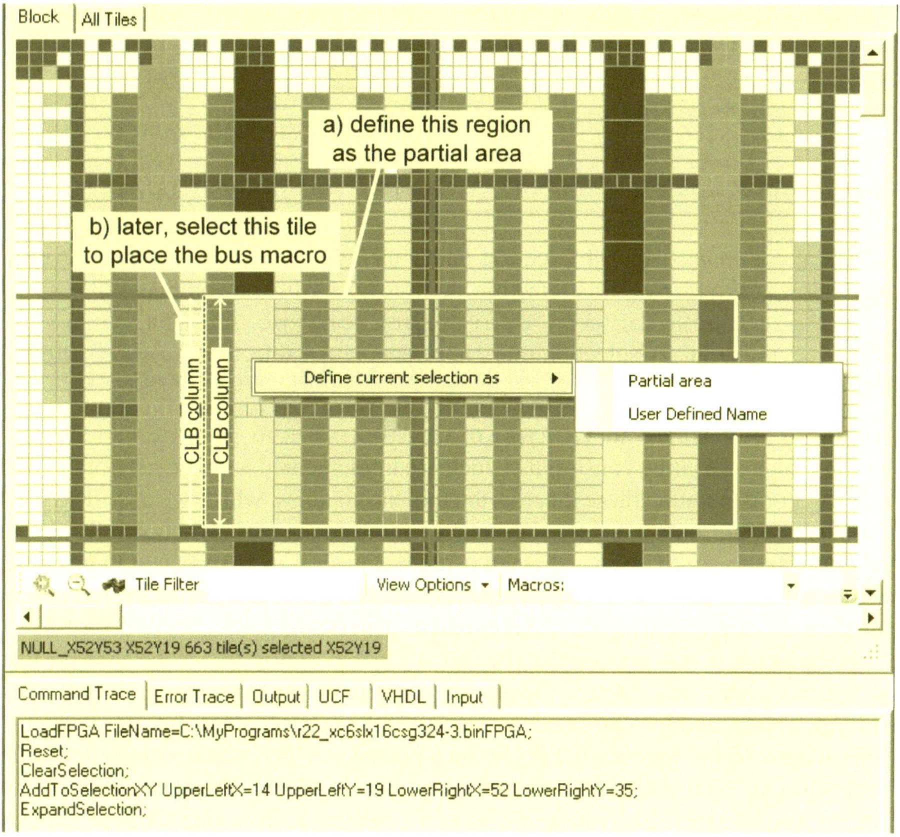

Fig. A.3 FPGA floorplanning. (**a**) in the beginning, we define the reconfigurable island as the `Partial area`; (**b**) later, we place the bus macro, as described in Sect. A.3 at the marked tile

island as shown in Fig. A.3. We select an area that spans the height of a clock region, because this is the height of a configuration frame. Clock regions are 16 CLBs, 4 BRAMs, or 4 DSP blocks in hight on Spartan-6 devices. The borders of the clock regions are marked in the GoAhead GUI, if activated in the View Options.

As all I/O pins are located at the four edges of the chip, it is recommended not to place a reconfigurable island too close to the edges of the fabric. For the same reason, it is not recommended (but possible) to use the clock regions at the very top or bottom of a Spartan-6 FPGA. A reconfigurable region can contain arbitrary resources, including logic (i.e., CLBs, BRAMs, or DSP48 tiles). However, for placing the bus macro in the next step, we have to ensure that CLBs are located at both sides of the left border of the reconfigurable island.

A partial area is defined by applying the GoAhead command `Define current selection as -> Partial area`, as shown in Fig. A.3. This results in a selection of resources that we can refer to by the symbolic name

PartialArea. When looking into the command trace, we will find the following commands that we could put into a GOA script file:

```
ClearSelection;
AddToSelectionXY UpperLeftX=14 UpperLeftY=19 ↩
    LowerRightX=52 LowerRightY=35;
StoreCurrentSelectionAs UserSelectionType=PartialArea;
```

These commands store a selection under the symbolic name PartialArea. This allows us to reselect this selection anytime later with the command: SelectUserSelection UserSelectionType=PartialArea;. Note that any arbitrary selection (e.g., non-rectangular selections) can be stored in this way.

Adding a Bus Macro in GoAhead GoAhead has to generate various constraints, such as routing constraints which are implemented with the help of blocker macros. For generating these macros, the tool has to know, which wires will be used for the communication between the static system and the partial modules. There are several ways to specify this and when using bus macros, it is sufficient to place the macro at the border of a reconfigurable region. The internal wires of the bus macro are than considered during the blocker macro generation.

For placing a bus macro, the Macro Placer dialog has to be opened that is located under the Macro menu. The macro placer provides a library with several macros, and for this example, we select the macro BM_S6_L4_R4_single. This macro is for Spartan-6 FPGAs and provides a horizontal connection with four wires in each direction between two adjacent CLBs. There is also a macro BM_S6_L4_R4_double, which provides a connection from one CLB to another one that is two CLBs far away. This macro can hop over a BRAM or DSP48 column or it can be used to link more wires by nesting two of these macros into each other (as shown in Fig. 2.16 on Page 70). Note that own macros can be used by adding the corresponding XDL description in the file dialog. GoAhead comes together with a library of horizontal bus macros for several device families. However, the zero logic overhead methods works in any arbitrary direction.

One interesting feature of GoAhead is the capability to cut out a module from any design and store it as a macro. This is possible by converting a design into an XDL netlist [BKT11]. With the Xilinx design language (XDL), Xilinx provides a very powerful interface to their tools which allows to perform any kind of modification to a system implemented on Xilinx FPGAs. A module that has been cut out from a design can then be placed to different positions, if the resource footprint of the FPGA fits the module requirements. This works even among different FPGAs within the same device family.

If activating in the View Options the option Possible Macro Placements, all possible placement positions (i.e., macro reference points or anchors) of the currently selected macro will be highlighted. A macros' placement position is defined in the tile view by selecting the target reference tile (typically a CLB). This fills the fields Anchor Tile and Anchor Slice in the Macro Placer.

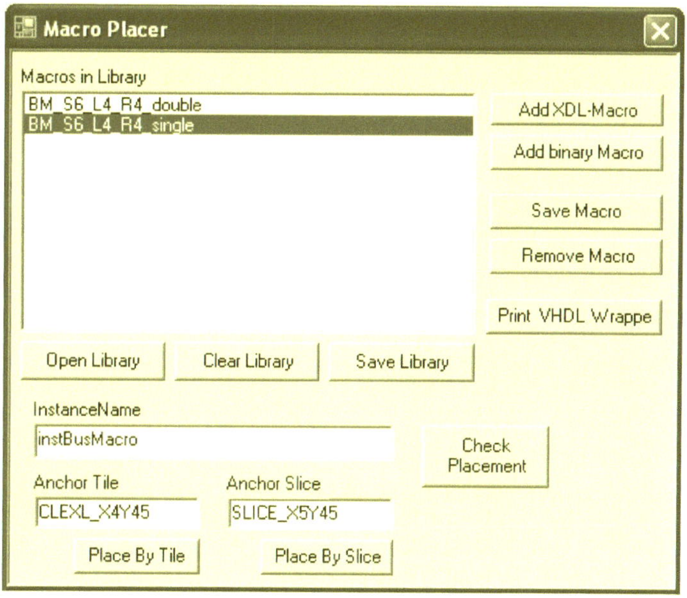

Fig. A.4 Bus macro instantiation using the Marco Placer

For placing the bus macro used in the demo system, a tile left outside the reconfigurable area has to be selected, as shown in Fig. A.3.

Next, we specify the instance name of the macro as `instBusMacro` inside the Macro Placer dialog. This name will be used for the VHDL and UCF template generation. We can compare our settings with Fig. A.4.

GoAhead can check if the selected macro of the library can be placed at the specified position by clicking the `Check Placement` button. The placement itself is accomplished by either clicking `Place by Tile` or `Place by Slice`. Both commands perform the same action and differ only in the coordinate system used to specify the macro reference point. While the tile grid is more intuitive to use, specifying slice coordinates has the longer tradition. The whole process results into the following log in the Command Trace:

```
ClearSelection;
AddToSelectionXY UpperLeftX=12 UpperLeftY=21 ↩
    LowerRightX=12 LowerRightY=21;
AddSingleMacroInstantiationBySlice MacroName=↩
    BM_S6_L4_R4_single InstanceName=instBusMacro ↩
    SliceName=SLICE_X5Y45;
```

Adding a Bus Macro in Xilinx ISE So far, a Bus Macro was added inside GoAhead and we have now to add the macro also in the ISE design project. The goal is to build the system as shown in Fig. A.2. In order to simplify this process,

GoAhead can generate VHDL instantiation templates for all macros instantiated within the tool. The VHDL declaration of the bus macro is generated by executing `Print VHDL Wrapper` within the Macro Placer. The result is printed in the VHDL text window tab in the GoAhead GUI and looks as follows:

```
component BM_S6_L4_R4_single is Port (
    RH0 : in STD_LOGIC;
    RH1 : in STD_LOGIC;
    RH2 : in STD_LOGIC;
    RH3 : in STD_LOGIC;
    LH0 : in STD_LOGIC;
    LH1 : in STD_LOGIC;
    LH2 : in STD_LOGIC;
    LH3 : in STD_LOGIC;
    RI0 : in STD_LOGIC;
    RI1 : in STD_LOGIC;
    RI2 : in STD_LOGIC;
    RI3 : in STD_LOGIC;
    RO0 : out STD_LOGIC;
    RO1 : out STD_LOGIC;
    RO2 : out STD_LOGIC;
    RO3 : out STD_LOGIC;
    LI0 : in STD_LOGIC;
    LI1 : in STD_LOGIC;
    LI2 : in STD_LOGIC;
    LI3 : in STD_LOGIC;
    LO0 : out STD_LOGIC;
    LO1 : out STD_LOGIC;
    LO2 : out STD_LOGIC;
    LO3 : out STD_LOGIC );
end component BM_S6_L4_R4_single;
```

This output is required in the architecture declaration block of the static top implementation file (assumed to be `static.vhd` in the following) in order to announce the bus macro. The port names are predefined within the XDL macro description and there exist at each side of the macro four inputs and four outputs that are connected to a corresponding bus macro outputs and inputs on the entire other side, as shown in Fig. A.2. The Figure hides the input port names RH and LH. These ports have to be set to `'1'` and are used internally by the macro. The reason for this is that the macro splits internally the 6-input LUT in two 5-input LUTs in order to connect two wires per LUT rather than only one wire like the bus macros provided by Xilinx. Splitting the LUTs requires one input to be connected to `'1'` and this cannot be hidden inside the macro.

GoAhead can also generate a macro instantiation template. The corresponding dialog is located under `Macro -> Macro Instantiations`. The new window provides a list containing information about all instantiated macros (in this example, only the bus macro). When applying `Print Macro Instantiations`, a macro instantiation template for the currently selected macro is printed into the VHDL output window. However, there are currently no signals connected to this

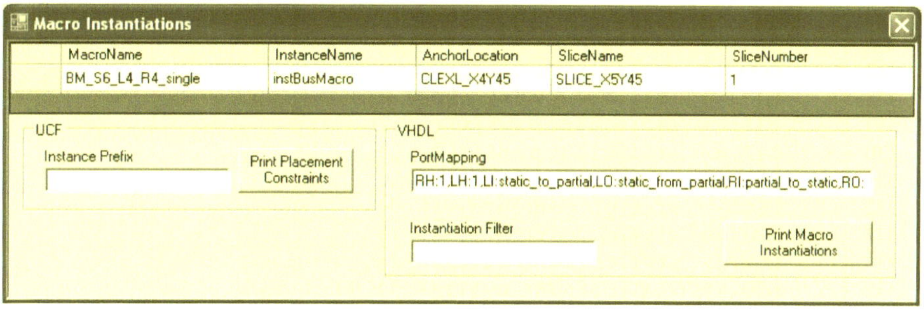

Fig. A.5 Macro instantiation window containing a port mapping string

instantiation. This can be performed by adding `Port Mappings`. A port mapping has the form:

```
MacroPortPrefix_a:{Signal_a|0|1|open}
          [,MacroPortPrefix_b:{Signal_b|0|1|open}]...
```

This means that we can specify the beginning of a macro port name (e.g., `RH`) in order to set the port to a constant `'1'` value by adding the port mapping `RH:1`. More mappings can be added by separating them with commas. If we connect a port prefix to a signal (e.g., `LI:static_to_partial`), vector connections will be generated for this signal. This will connect the `LH` and `RH` inputs to `'1'` and the remaining ports according to the signal names listed in Fig. A.2. The corresponding port mapping is:

```
RH:1,LH:1,LI:static_to_partial,LO:static_from_partial,
          RI:partial_to_static,RO:partial_from_static
```

For the `Hello World` system, the Macro Instantiations window looks as follows (Fig. A.5):

After applying `Print Macro Instantiations`, the VHDL output window contains now:

```
instBusMacro : BM_S6_L4_R4_single
Port Map (
     RH0 => '1',
     RH1 => '1',
     RH2 => '1',
     RH3 => '1',
     LH0 => '1',
     LH1 => '1',
     LH2 => '1',
     LH3 => '1',
     RI0 => partial_to_static(0),
     RI1 => partial_to_static(1),
     RI2 => partial_to_static(2),
     RI3 => partial_to_static(3),
     RO0 => partial_from_static(0),
```

```
        RO1 => partial_from_static(1),
        RO2 => partial_from_static(2),
        RO3 => partial_from_static(3),
        LI0 => static_to_partial(0),
        LI1 => static_to_partial(1),
        LI2 => static_to_partial(2),
        LI3 => static_to_partial(3),
        LO0 => static_from_partial(0),
        LO1 => static_from_partial(1),
        LO2 => static_from_partial(2),
        LO3 => static_from_partial(3) );
```

This instantiation template can be added to architecture of the top implementation file (`static.vhd`). This requires the following signal declarations:

```
signal partial_from_static  : std_logic_vector(3 downto 0) := (others=>'1');
signal partial_to_static    : std_logic_vector(3 downto 0) := (others=>'1');
signal static_from_partial  : std_logic_vector(3 downto 0) := (others=>'1');
signal static_to_partial    : std_logic_vector(3 downto 0) := (others=>'1');

  attribute keep of partial_from_static  : signal is "true";
  attribute keep of partial_to_static    : signal is "true";
  attribute keep of static_from_partial  : signal is "true";
  attribute keep of static_to_partial    : signal is "true";
```

Note that all signals are initialized to `'1'`. This is done as each Slice, BRAM, or DSP48 primitive input can be set to `'1'` without additional logic (which is not the case for the default `'0'` value). In addition, a keep attribute is set for each signal. Together with the `map -u` option (do not truncate unconnected signals), this allows to implement a module without the need of external signals. As the last modification to our static design, we connect the bus macro with the push button and LED signals:

```
LEDs <= static_from_partial;
static_to_partial <= BUTTONs;
```

Despite that macros can contain vectorized entities, single bit signals have been used for the entity signals of all provided macros in the library. This is required to be compatible to the Xilinx ISE tool chain. However, the instantiation templates are connected more convenient to signal vectors. Several Xilinx ISE Versions have had problems when instantiating NMC hardmacros in a sub-hierarchy of a design.

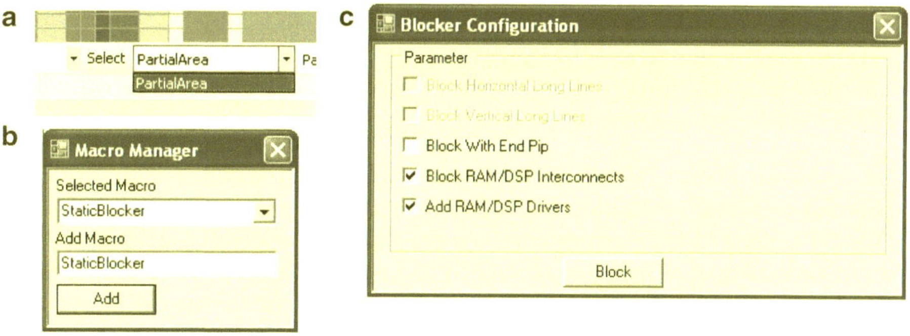

Fig. A.6 (**a**) `PartialArea` selection. (**b**) Adding a `StaticBlocker` macro. (**c**) Blocker settings

In order to overcome this issue, macros should only be instantiated at the top level of a design. All instantiation templates are generated by just two GoAhead commands:

```
PrintWrapperForMacro MacroName=BM_S6_L4_R4_single;
PrintVHDLMacroInstantiation PortMapping=RH:1,LH:1,LI:↩
    static_to_partial,LO:static_from_partial,RI:↩
    partial_to_static,RO:partial_from_static
```

Generating Physical Place and Route Constraints By instantiating a macro in GoAhead, its resources are announced to the tool and can be considered when generating implementation constraints. This is used to generate UCF placement constraints for the `static.ucf` file of the example ISE project. The macro placement constraints are generated with the command `Print Placement Constraints` inside the Macro Instantiations dialog. This will generate the following output in the UCF output window:

```
INST "instBusMacro" LOC = "SLICE_X5Y45"; # by_GoAhead
```

GoAhead can generate placement prohibit constraints for the reconfigurable island in order to prevent the static system from instantiating resources inside this region. This is done by selecting the `PartialArea` group in the drop down menu of the select and view option bar (see also Fig. A.6a). For this selection, the command `UCF -> Prohibit` has to be applied. By leaving the `Exclude used Slices` checkbox activated, GoAhead will not generate prohibit statements for the slices used by the bus macro. The result of the prohibit command is a list of UCF constraints containing primitives of the FPGA fabric (e.g., slices, BRAMs, DSP primitives) that are not allowed to be used by the placer:

```
CONFIG PROHIBIT = "SLICE_X6Y47"; #generated_by_GoAhead
CONFIG PROHIBIT = "SLICE_X7Y47"; #generated_by_GoAhead
CONFIG PROHIBIT = "SLICE_X6Y46"; #generated_by_GoAhead
...
```

The result is printed into the GOAHEAD UCF output window and optionally appended to a file, when entering a file name (in our example: `static.ucf`). With the generated UCF constraints, we run the Xilinx vendor placer in the ISE design environment or from the command line by calling MAP. Note that opposed to earlier versions of ISE, the placement is now performed during the map phase. Despite that the route tool is called *par* for place and route, it now only provides routing functionality.

Prohibit statements have no impact on the router. For prohibiting the router from using resources within the reconfigurable island, GOAHEAD generates *blocker macros* that occupy all routing resources within a definable region, as introduced in Sect. 3.2.4. The blocker macro generation starts by adding a new macro inside the Macro Manager (see Fig. A.6b). The Macro Manager is opened from the menu with `Macro -> Macro Manager`. The example is shown for the name `StaticBlocker`. After this, the partial area (`PartialArea`) has to be selected (Fig. A.6a). Next, we run from the menu `Commands -> Block Selection`. The result is a macro stored inside the tool. By applying the command `Macro -> Generate XDL Code`, the blocker macro is written to a file.

The following GOAHEAD commands have been executed (and logged in the Command Trace) when generating the place and route constraints:

```
PrintLocationConstraintsForPlacedMacros;
ClearSelection;
SelectUserSelection UserSelectionType=PartialArea;
PrintProhibitStatementsForSelection ExcludeUsedSlices=↩
    True FileName=Static.ucf Append=True;
AddMacro MacroName=StaticBlocker;
SelectMacro MacroName=StaticBlocker;
BlockSelection SliceNumber=0 BlockArcsInRAMTiles=True ↩
    BlockWithEndPips=False;
GenerateXDL FileName=StaticBlocker.xdl;
```

A.3.1 Routing and Bitstream Generation

In the last steps, we generated a placed design and all constraints required for the physical implementation of the static system. The missing routing and bitstream generation is performed by the script `static.bat`. The script is called as follows:

```
>static InputProject  [Blocker_XDL_FILE ]  ↩
                 [Blocker_XDL_FILE BUFGInstanceName] ...
```

The options are:

`InputProject`: the prefix of the `MyProject_map.ncd` file. (specified without the `_map.ncd` at the end)

`Blocker_XDL_FILE`: a file generated by GoAhead to constrain the routing

`BUFGInstanceName`: an optional instance name used to connect blocker primitives to a global clock net (the net is specified by its BUFG driver instance)

For the here presented example, the call is:

```
>static top StaticBlocker.xdl instBUFG25
```

The operation of the script is illustrated in Fig. A.7. It takes the placed design, and converts it into the XDL netlist format. Into this netlist, GoAhead adds the blockers from the options list. If an optional BUFG driver instance name is specified for a blocker, the blocker will be connected to the corresponding global clock net that is driven by the BUFG. Specifying a clock net by its BUFG instance name prevents name mismatches, because net names might get changed by the synthesis tool. By connecting blockers to some global clock signals, the corresponding clock tree drivers of the FPGA fabric will be activated. This allows the partial modules (that will replace the blocker primitives) to automatically access/use the specified clock nets at runtime.

After integrating optionally blockers, the netlist is converted back to the Xilinx binary netlist format NCD. Next, the script calls the Xilinx vendor router through the tool `FPGA_editor`. After this, a timing report is generated as well as the static configuration bitstream.

A.3.2 Resulting Static System

As shown in the FPGA Editor screen shot in Fig. A.8, the result of the static implementation is a system that provides a reconfigurable island. This island provides access to the global `clk25` clock network. Furthermore, communication between the static and partial part of the system can be carried out with the help of a bus macro.

A.3.3 Remarks

For the `''Hello World''` system, no timing constraints have been specified. Xilinx provides a constraint called `tpsync`. With this constraint, the time slack to and from the bus macro can be specified. Consequently, we can define the time slack

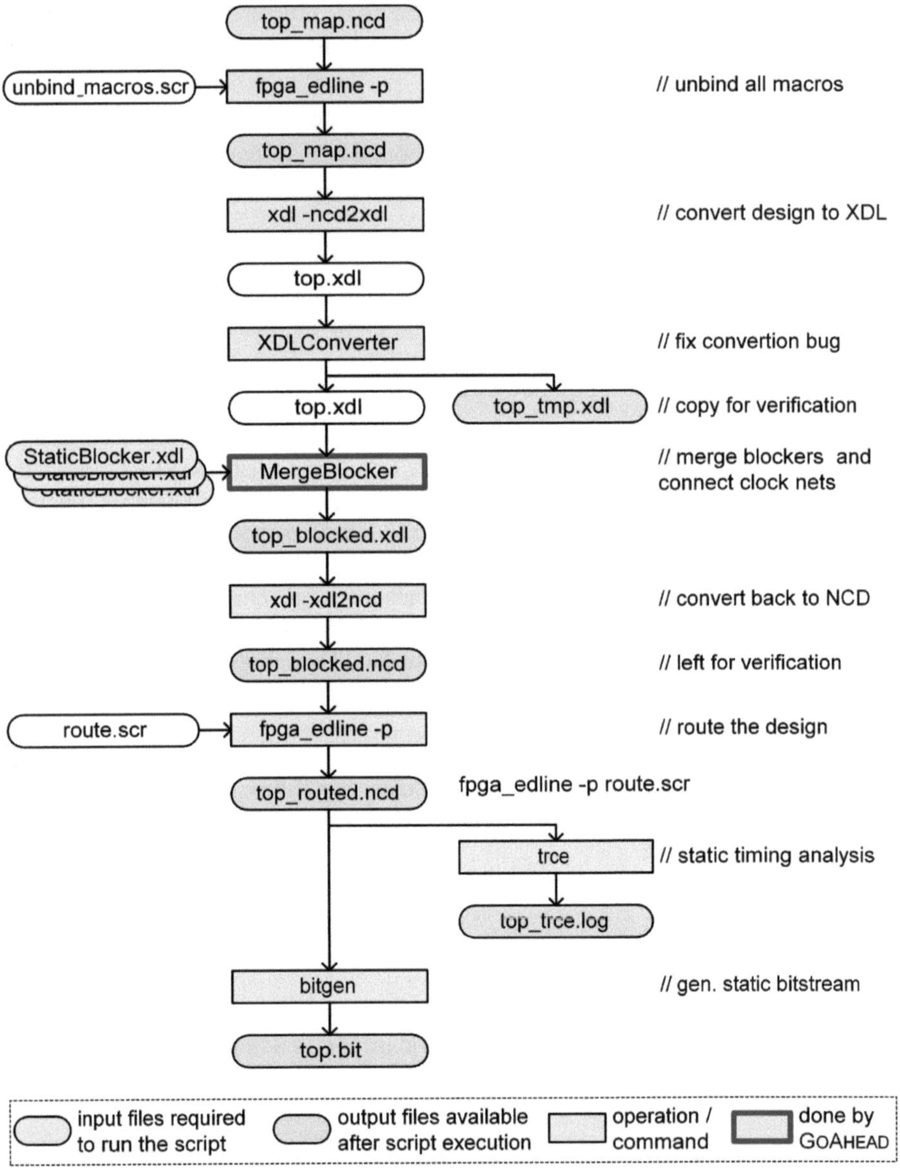

Fig. A.7 Static routing script. The script requires a placed input netlist and optional blockers. The script generates further scripts on-the-fly in order to call the Xilinx vendor tool `fpga_editor`. The output is a configuration bitsrteam and timing report. The example is for `>static top StaticBlocker.xdl`

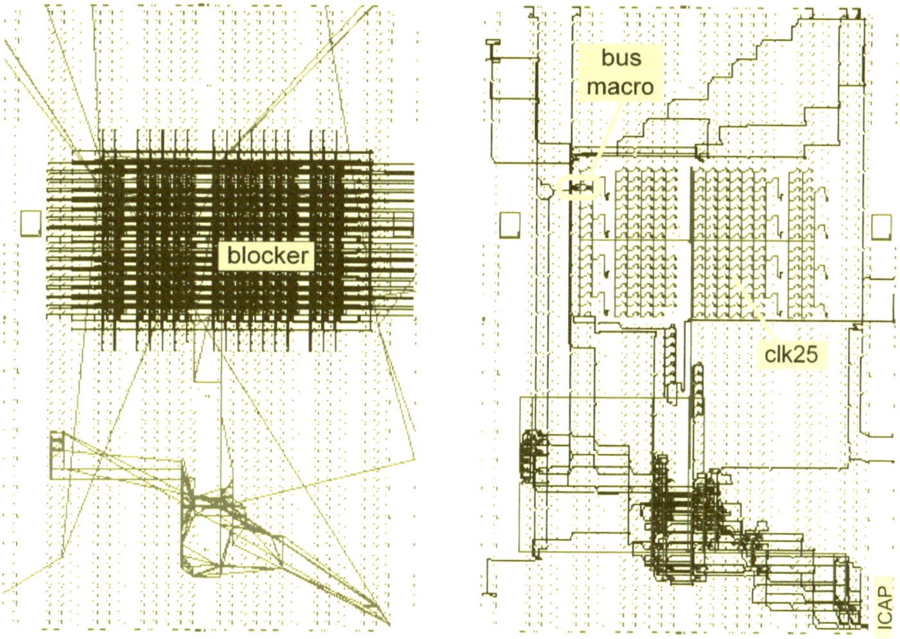

Fig. A.8 Implemented ``Hello World'' system; *left*: static design before routing, and *right*: after routing (with deleted blocker)

of the signals crossing the reconfigurable border individually for the static and the partial part of the system. Hence, the partial module timing can be fully decoupled from the static system routing. This is in particular required when implementing both parts in parallel design teams.

The configuration via ICAP is rather straight and it is sufficient to stream the generated bitstream data to the configuration port. This can be controlled either by clock gating or by an enable signal. Using the ICAP port requires to instantiate the ICAP primitive in the top level design file. One error source to be avoided is that the configuration data has to be bit-wise permuted within each byte of the bitstream ($d_0 <= d_7$, $d_1 <= d_6$, ...). In the example in Fig. A.8, the configuration data is supplied via an UART (located on the FPGA) by a host PC connected to the FPGA board.

A.4 Implementing the Partial Modules

As depicted in Fig. A.2b, the partial modules will be implemented independently from the static system. In the following sections, it will be shown from scratch, how to implement a partial module for the introduced `Hello World` example.

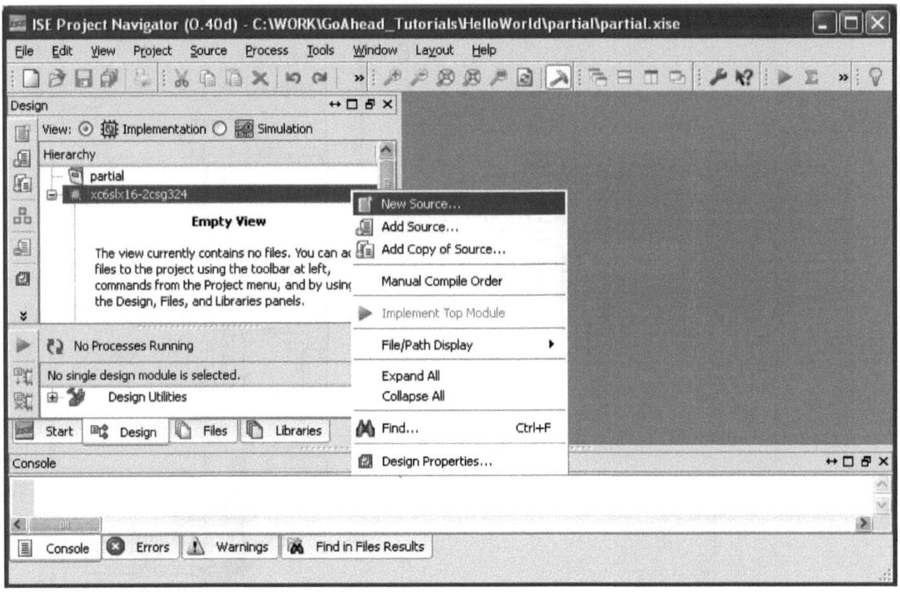

Fig. A.9 Adding `partial.vhd` and `partial.ucf` to an empty ISE project

A.4.1 Creating an Empty Partial Module Project

The implementation of the partial module starts with the creation of a new empty ISE project. For this project, we select the same FPGA that has been used before for the static system. We name the project `partial`. After this, we add two new (empty) source files to the project (see also Fig. A.9):

`partial.vhd` : a VHDL module file containing all logic. As a partial module has typically only internal signals, the top level entity will only contain a dummy clock input pin (named `"CLKin"` in the example).

`partial.ucf` : an implementation constraints file

Next, we copy the bus macro NMC file (`BM_S6_L4_R4_single.nmc`) to the partial project directory. This file is provided with GOAHEAD. After this, we copy from the static implementation file (`static.vhd`) the bus macro component declaration, the declaration of the bus macro connecting signals (e.g., `static_to_partial`), including the keep attribute statements, and the bus macro instantiation to our new partial implementation file (`partial.vhd`). In addition, we add a component instantiation of a BUFG clock driver to `partial.vhd` (the BUFG declaration is provided by Xilinx in the UNISIM library). The input will be connected a the top level interface signal `CLKin` and the output has to be connected to the signal `clk25`. By deactivating the "ADD I/O Buffers" synthesize option, as shown in Fig. A.10, no input pin or routing towards the BUFG primitive will be generated. With this option, an *internal primitive input* (here a BUFG input) will act as the top level input connection. This fits best the partial module design, because

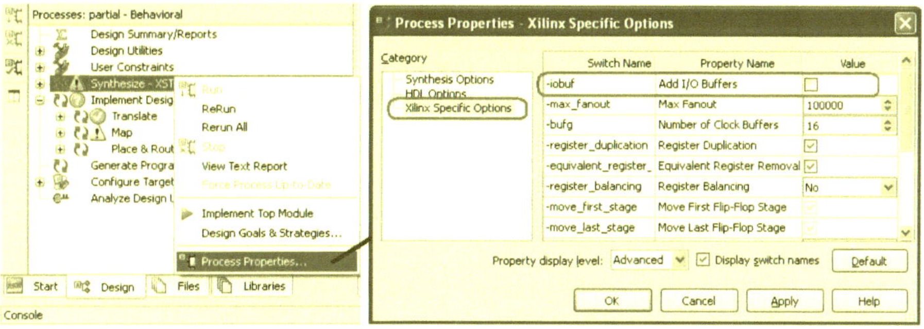

Fig. A.10 Synthesize options for the partial module implementation

the routing to the BUFG is part of the static design. As the final VHDL code, we implement the module logic according to Fig. A.2b. The final `partial.vhd` file looks as follows:

```
-------------------------------------------------
-- HELLO WORLD -- Partial
-------------------------------------------------
library IEEE,UNISIM;
use IEEE.STD_LOGIC_1164.ALL;
use UNISIM.VComponents.all;

entity partial is
    Port ( CLKin : in  STD_LOGIC );
end partial;

architecture Behavioral of partial is

component BM_S6_L4_R4_single is
    Port ( LH0 : in  STD_LOGIC;
           LH1 : in  STD_LOGIC;
           LH2 : in  STD_LOGIC;
           LH3 : in  STD_LOGIC;
           RH0 : in  STD_LOGIC;
           RH1 : in  STD_LOGIC;
           RH2 : in  STD_LOGIC;
           RH3 : in  STD_LOGIC;
           LI0 : in  STD_LOGIC;
           LI1 : in  STD_LOGIC;
           LI2 : in  STD_LOGIC;
           LI3 : in  STD_LOGIC;
           LO0 : out  STD_LOGIC;
           LO1 : out  STD_LOGIC;
           LO2 : out  STD_LOGIC;
           LO3 : out  STD_LOGIC;
           RI0 : in  STD_LOGIC;
           RI1 : in  STD_LOGIC;
```

```
              RI2 : in   STD_LOGIC;
              RI3 : in   STD_LOGIC;
              RO0 : out   STD_LOGIC;
              RO1 : out   STD_LOGIC;
              RO2 : out   STD_LOGIC;
              RO3 : out   STD_LOGIC    );
end component BM_S6_L4_R4_single;

 attribute KEEP : string;

signal partial_from_static  : std_logic_vector(3 downto 0) := (others=>'1');
signal partial_to_static    : std_logic_vector(3 downto 0) := (others=>'1');
signal static_from_partial  : std_logic_vector(3 downto 0) := (others=>'1');
signal static_to_partial    : std_logic_vector(3 downto 0) := (others=>'1');

  attribute keep of partial_from_static : signal is ``true'';
  attribute keep of partial_to_static   : signal is ``true'';
  attribute keep of static_from_partial : signal is ``true'';
  attribute keep of static_to_partial   : signal is ``true'';

signal CLK25 : std_logic := '1';

begin

instBUFG25 : BUFG
  port map ( I => CLKin,
             O => CLK25 );

instBusMacro : BM_S6_L4_R4_single
Port Map (
    RH0 => '1',
    RH1 => '1',
    RH2 => '1',
    RH3 => '1',
    LH0 => '1',
    LH1 => '1',
    LH2 => '1',
    LH3 => '1',
    RI0 => partial_to_static(0),
    RI1 => partial_to_static(1),
    RI2 => partial_to_static(2),
    RI3 => partial_to_static(3),
    RO0 => partial_from_static(0),
    RO1 => partial_from_static(1),
    RO2 => partial_from_static(2),
```

```
          RO3 => partial_from_static(3),
          LI0 => static_to_partial(0),
          LI1 => static_to_partial(1),
          LI2 => static_to_partial(2),
          LI3 => static_to_partial(3),
          LO0 => static_from_partial(0),
          LO1 => static_from_partial(1),
          LO2 => static_from_partial(2),
          LO3 => static_from_partial(3)   );

-------- The OR/NOR Gate Module --------
partial_to_static(0) <= partial_from_static(0) OR partial_from_static(1) OR partial_from_static(2) OR partial_from_static(3);
partial_to_static(2) <= NOT (partial_from_static(0) OR partial_from_static(1) OR partial_from_static(2) OR partial_from_static(3));

process (clk25)
begin
    if clk25'event and clk25='1' then
        partial_to_static(1) <= partial_from_static(0) OR partial_from_static(1) OR partial_from_static(2) OR partial_from_static(3);
        partial_to_static(3) <= NOT (partial_from_static(0) OR partial_from_static(1) OR partial_from_static(2) OR partial_from_static(3));
    end if;
end process;

end Behavioral;
```

Most of the lines have been copied from the static implementation file. This code was generated by GoAhead throughout Sect. A.3. Consequently, the partial module implementation can be carried out *before and independent to* the implementation of the static system. GoAhead will assist this by providing implementation and constraint templates. The main things that have to be done beforehand for both, the static system and the partial modules is the partial region allocation and the placement of the bus macros. For the latter one, it is required that the relative bus macro placement at the static to partial border is identical for all islands that will host the same set of partial modules at runtime.

The listing of `partial.vhd` gives the implementation of the OR/NOR module according to Fig. A.2b. Note that this time, we leave the two signals located in the static part of the system (`static_to_partial` and `static_from_partial`) unconnected. These signals will be set to `'1'` because of the initial value specification in the signal declarations. Again, only a logical `'1'` level can be created at any logic primitive input without extra implementation cost (in terms of LUTs).

A.4.2 Partial Module Implementation Constraints

After the VHDL implementation, we will add some implementation constraints to the `partial.ucf` file, as described in the following sections. The first two entries can be copied from the static system (`static.ucf`):

```
INST "instBUFG25" LOC = "BUFGMUX_X2Y2"; # GCLK1
INST "instBusMacro" LOC = "SLICE_X5Y45"; # ↩
    generated_by_GoAhead
```

The first line specifies the BUFG clock driver primitive of the FPGA fabric that will drive the `clk25` net. By specifying the BUFG primitive, it is defined which global clock network will be used for `clk25`. Note that there are 16 global clock networks and corresponding BUFG primitives available on Spartan-6 FPGAs. All these networks are basically identical and the corresponding BUFG primitive names are available in GoAhead under `File -> Device Info`. However, we have to ensure that the same BUFG constraint is used than in the static system in order to guarantee that the partial modules connect to the same global clock network that has been activated during the implementation of the static system (see Fig. A.8 for an example of an activated clock network).

The second line in the UCF file specifies the placement of the bus macro. The placement is identical to the placement position used in the static system. In general, it is required that the relative position of the bus macros is identical for each island that should host the same set of partial modules.

Before we can run the physical implementation, prohibit constraints and blocker macros have to be generated. As opposed to the prohibit constraints used for the static system, we have now to prohibit everything around the reconfigurable island. This is required to prevent the partial modules from taking resources outside the island. The implementation of this in GoAhead is very similar to the static system constraints generation. It is possible to select the `PartialArea` and invert the selection with the command `Edit -> Invert Selection`, as shown in Fig. A.11. This will create a selection containing all tiles except the ones from the reconfigurable region. For this selection, the prohibit statements are generated by applying `UCF -> Prohibit`. The prohibit statements can be directly appended to `partial.ucf`.

For the same selection, a blocker macro has to be generated. This starts with adding a new macro in the Macro Manager (located under the `Macro` menu). In this example, we give this macro the name `PartialBlocker`. After this, we open the blocker dialog with `Commands -> Block Selection`. We use the same settings as being used before for generating the static blocker (see Fig. A.6c for an example). As the last constraints generation step in GoAhead, we store the `PartialBlocker` in the project directory of the partial system via the menu with `Macro -> Generate XDL Code`.

We give this blocker the filename `PartialBlocker.xdl`.

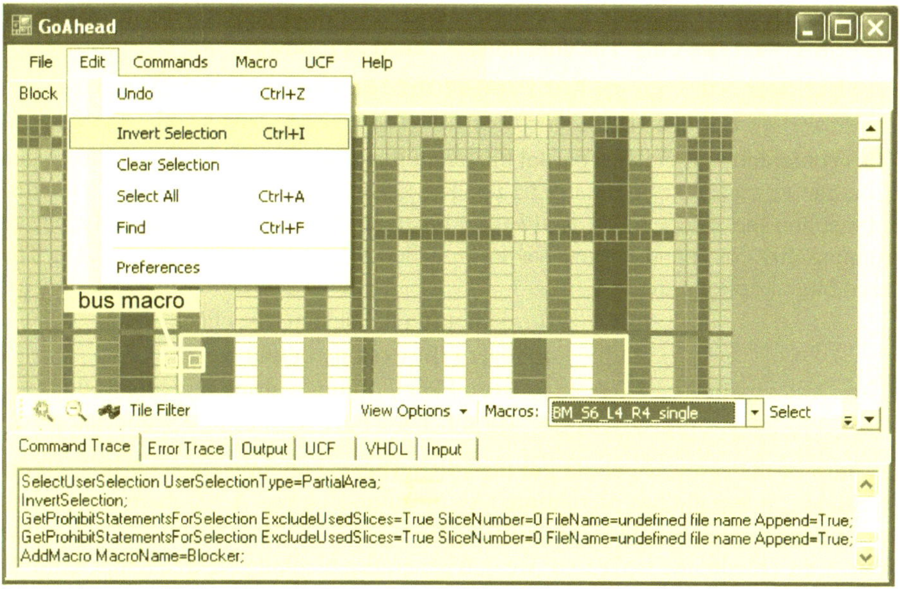

Fig. A.11 Inverting the selection in GoAhead to generate prohibit statements and a blocker macro *around* the reconfigurable region

A.4.3 Routing and Partial Bitstream Generation

With the prohibit and the placement constraints for the reconfigurable island, logic synthesis and placement can be performed (identical to what has been done during the implementation of the static system). Consequently, we run the implementation until map has finished (either in ISE or on the command line). The routing is carried out by the batch script `partial.bat` that is very similar to `static.bat`. For this example, the call is:

```
>partial partial PartialBlocker.xdl
```

For the partial modules, it is not required to connect the blocker primitives to a clock network. This is because the clock routing is part of the static system implementation. The flow goes the same way via an XDL netlist in order to include the blocker (here `PartialBlocker.xdl`) before running the router in the FPGA_editor. The main difference is the generation of the partial bitstream.

A.4.3.1 Partial Bitstream Generation

In previous versions of the Xilinx bitstream generation tools, there was an option for generating a partial bitstream for a definable region of the FPGA fabric (a bounding box). This option is unfortunately not available for recent FPGA families. However, there is an option to generate differential bitstreams using the `bitgen -r switch`. The command is called as follows:

```
>bitgen -d -w -g ActiveReconfig:Yes -g CRC:Disable
        to_file.ncd -r from_file.bit partial_diff.bit
```

This command will create a partial bitstream that contains the differences to switch from the `from_file` to the `to_file`. The `to_file` is always in the NCD netlist format and the `from_file` is always a bitstream (`*.bit`). Consequently, before running this command, bitgen has to be called to generate the `from_file`. The used bitgen options are:

`-d`:	do not run the design rule check DRC
`-w`:	overwrite existing output files
`-g ActiveReconfig:Yes`:	keep FPGA running during reconfiguration
`-g CRC:Disable`:	do not check the CRC during configuration

In most of the cases, we could use the CRC. This is only prevented when applying bitstream manipulations at runtime. As illustrated in Fig. A.12, there are several ways to generate partial bitstreams using `bitgen -r`. Note that by swapping the `from_file` with the `to_file`, we will generate a blanking bitstream that removes the partial module from the system (if loaded to the FPGA). It is recommended to use module blanking rather than overwriting one module directly by another one. This reduces power consumption during reconfiguration [BKT10].

As shown in Fig. A.12, differential partial bitstreams can be generated using three different methods:

(a) *Incremental*: here, the partial module is generated as an increment to an existing design (typically a static system).
(b) *Separated*: here, the partial module is strictly separated from any other parts of the system in order to generate a differential bitstream among a completely empty system. This is supported by GoAhead with the command `CutOff-FromDesign`. The command can cut away everything from a design that is not selected. This bitstream generation method is used in the partial routing script.
(c) *Remove*: here, the GoAhead `CutOffFromDesign` command is used to delete the part of the design a bitstream is going to be generated for. By taking the original bitstream as the target (`to_file`), a partial bitstream containing only the deleted parts is generated.

As long as resource columns span a full clock region height, these methods can be applied to any arbitrary selection of logic, BRAM, or DSP48 resource columns.

In the bitgen call options, it is specified not to use the design rule check DRC (`-d`). However, this is only required when cutting out modules using the methods illustrated in Fig. A.12b, c. In that examples, there will be input and output antennas (i.e., signals that have either only an input or only a driver, but not both). These antennas are created by cutting out the module and consequently by cutting the signal paths crossing the static to partial border (for example the internal wires of a bus macro). Consequently, the DRC can be ignored. As an alternative, GoAhead

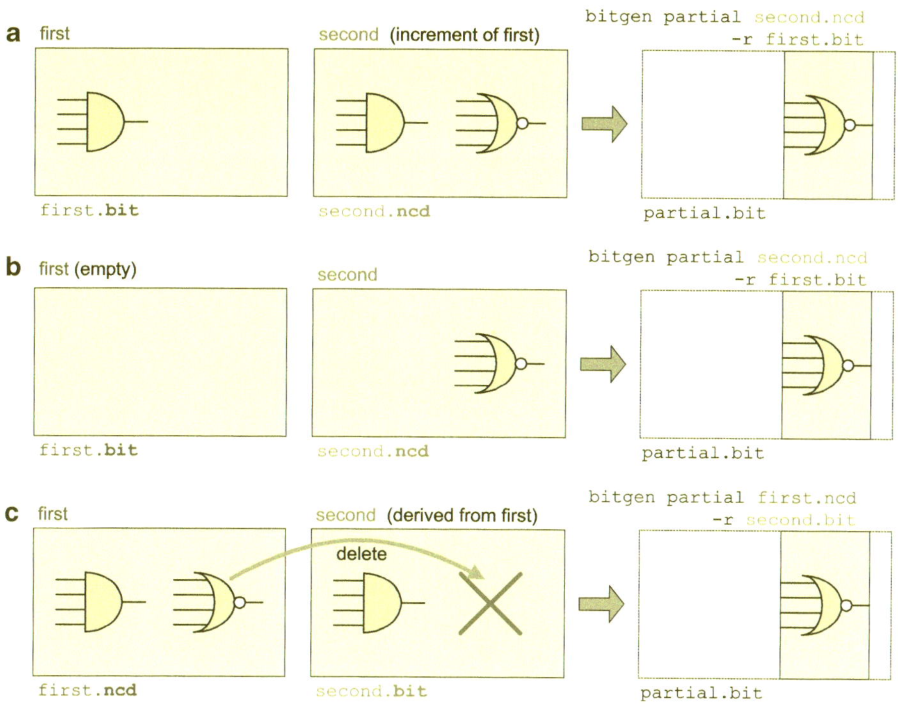

Fig. A.12 Partial bitstream generation using the differential bitstream option (`bitgen -r`). (**a**) In the case the partial part is implemented as an increment to another (typically static) system. (**b**) If the partial module is completely separated. (**c**) As an alternative to (**b**), we can delete the parts that should be added to the bitstream. Note, by changing first with second, a blanking bitstream can be generated

can create netlists at any configuration state by merging modules on the netlist level into a static system. In other words, GOAHEAD can cut out modules directly from a design and store them as partial netlists (using the GOAHEAD command `CutOffFromDesign`). These netlists can be used to create incremental systems such that the bitstream generation method from Fig. A.12a can be used. In this case, partial bitstreams can be generated that pass the bitgen DRC check.

All partial modules can be implemented the same way and they can share the same generated constraints.

References

[ABD+05] Ahmadinia A, Bobda C, Ding J, Majer M, Teich J, Fekete S, van der Veen J (2005) A practical approach for circuit routing on dynamic reconfigurable devices. In: Proceedings of the 16th IEEE international workshop on rapid system prototyping (RSP), June 2005, pp 84–90

[ABD+07] Athanas P, Bowen J, Dunham T, Patterson C, Rice J, Shelburne M, Suris J, Bucciero M, Graf J (2007) Wires on demand: run-time communication synthesis for reconfigurable computing. In: International conference on field programmable logic and applications (FPL), Aug 2007. IEEE, pp 513–516.

[Abe11] Abel N (2011) Design and implementation of an object-oriented framework for dynamic partial reconfiguration. PhD thesis, University of Heidelberg, Heidelberg

[ABM+05] Ahmadinia A, Bobda C, Majer M, Teich J, Fekete S, van der Veen J (2005) DyNoC: a dynamic infrastructure for communication in dynamically reconfigurable devices. In: Proceedings of the international conference on field-programmable logic and applications (FPL), Aug 2005, pp 153–158

[Abo09] Abound Logic, Inc (2009) The raptor family of FPGAs (V1.0)

[Alt07] Altera Corporation (2007) Avalon memory-mapped interface specification V3.3. Available online: www.altera.com/literature/manual/mnl_avalon_spec.pdf

[Alt09] Altera Inc (2009) Altera devices. Available online: www.altera.com/products/devices/dev-index.jsp

[AS93] Athanas PM, Silverman HF (1993) Processor reconfiguration through instruction-set metamorphosis: compiler and architectures. IEEE Comput 26(3):11–18

[Ash08] Ashenden PJ (2008) The designer's guide to VHDL, 3rd edn. Morgan Kaufmann

[AT05] Asadi G-H, Tahoori MB (2005) Soft error mitigation for SRAM-based FPGAs. In: Proceedings of the 23rd IEEE VLSI test symposium (VTS), IEEE Computer Society, pp 207–212

[BBH+06] Bieser C, Bahlinger M, Heinz M, Stops C, Müller-Glaser KD (2006) A novel partial bitstream merging methodology accelerating Xilinx Virtex-II FPGA based RP system setup. In: Proceedings of the international conference on field programmable logic and applications (FPL), pp 1–4

[BD01] Brebner GJ, Diessel O (2001) Chip-based reconfigurable task management. In: Proceedings of the 11th international conference on field programmable logic and application (FPL), Springer, pp 182–191

[Bec07] Beckhoff C (2007) Entwurf und Implementierung von Hardwaremodulen zur Dekompression von FPGA-Konfigurationsdaten. In: Studienarbeit, Lehrstuhl für Hardware-Software-Co-Design. Universtät Erlangen-Nürnberg, Erlangen

D. Koch, *Partial Reconfiguration on FPGAs: Architectures, Tools and Applications*, Lecture Notes in Electrical Engineering 153, DOI 10.1007/978-1-4614-1225-0,

[BHB+07] Braun L, Hübner M, Becker J, Perschke T, Schatz V, Bach S (2007) Circuit switched run-time adaptive network-on-chip for image processing applications. In: International conference on field programmable logic and applications (FPL), Aug 2007. IEEE, pp 688–691

[BJRK+03] Blodget B, James-Roxby P, Keller E, McMilla S, Sundararajan P (2003) A self-reconfiguring platform. In: Proceedings of international conference on field-programmable logic and applications (FPL), pp 565–574

[BKT10] Beckhoff C, Koch D, Torresen J (2010) Short-circuits on FPGAs caused by partial runtime reconfiguration. In: Proceedings of the international conference on field programmable logic and applications (FPL), Aug 2010, pp 596–601

[BKT11] Beckhoff C, Koch D, Torresen J (2011) The Xilinx Design Language (XDL): tutorial and use cases. In: Proceedings of the 6th international workshop on reconfigurable communication-centric systems-on-chip (ReCoSoC), pp 1–8

[BLC07] Becker T, Luk W, Cheung PYK (2007) Enhancing relocatability of partial bitstreams for run-time reconfiguration. In: Proceedings of the 15th annual IEEE symposium on Field-programmable Custom Computing Machines (FCCM), IEEE Computer Society, pp 35–44

[BM02] Benini L, De Micheli G (2002) Networks on chips: a new SoC paradigm. Computer 35(1):70–78

[BMG06] Bieser C, Mueller-Glaser K-D (2006) Rapid prototyping design acceleration using a novel merging methodology for partial configuration streams of Xilinx Virtex-II FPGAs. In: Proceedings of the 17th IEEE international workshop on Rapid System Prototyping (RSP). IEEE Computer Society, pp 193–199

[BMK+04] Bobda C, Majer M, Koch D, Ahmadinia A, Teich J (2004) A dynamic NoC approach for communication in reconfigurable devices. In: Proceedings of international conference on field-programmable logic and applications (FPL). Volume 3203 of lecture notes in computer science (LNCS). Springer, pp 1032–1036

[BMMAAS09] Bruneel K, Abouelella FMMA, Stroobandt D (2009) TMAP: s reconfigurability-aware technology mapper. In: Jacquemod G, Luxey C, Damiano J-P (eds) Design, automation and test Europe: university booth (DATE), Apr 2009

[BMN+01] Baumgarte V, May F, Nückel A, Vorbach M, Weinhardt M (2001) PACT XPP – A self-reconfigurable data processing architecture. In: Proceedings of the international conference on Engineering of Reconfigurable Systems and Algorithms (ERSA), June 2001, pp 64–70

[BR97] Betz V, Rose J (1997) VPR: a new packing, placement and routing tool for FPGA research. In: Proceedings of the 7th international workshop on field-programmable logic and applications (FPL). Springer, pp 213–222

[Bre97] Brebner G (1997) The swappable logic unit: a paradigm for virtual hardware. In: Pocek KL, Arnold J (eds) IEEE symposium on FPGAs for custom computing machines (FPGA), Apr 1997, pp 77–86

[BRM99] Betz V, Rose J, Marquardt A (eds) (1999) Architecture and CAD for deep-submicron FPGAs. Kluwer, Norwell

[BT93] Babb J, Tessier R (1993) Virtual wires: overcoming pin limitations in FPGA-based logic emulators. In: IEEE workshop on FPGAs for custom computing machines, pp 142–151

[BTD+97] Babb J, Tessier R, Dahl M, Hanono S, Hoki D, Agarwal A (1997) Logic emulation with virtual wires. IEEE trans Comput Aided Design 16:609–626

[CKL+] Curt D, Kalara P, Leblanc R, Eck V, Trynosky S, Lindholm J, Bauer T, Blodget B, McMillan S, Philip J, Prasanna S, Keller E (2004) Reconfiguration of the programmable logic of an integrated circuit. WO Patent WO002004055986A3, issued 25, Nov 2004

[CLC+02] Compton K, Li Z, Cooley J, Knol S, Hauck S (2002) Configuration relocation and defragmentation for run-time reconfigurable computing. IEEE Trans Very Large Scale Integ Syst 10:209–220

[CMS06] Claus C, Müller FH, Stechele W (2006) Combitgen: a new approach for creating partial bitstreams in Virtex-II Pro. In: Karl W, Becker J, Großpietsch K-E, Hochberger C, Maehle E (eds) Workshops proceedings of the 19th international conference on Architecture of Computing Systems (ARCS). Volume 81 of lecture notes in informatics. GI, pp 122–131

[CMZS07] Claus C, Müller FH, Zeppenfeld J, Stechele W (2007) A new framework to accelerate Virtex-II Pro dynamic partial self-reconfiguration. In: Proceedings of the IEEE 21th International Parallel and Distributed Processing Symposium (IPDPS), Mar 2007, pp 1–7

[CPF09] Carver JM, Pittman RN, Forin A (2009) Automatic bus macro placement for partially reconfigurable FPGA designs. In: Proceeding of the ACM/SIGDA international symposium on Field Programmable Gate Arrays (FPGA). ACM, pp 269–272

[CS1] Department of Computer Science 12, ReCoNets-Project website, University of Erlangen Nuremberg, Germany. www.reconets.de

[Deh99] Dehon A (1999) Balancing interconnect and computation in a reconfigurable computing array (or, why you don't really want 100% LUT utilization. In: Proceedings of the international symposium on Field Programmable Gate Arrays (FPGA), pp 69–78

[DF07] Dittmann F, Frank S (2007) Hard real-time reconfiguration port scheduling. In: Proceedings of the conference on design, automation and test in Europe (DATE). EDA Consortium, pp 123–128

[Dij59] Dijkstra EW (1959) A note on two problems in connexion with graphs. Numer Math 1:269–271

[DL09] Demirsoy SS, Langhammer M (2009) Cholesky decomposition using fused datapath synthesis. In: Proceeding of the ACM/SIGDA international symposium on Field-Programmable Gate Arrays (FPGA). ACM, pp 241–244

[DT01] Dally WJ, Towles B (2001) Route packets, not wires: on-chip interconnection networks. In: Proceedings of the 38th conference on design automation (DAC), pp 684–689

[EAWJ02] Elnozahy EN, Alvisi L, Wang Y-M, Johnson D (2002) A survey of rollback-recovery protocols in message-passing systems. ACM Comput Surv 34(3): 375–408

[epc07] Altera Inc. (2007) Enhanced configuration devices (EPC4, EPC8 & EPC16) data sheet. www.altera.com/literature/hb/cfg/ch_14_vol_2.pdf

[EW0] Can programmables break out of $3.6bn niche. ElectronicWeekly, 26.06.2009

[Fid88] Fidge C (1988) Timestamps in message-passing systems that preserve the partial ordering. In: Proceedings of the 11th Australian Computer Science Conference (ACSC), pp 56–66

[FKS+08] Fekete SP, Kamphans T, Schweer N, Tessars C, van der Veen JC, Angermeier J, Koch D, Teich J (2008) No-break dynamic defragmentation of reconfigurable devices. In: FPL 2008, international conference on field programmable logic and applications, Heidelberg, pp 113–118

[Fly05] Flynn M (2005) Area – time – power and design effort: the basic tradeoffs in application specific systems. In: Proceedings of the IEEE international conference on Application-Specific Systems, Architecture Processors (ASAP). IEEE Computer Society, pp 3–6

[FvdVA+07] Fekete SP, van der Veen J, Angermeier J, Göhringer D, Majer M, Teich J (2007) Scheduling and communication-aware mapping of HW/SW modules for dynamically and partially reconfigurable SoC architectures. In: Proceedings of the 20th international conference on Architecture of Computing Systems (ARCS), Zurich. VDE-Verlag, pp 151–160

[FvdVMT06] Fekete S, van der Veen J, Majer M, Teich J (2006) Minimizing communication cost for reconfigurable slot modules. In: Proceedings of 16th international conference on field programmable logic and applications (FPL), Aug 2006, pp 535–540

[GLS99] Guccione S, Levi D, Sundararajan P (1999) JBits: Java based interface for reconfigurable computing. In: Proceedings of the 2nd annual military and aerospace applications of programmable devices and technologies (MAPLD)

[GSS09] Gao S, Schmidt A, Sass R (2009) Hardware implementation of mpi barrier on an fpga cluster. In: Proceedings of the international conference on field programmable logic and applications (FPL), pp 12–17

[GZ97] Gupta RK, Zorian Y (1997) Introducing core-based system design. IEEE Des Test 14(4):15–25

[Hal07] Halfhill TR (2007) Tabulas time machine – Rapidly reconfigurable chips will challenge conventional FPGAs. Microprocessor Report, Issue 032910. Available online: www.tabula.com/news/M11_Tabula_Reprint.pdf.

[Hau98] Hauck S (1998) Configuration prefetch for single context reconfigurable coprocessors. In: Proceedings of the sixth ACM/SIGDA international symposium on Field Programmable Gate Arrays (FPGA). ACM, pp 65–74

[Heu03] Heun V (2003) Grundlegende algorithmen. Vieweg

[HHK10] Hempel G, Hochberger C, Koch A (2010) A comparison of hardware acceleration interfaces in a customizable soft core processor. In: Proceedings of the international conference on field programmable logic and applications (FPL). IEEE Computer Society, pp 469–474

[HKKP07a] Hagemeyer J, Kettelhoit B, Koester M, Porrmann M (2007) A design methodology for communication infrastructures on partially reconfigurable FPGAs. In: International conference on field programmable logic and applications (FPL), Aug 2007. IEEE, pp 331–338

[HKKP07b] Hagemeyer J, Kettelhoit B, Koester M, Porrmann M (2007) Design of homogeneous communication infrastructures for partially reconfigurable FPGAs. In: Proceedings of the international conference on Engineering of Reconfigurable Systems and Algorithms (ERSA), Jun 2007

[HKP06] Hagemeyer J, Kettelhoit B, Porrmann M (2006) Dedicated module access in dynamically reconfigurable systems. In: Proceedings of the 20th International Parallel and Distributed Processing Symposium (IPDPS)

[HKT03] Haubelt C, Koch D, Teich J (2003) Basic OS support for distributed reconfigurable hardware. In: Proceedings of the third international workshop on systems, architectures, modeling, and simulation, July 2003, pp 18–22

[HKT11] Hansen SG, Koch D, Torresen J (2011) High speed partial run-time reconfiguration using enhanced ICAP hard macro. In: Proceedings of the 18th reconfigurable architectures workshop (RAW). IEEE

[HLK02] Horta EL, Lockwood JW, Kofuji ST (2002) Using PARBIT to implement partial run-time reconfigurable systems. In: Proceedings of the 12th international conference on field-programmable logic and applications (FPL). Springer, pp 182–191

[HLTP02] Horta EL, Lockwood JW, Taylor DE, Parlour D (2002) Dynamic hardware plugins in an FPGA with partial run-time reconfiguration. In: Proceedings of the 39th conference on design automation (DAC). ACM, pp 343–348

[HM01] Huang W-J, McCluskey EJ (2001) Column-based precompiled configuration techniques for FPGA. In: Proceedings of the the 9th annual IEEE symposium on field-programmable custom computing machines (FCCM). IEEE Computer Society, pp 137–146

[HSKB06] Hübner M, Schuck C, Kühnle M, Becker J (2006) New 2-dimensional partial dynamic reconfiguration techniques for real-time adaptive microelectronic circuits. In: Proceedings of the IEEE computer society annual symposium on emerging VLSI technologies and architectures (ISVLSI), p 97

[Huf52] Huffman DA (1952) A method for the construction of minimum-redundancy codes. Proc IRE 40(9):1098–1101

[HW99] Hauck S, Wilson WD (1999) Runlength compression techniques for FPGA configurations. In: Proceedings of the seventh annual IEEE symposium on field-programmable custom computing machines (FCCM99). IEEE Computer Society, pp 286–287

[Inc02] Xilinx Inc (2002) Two flows for partial reconfiguration: module based or difference based. Available online: www.xilinx.com/bvdocs/appnotes/xapp290.pdf

[Inc03] Atmel Inc (2003) AT40K series configuration user guide. Available online: http://www.atmel.com/dyn/resources/prod_documents/DOC1009.PDF

[Inc05a] Xilinx Inc (2005) Using look-up tables as shift registers (SRL16) in Spartan-3 generation FPGAs. Available online: www.xilinx.com/bvdocs/appnotes/xapp465.pdf

[Inc05b] Xilinx Inc (2005) Virtex-II platform FPGAs: complete data sheet. Available online: http://www.xilinx.com/support/documentation/data_sheets/ds031.pdf

[Inc07] Xilinx Inc (2007) Xilinx Virtex-II platform FPGA user guide. Available online: http://www.xilinx.com/support/documentation/user_guides/ug002.pdf

[Inc08a] Altera Inc (2008) Logic array blocks and adaptive logic modules in Stratix IV devices. Available online: http://www.altera.com/literature/hb/stratix-iv/

[Inc08b] Xilinx Inc (2008) CLB white papers. Available online: http://www.xilinx.com/support/documentation/wpfpgafeaturedesign_clb.htm/

[Inc08c] Xilinx Inc (2008) PlanAhead design analysis tool. Available online: http://www.xilinx.com/ise/optional_prod/planahead.htm

[Int99] International Business Machines corporation (IBM) (1999) The CoreConnect bus architecture. Available online: http://www-03.ibm.com/chips/products/coreconnect/

[JS90] Jones G, Sheeran M (1990) Circuit design in Ruby. In Staunstrup, J (ed) Formal Methods for VLSI Design. North-Holland

[JW05] Jerraya AA, Wolf W (2005) Hardware/software interface codesign for embedded systems. Computer 38(2):63–69

[KBAT07] Koch D, Bobda C, Ahmadinia A, Teich J (2007) FPGA architecture extensions for preemptive multitasking and hardware defragmentation. In: Proceedings of International Conference on Field-Programmable Technology 2007 (ICFPT '07), Dec 2007. IEEE, pp 433–436

[KBT07] Koch D, Beckhoff C, Teich J (2007) Bitstream decompression for high speed FPGA configuration from slow memories. In: Proceedings of international conference on field-programmable technology (ICFPT), Dec 2007. IEEE, pp 161–168

[KBT08] Koch D, Beckhoff C, Teich J (2008) ReCoBus-builder – A novel tool and technique to build statically and dynamically reconfigurable systems for FPGAs. In: Proceedings of international conference on field-programmable logic and applications (FPL 08), Sept 2008, pp 119–124

[KBT09a] Koch D, Beckhoff C, Teich J (2009) A communication architecture for complex runtime reconfigurable systems and its implementation on Spartan-3 FPGAs. In: Proceedings of the 17th ACM/SIGDA international symposium on field-programmable gate arrays (FPGA 2009), Feb 2009. ACM, pp 233–236

[KBT09b] Koch D, Beckhoff C, Teich J (2009) Hardware decompression techniques for FPGA-based embedded systems. ACM Trans Reconfig Technol Syst 2:9:1–9:23

[KBT09c] Koch D, Beckhoff C, Teich J (2009) Minimizing internal fragmentation by fine-grained two-dimensional module placement for runtime reconfigurable systems. In: 17th annual IEEE symposium on field-programmable custom computing machines (FCCM 2009), Apr 2009. IEEE Computer Society, pp 251–254

[KBT10] Koch D, Beckhoff C, Torresen J (2010) Obstacle-free two-dimensional online-routing for run-time reconfigurable FPGA-based systems. In: Proceedings of international conference on field-programmable technology (ICFPT10́). IEEE, pp 208–215

[KD06a] Koh S, Diessel O (2006) COMMA: a communications methodology for dynamic module-based reconfiguration of FPGAs. In: Workshops proceedings of the 19th international conference on architecture of computing systems (ARCS), Mar 2006, pp 173–182

[KD06b] Koh S, Diessel O (2006) COMMA: a communications methodology for dynamic module reconfiguration in FPGAs. In: Proceedings of the 14th annual IEEE symposium on field-programmable custom computing machines (FCCM). IEEE Computer Society, pp 273–274

[KdlTRJ06] Krasteva YE, de la Torre E, Riesgo T, Joly D (2006) Virtex II FPGA bitstream manipulation: application to reconfiguration control systems. In: Proceedings of the 16th international conference on field programmable logic and applications (FPL). IEEE, pp 1–8

[KHST07] Koch D, Haubelt C, Streichert T, Teich J (2007) Modeling and synthesis of hardware-software morphing. In: Proceedings of the international symposium on circuits and systems (ISCAS 2007), May 2007, pp 2746–2749

[KHT07] Koch D, Haubelt C, Teich J (2007) Efficient hardware checkpointing – Concepts, overhead analysis, and implementation. In: Proceedings of the 15th ACM/SIGDA international symposium on field-programmable gate arrays (FPGA 2007), Feb 2007. ACM, pp 188–196

[KHT08] Koch D, Haubelt C, Teich J (2008) Efficient reconfigurable on-chip buses for FPGAs. In: 16th annual IEEE symposium on field-programmable custom computing machines (FCCM 2008), Apr 2008. IEEE Computer Society, pp 287–290

[KJdlTR05] Krasteva YE, Jimeno AB, de la Torre E, Riesgo T (2005) Straight method for reallocation of complex cores by dynamic reconfiguration in Virtex II FPGAs. In: Proceedings of the 16th IEEE international workshop on rapid system prototyping (RSP), pp 77–83

[KKH+06] Kissler D, Kupriyanov A, Hannig F, Koch D, Teich J (2006) A generic framework for rapid prototyping of system-on-chip designs. In: Proceedings of international conference on computer design (CDES), June 2006, pp 189–195

[KKT06] Koch D, Körber M, Teich J (2006) Searching RC5-keys with distributed reconfigurable computing. In: Proceedings of international conference on engineering of reconfigurable systems and algorithms (ERSA 2006), June 2006. CSREA Press, pp 42–48

[KLPR05] Kalte H, Lee G, Porrmann M, Rückert U (2005) REPLICA: a bitstream manipulation filter for module relocation in partial reconfigurable systems. In: Proceedings of the 19th International Parallel and Distributed Processing Symposium – Reconfigurable architectures workshop (IPDPS). IEEE Computer Society

[KMD+06] Kapre N, Mehta N, Delorimier M, Rubin R, Barnor H, Wilson MJ, Wrighton M, Dehon A (2006) Packet-switched vs. time-multiplexed FPGA overlay networks. In: Proceedings of the IEEE symposium on Field-programmable Custom Computing Machines (FCCM). IEEE, pp 205–216

[Kot] Kotolly A (2006) The economics of structured- and standard-cell-ASIC designs. Electronic News, 16.03.2006

[KP05] Kalte H, Porrmann M (2005) Context saving and restoring for multitasking in reconfigurable systems. In: Proceedings of the 15th international conference on field programmable logic and applications (FPL), Aug 2005, pp 223–228

[KP06] Kalte H, Porrmann M (2006) REPLICA2Pro: task relocation by bitstream manipulation in Virtex-II/Pro FPGAs. In: Proceedings of the 3rd conference on computing frontiers (CF). ACM, pp 403–412

[KPR02] Kalte H, Porrmann M, Rückert U (2002) A prototyping platform for dynamically reconfigurable system on chip designs. In: Proceedings of the IEEE workshop heterogeneous reconfigurable systems on chip (SoC)

[KPR04a] Kalte H, Porrmann M, Rückert U (2004) Study on column wise design compaction for reconfigurable systems. In: Proceedings of the IEEE international conference on field programmable technology (FPT), Dec 2004

[KPR04b] Kalte H, Porrmann M, Rückert U (2004) System-on-programmable-chip approach enabling online fine-grained 1D-placement. In: Proceedings of the 11th reconfigurable architectures workshop (RAW), pp 141–146

[KR07] Kuon I, Rose J (2007) Measuring the gap between FPGAs and ASICs. Tran Comput-Aided Des Integr Circ Syst 26(2):203–215

[KRS+10] Koch D, Reimann F, Streichert T, Haubelt C, Teich J (2010) ReCoNets – design methodology for embedded systems consisting of small networks of reconfigurable nodes and connections. In: Platzner M, Teich J, Wehn N (eds) Dynamically reconfigurable systems. Springer, pp 223–244

[KSHT08a] Koch D, Streichert T, Haubelt C, Teich J (2008) Logic chip, logic system and method for designing a logic chip. Patent PCT/EP2008/007342

[KSHT08b] Koch D, Streichert T, Haubelt C, Teich J (2008) Logic chip, method and computer program for providing a configuration information for a configurable logic chip. Patent PCT/EP2008/007343

[KSHT09] Kissler D, Strawetz A, Hannig F, Teich J (2009) Power-efficient reconfiguration control in coarse-grained dynamically reconfigurable architectures. In: Proceedings of the 18th international workshop on power and timing modeling, optimization, and simulation (PATMOS), Sept 2009. Lecture notes in computer science (LNCS, vol 5349). Springer, pp 307–317

[KT04] Koch D, Teich J (2004) Platform-independent methodology for partial reconfiguration. In: Proceedings of the 1st conference on computing frontiers (CF'04). ACM, pp 398–403

[KT10] Koch D, Torresen J (2010) Routing optimizations for component-based system design and partial run-time reconfiguration on FPGAs. In: Proceedings of international conference on field-programmable technology (ICFPT). IEEE, pp 460–464

[KT11a] Koch D, Torresen J (2011) A routing architecture for mapping dataflow graphs at run-time. In: Proceedings of international conference on field-programmable logic and applications (FPL 11), Sept 2011, pp 286–290

[KT11b] Koch D, Torresen J (2011) FPGASort: a high performance sorting architecture exploiting run-time reconfiguration on FPGAs for large problem sorting. In: Proceedings of the 19th ACM/SIGDA international symposium on field-programmable gate arrays (FPGA 2011), Feb 2011. ACM, pp 45–54

[KV06] Kachris C, Vassiliadis S (2006) Performance evaluation of an adaptive FPGA for network applications. In: Proceedings of the seventeenth IEEE international workshop on rapid system prototyping (RSP). IEEE Computer Society, pp 54–62

[LAB+05] Lewis D, Ahmed E, Baeckler G, Betz V, Bourgeault M, Galloway D, Hutton M, Lane C, Lee A, Leventis P, Mcclintock C, Padalia K, Pedersen B, Powell G, Ratchev B, Reddy S, Schleicher J, Stevens K, Yuan R, Cliff R, Rose J (2005) The Stratix II logic and routing architecture. In: Proceedings of the ACM/SIGDA 13th international symposium on field-programmable gate arrays (FPGA). ACM, pp 14–20

[LBM+06] Lysaght P, Blodget B, Mason J, Young J, Bridgford B (2006) Invited paper: enhanced architecture, design methodologies and CAD tools for dynamic reconfiguration of Xilinx FPGAs. In: Proceedings of the 16th international conference on field programmable logic and application (FPL), Aug 2006, pp 1–6

[LCK07] Lu R, Cao A, Koh C-K (2007) SAMBA-Bus: a high performance bus architecture for system-on-chips. IEEE Trans VLSI Syst 15(1):69–79

[LH99] Li Z, Hauck S (1999) Don't care discovery for FPGA configuration compression. In: Proceedings of the 7th ACM/SIGDA international symposium on field programmable gate arrays (FPGA). ACM, pp 91–98

[LH01] Li Z, Hauck S (2001) Configuration compression for Virtex FPGAs. In: Proceedings of the 9th annual IEEE symposium on field-programmable custom computing machines (FCCM). IEEE Computer Society, pp 147–159

[LK10] Lange H, Koch A (2010) Architectures and execution models for hardware/-software compilation and their system-level realization. IEEE Trans Comput 59:1363–1377

[LLTY04] Lemieux G, Lee E, Tom M, Yu A (2004) Directional and single-driver wires in FPGA interconnect. In: IEEE international conference on field-programmable technology (FPT)

[LNTT01] Lockwood JW, Naufel N, Turner JS, Taylor DE (2001) Reprogrammable network packet processing on the field programmable port extender (FPX). In: Proceedings of the 1998 ACM/SIGDA sixth international symposium on field programmable gate arrays (FPGA). ACM, pp 87–93

[LP08] Lübbers E, Platzner M (2008) Communication and synchronization in multi-threaded reconfigurable computing systems. In: Proceedings of the international conference on engineering of reconfigurable systems and algorithms (ERSA), Jun 2008, pp 83–89

[LSC96] Luk W, Shirazi N, Cheung PYK (1996) Modelling and optimising run-time reconfigurable systems. In: IEEE symposium on FPGAs for custom computing machines. IEEE Computer Society Press, pp 167–176

[LSC97] Luk W, Shirazi N, Cheung PYK (1997) Compilation tools for run-time reconfigurable designs. In: Proceedings of the 5th IEEE symposium on FPGA-based custom computing machines (FCCM). IEEE Computer Society, pp 56–65

[Ltd07] Advanced RISC Machines Ltd (2007) AMBA system architecture. Available online: http://www.arm.com/products/solutions/AMBAHomePage.html

[LWH02] Landaker WJ, Wirthlin MJ, Hutchings B (2002) Multitasking hardware on the SLAAC1-V reconfigurable computing system. In: Proceedings of the 12th international conference on field-programmable logic and applications (FPL), pp 806–815

[Mar11] Bourgeault M (2011) Alteras partial reconfiguration flow. Available online: http://www.eecg.utoronto.ca/ ~jayar/FPGAseminar/FPGA_Bourgeault_June23_2011.pdf

[MBV+02] Marescaux T, Bartic A, Verkest D, Vernalde S, Lauwereins R (2002) Interconnection networks enable fine-grain dynamic multi-tasking on FPGAs. In: Proceedings of the 12th international conference on field-programmable logic and applications (FPL), pp 795–805

[ME95] Mcmurchie L, Ebeling C (1995) Pathfinder: a negotiation-based performance-driven router for FPGAs. In: Proceedings of the 1995 ACM third international symposium on field-programmable gate arrays (FPGA), pp 111–117

[Met04] Metzgen P (2004) A high performance 32-bit ALU for programmable logic. In: Proceedings of the 2004 ACM/SIGDA 12th international symposium on field programmable gate arrays (FPGA). ACM, pp 61–70

[MMP+03] Mesquita D, Moraes F, Palma J, Möller L, Calazans N (2003) Remote and partial reconfiguration of FPGAs: tools and trends. In: Proceedings of the 17th international symposium on parallel and distributed processing (IPDPS). IEEE Computer Society, p 177.1

[MNC+03] Mignolet J-Y, Nollet V, Coene P, Verkest D, Vernalde S, Lauwereins R (2003) Infrastructure for design and management of relocatable tasks in a heterogeneous reconfigurable system-on-chip. In: Proceedings of the conference on design, automation and test in Europe (DATE). IEEE Computer Society, pp 986–991

[MNM+04] Marescaux T, Nollet V, Mignolet J-Y, Bartic A, Moffat W, Avasare P, Coene P, Verkest D, Vernalde S, Lauwereins R (2004) Run-time support for heterogeneous multitasking on reconfigurable SoCs. Integr VLSI J 38(1):107–130

[MSCL06] Mak TST, Sedcole NP, Cheung PYK, Luk W (2006) On-FPGA communication architectures and design factors. In: Proceedings of the 16th international conference on field programmable logic and applications (FPL). IEEE, pp 1–8

[MSCL08] Mak T, Sedcole P, Cheung PYK, Luk W (2008) Wave-pipelined signalling for On-FPGA communication. In: IEEE international conference on field-programmable technology (FPT)

[MTAB07] Majer M, Teich J, Ahmadinia A, Bobda C (2007) The erlangen slot machine: a dynamically reconfigurable FPGA-based computer. J VLSI Sig Process Syst 47(1):15–31

[OM07] Oliver TF, Maskell DL (2007) Prerouted FPGA cores for rapid system construction in a dynamic reconfigurable system. EURASIP J Embed Syst 2007(1):7 [ope] opencores http://www.opencores.org

[PdMM+02] Palma JC, de Mello AV, Möller L, Moraes F, Calazans N (2002) Core communication interface for FPGAs. In: Proceedings of the 15th symposium on integrated circuits and systems design (SBCCI), p 183

[PMW04] Pan JH, Mitra T, Wong W-F (2004) Configuration bitstream compression for dynamically reconfigurable FPGAs. In: Proceedings of the IEEE/ACM international conference on computer-aided design (ICCAD), pp 766–773

[Sal04] Salomon D (2004) Data compression – The complete reference. Springer

[SAP08] Suris J, Athanas P, Patterson C (2008) An efficient run-time router for connecting modules in FPGAs. In: Proceedings of international conference on field-programmable logic and applications (FPL), Sept 2008, pp 125–130

[SBB+06] Sedcole P, Blodget B, Becker T, Anderson J, Lysaght P (2006) Modular dynamic reconfiguration in Virtex FPGAs. IEE 153(3):157–164

[SBKV05] Sethuraman B, Bhattacharya P, Khan J, Vemuri R (2005) LiPaR: a light-weight parallel router for FPGA-based networks-on-chip. In: Proceedings of the 15th ACM great lakes symposium on VLSI (GLSVSLI), New York. ACM, pp 452–457

[SCCL04] Sedcole P, Cheung PYK, Constantinides GA, Luk W (2004) A structured methodology for system-on-an-FPGA design. In: 14th international conference on field programmable logic and application (FPL), Aug 2004. Volume 3203 of lecture notes in computer science, pp 1047–1051

[SF08] Silvia M, Ferreira JC (2008) Generation of partial FPGA configurations at run-time. In: Proceedings of international conference on field-programmable logic and applications (FPL), Sept 2008, pp 367–372

[SGS05] Sunwoo J, Garimella S, Stroud C (2005) On embedded processor reconfiguration of logic BIST for FPGA cores in SoCs. In: Proceedings of the 14th IEEE North Atlantic test workshop (NATW), May 2005

[Sil02] Silicore™ (2002) WISHBONE System-on-Chip (SoC)interconnection architecture for portable IP cores Rev. B.3. Available online: http://www.opencores.org/projects.cgi/web/wishbone/wbspec_b3.pdf

[SKHT06] Streichert T, Koch D, Haubelt C, Teich J (2006) Modeling and design of fault-tolerant and self-adaptive reconfigurable networked embedded systems. EURASIP J Embed Syst 2006:1–15. Article ID 42168, doi:10.1155/ES/2006/42168

[SLM00] Simmler H, Levinson L, Manner R (2000) Multitasking on FPGA coprocessors. In: Proceedings of the 10rd international conference on field programmable logic and application (FPL), pp 121–130

[SLV03] Stitt G, Lysecky R, Vahid F (2003) Dynamic hardware/software partitioning: a first approach. In: Proceedings of the 40th conference on design automation (DAC). ACM, pp 250–255

[SNH+10] Schallenberg A, Nebel W, Herrholz A, Hartmann P, Grüttner K, Oppenheimer F (2010) POLYDYN – object-oriented modelling and synthesis targeting dynamically reconfigurable FPGAs. In: Platzner M, Teich J, Wehn N (eds) Dynamically reconfigurable systems. Springer, pp 139–158

[spi07] The Spirit Consortium (2007) http://www.spiritconsortium.org/

[SPLC12] Saldana M, Patel A, Liu HJ, Chow P (2012) Using partial reconfiguration and message passing to enable FPGA-based generic computing platforms. EURASIP J Embed Syst 2012(127302):10

[SS82] Storer JA, Szymanski TG (1982) Data compression via textual substitution. J ACM 29(4):928–951

[SSK+07] Streichert T, Strengert C, Koch D, Haubelt C, Teich J (2007) Communication aware optimization of the task binding in hardware/software reconfigurable networks. Int J Circ Syst 2(1):29–36

[SV98] Scalera SM, Vázquez JR (1998) The design and implementation of a context switching FPGA. In: Proceedings of the IEEE symposium on FPGAs for custom computing machines (FCCM). IEEE Computer Society, p 78

[SW07] Snider GS, Williams S (2007) Nano/CMOS architectures using a field-programmable nanowire interconnect. Nanotechnology 18(3):1–11

[SWT09] Sim JE, Wong W-F, Teich J (2009) Optimal placement-aware trace-based scheduling of hardware reconfigurations for FPGA accelerators. In: Proceedings of the 17th annual IEEE symposium on field-programmable custom computing machines (FCCM), Apr 2009. IEEE Computer Society (to appear)

[Syn08] Synopsys Inc (2008) DesignWare GTECH library. Available online: www.synopsys.com/dw/doc.php/doc/dwf/manuals/dw_gtech.pdf

[TCJW97] Trimberger S, Carberry D, Johnson A, Wong J (1997) A time-multiplexed FPGA. In: Proceedings of the 5th IEEE symposium on FPGA-based custom computing machines (FCCM). IEEE Computer Society, pp 22–28

[Tei07] Teich J (2007) Lecture on reconfigurable computing: Chapter 1. inroduction, University of Erlangen-Nuremberg. Available online: http://www12.informatik.uni-erlangen.de/edu/rc/slides/1_RC_introduction.pdf

[Tes98] Tessier R (1998) Negotiated A* routing for FPGAs. In: Proceedings of the fifth canadian workshop on field-programmable devices (FPD)

[TT03] Tiwari A, Tomko KA (2003) Scan-chain based watch-points for efficient run-time debugging and verification of FPGA designs. In: Proceedings of the conference on Asia South Pacific design automation (ASPDAC). ACM, pp 705–711

[vsi07] Virtual Socket Interface Alliance (2007) Legacy documents of the VSI alliance. Available online: http://vsi.org/

[VSV+01] Vanmeerbeeck G, Schaumont P, Vernalde S, Engels M, Bolsens I (2001) Hardware/software partitioning of embedded system in OCAPI-xl. In: Proceedings of the ninth international symposium on hardware/software codesign. ACM, pp 30–35

[WAL+93] Wazlowski M, Agarwal L, Lee T, Smith A, Lam E, Athanas P, Silverman H, Ghosh S (1993) PRISM-II compiler and architecture. In: Buell DA, Pocek KL (eds) IEEE workshop on FPGAs for custom computing machines (FCCM). IEEE Computer Society Press, pp 9–16

[WGNH01] Wheeler T, Graham P, Nelson BE, Hutchings B (2001) Using design-level scan to improve FPGA design observability and controllability for functional verification. In: Proceedings of the 11th international conference on field programmable logic and application (FPL), pp 483–492

[WH95] Wirthlin M, Hutchings BL (1995) DISC: the dynamic instruction set computer. In: Schewel J (ed) Proceedings on Field Programmable Gate Arrays (FPGAs) for fast board development and reconfigurable computing (SPIE) 2607. SPIE – The International Society for Optical Engineering, pp 92–103

[WH96] Wirthlin MJ, Hutchings BL (1996) Sequencing run-time reconfigured hardware with software. In: Proceedings of the 1996 ACM fourth international symposium on field-programmable gate arrays (FPGA). ACM, pp 122–128

[WJ06] Wang F, Jean J (2006) Architectural support for runtime 2D partial reconfiguration. In: Proceedings of international conference on engineering of reconfigurable systems and algorithms (ERSA), June 2006. CSREA Press, pp 231–232

[WL05] Wetekam G, Lutz B(2005) Hardware-Implementierung einer 3D-Huffman-Decodierung für dynamische Volumendaten. In: Hardware for visual computing workshop, Universität Tübingen

[WP02] Walder H, Platzner M (2002) Non-preemptive multitasking on FPGA: task placement and footprint transform. In: Proceedings of the international conference on engineering of reconfigurable systems and algorithms (ERSA), June 2002. CSREA Press, pp 24–30

[WP04] Walder H, Platzner M (2004) A runtime environment for reconfigurable operating systems. In: Proceedings of the 14th international conference on field programmable logic and application (FPL), pp 831–835

[xcf07] Xilinx Inc (2007) Platform flash in-system programmable configuration PROMs. Available online: www.xilinx.com/support/documentation/data_sheets/ds123.pdf

[Xil07] Xilinx Inc (2007) Xilinx: silicon devices. Available online: www.xilinx.com/products/silicon_solutions/

[Xil11] Xilinx Inc (2011) Partial reconfiguration user guide (Rel 13.2). Available online: www.xilinx.com/support/documentation/sw_manuals/xilinx13_2/ug702.pdf

[XRT03] Xu W, Ramanarayanan R, Tessier R (2003) Adaptive fault recovery for networked reconfigurable systems. In: Proceedings of the 11th annual IEEE symposium on field-programmable custom computing machines (FCCM). IEEE Computer Society, p 143

[ZKS04] eferino CA, Kreutz ME, Susin AA (2004) RASoC: a router soft-core for networks-on-chip. In: Proceedings of the conference on design, automation and test in Europe (DATE). IEEE Computer Society, p 30198

[ZL77] Ziv J, Lempel A (1977) A universal algorithm for sequential data compression. IEEE Trans Inform Theory 23(3):337–343

About the Author

Since his childhood, it was clear that Dirk Koch would become an engineer. There was virtually no technical device safe from getting investigated to its core within the household. During his schooling, he played around with micro electronics, mechanics, and even lasers. He built his own 6502 microcomputer and used his Commodore 64 as an EPROM programmer. Dirks' first contact with reconfigurable logic was when building a digital alarm clock in a Lattice ispLSI 1016 CPLD.

During his time at the University of Paderborn, Germany, Dirk focussed on power electronics, renewable energy, and engine control. He worked on his diploma thesis at the Queensland University of Technology (QUT), Brisbane, Australia. There, he implemented a special radial basis function neural network for function approximation on the reconfigurable course grained PACT XPP architecture.

After this, Dirk joined the group of Prof. Jürgen Teich at the University of Erlangen-Nuremberg, Germany. He worked on the *ReCoNets* and *ReCoNodes* projects, which were funded by the German Science Foundation (DFG) within the Priority Programm SPP 1148 "Rekonfigurierbare Rechensysteme". This was the time Dirk got in touch with FPGAs and partial runtime reconfiguration. Throughout his time in Erlangen, he worked on distributed control systems that exploit runtime reconfiguration of FPGAs for increasing system reliability and flexibility. He graduated in Erlangen with the PhD thesis "Architectures, Methods, and Tools for Distributed Runtime Reconfigurable FPGA-based Systems". His thesis won the "EDAA Outstanding Dissertations Award 2010" in the category "New directions in physical design, design for manufacturing and CAD for analogue circuits and MEMS". The prize is awarded by the European Design and Automation Association.

Having had the experience of a summer in Australia, Dirk decided to go for some winters in Norway; and since the end of 2009, he works as a postdoctoral researcher at the University of Oslo. There, he continues his work on the project "Context Switching Reconfigurable Hardware for Communication Systems"

D. Koch, *Partial Reconfiguration on FPGAs: Architectures, Tools and Applications*, Lecture Notes in Electrical Engineering 153, DOI 10.1007/978-1-4614-1225-0,

(COSRECOS), which is funded by the Norwegian Research Council under grant 191156V30. The overall goal of the COSRECOS project is – as the overall goal of this book – to make partial runtime reconfiguration more accessible to real-world applications.

Index

D. Koch, *Partial Reconfiguration on FPGAs: Architectures, Tools and Applications*,
Lecture Notes in Electrical Engineering 153, DOI 10.1007/978-1-4614-1225-0,

MIX
Papier aus verantwortungsvollen Quellen
Paper from responsible sources
FSC® C105338

If you have any concerns about our products,
you can contact us on
ProductSafety@springernature.com

In case Publisher is established outside the EU,
the EU authorized representative is:
Springer Nature Customer Service Center GmbH
Europaplatz 3, 69115 Heidelberg, Germany

Printed by Libri Plureos GmbH
in Hamburg, Germany